FIBER BUNDLES
AND HOMOTOPY

FIBER BUNDLES AND HOMOTOPY

Dai Tamaki
Shinshu University, Japan

World Scientific

EW JERSEY · LONDON · SINGAPORE · BEIJING · SHANGHAI · HONG KONG · TAIPEI · CHENNAI · TOKYO

Published by

World Scientific Publishing Co. Pte. Ltd.

5 Toh Tuck Link, Singapore 596224

USA office: 27 Warren Street, Suite 401-402, Hackensack, NJ 07601

UK office: 57 Shelton Street, Covent Garden, London WC2H 9HE

Library of Congress Control Number: 2021021137

British Library Cataloguing-in-Publication Data
A catalogue record for this book is available from the British Library.

FIBER BUNDLES AND HOMOTOPY

ISBN 978-981-123-799-7 (hardcover)
ISBN 978-981-123-809-3 (ebook for institutions)
ISBN 978-981-123-810-9 (ebook for individuals)

For any available supplementary material, please visit
https://www.worldscientific.com/worldscibooks/10.1142/12308#t=suppl

Printed in Singapore

Preface

In this book, we first study basic properties of fiber bundles and then, by abstracting their properties, we introduce fundamental concepts and methodologies in homotopy theory.

Homotopy theory is a relatively new research area in the long history of mathematics. The idea of continuous deformation was made precise around 1920s or 1930s in the attempts to make the idea of Poincaré rigorous. After the World War II, homotopy theory has been developed rapidly and became a matured field utilizing abstract and complex concepts.

The notion of homotopy, i.e. continuous deformation, naturally appears in many fields, especially in geometry. One of first applications of homotopy theory is the study of manifolds, and related geometric objects. Homotopy theory, then, grew up as an independent field of "theory of continuous deformations". The modern homotopy theory does not look like a kind of geometry.

The aim of this book is to give an introduction to this highly abstract modern homotopy theory by showing how fundamental concepts in homotopy theory arise from geometric problems. We choose fiber bundles as a main example.

This book consists of six chapters and an appendix:

- Chapter 1: How to Bundle Fibers.
  The aim of this chapter is to give an intuition of fiber bundles by simple examples before a precise definition is introduced.
- Chapter 2: Covering Spaces as a Toy Model.
  Basic properties of covering spaces are reviewed as a toy model of relations between fiber bundles and homotopy studied in Chapter 4.

- Chapter 3: Basic Properties of Fiber Bundles.
  We begin with one of the most primitive forms of fiber bundles by using only local trivializations and then introduce structure groups by using coordinate transformations, including principal bundles.
- Chapter 4: Classification of Fiber Bundles.
  The aim of this chapter, and the first half of this book, is to show that, for a "good group" G and a "good space" X, the set $P_G(X)$ of isomorphism classes of principal G-bundles over X can be identified with the homotopy set $[X, BG]$ from X to the classifying space BG of G.
  After preparing necessary tools such as the covering homotopy theorem, CW complexes, and homotopy groups, the classification theorem is proved under the hypothesis that a universal bundle exists. And then two kinds of constructions of universal bundles are explained.
  At the end of this chapter, we review properties of covering spaces from the viewpoint of fiber bundles.
- Chapter 5: Fibrations.
  We introduce the notion of fibration, more precisely, Hurewicz fibration and Serre fibration, as a generalization of fiber bundle, and then study their basic properties. We also introduce a further generalization, called quasifibration.
- Chapter 6: Postscript.
  The meaning and role of homotopy theory is explained from the author's personal viewpoint.
- Appendix A.
  The appendix consists of three sections, whose topics are the meaning of compact-open topology, vector bundles, and simplicial techniques, respectively.

Fiber bundles themselves have been one of most fundamental structures used in many kinds of geometry, especially in differential geometry. In this book, however, we will not discuss such applications, mainly because of the author's limited knowledge. The reader is encouraged to take a look at other books for geometric applications of fiber bundles.

Acknowledgments

I began to lecture fiber bundles in April, 1993. It was just two months after I finished my Ph.D. at the University of Rochester under the guidance of Professor Fred Cohen. It was my first experience to teach Japanese students in Japanese. I have taught calculus and linear algebra in the united states, but have never taught such advanced topics as fiber bundles. To make things worse, I found most students are juniors on the contrary to my plan to teach seniors and graduate students,

I am very grateful to the students who attended and listened to me in the class. I was really encouraged to see them listening to me and taking notes.

At the end of my 1993 class I distributed the first version of this note. In 1996, I had the second chance to teach fiber bundles. I distributed a revised version at the end of the class again. I was happy that several students found mistakes and typos in the notes. Juno Mukai was generous to offer me his grant to print this note in 2001. I used the printed version as a textbook for seniors reading course in 2001, during which lots of mistakes and typos were found. Their work made this note accurate enough to be publicized. I owe a lot to the members, Aoki, Ohtou, Shimada, Tsuruta, and Shiina.

I realized the importance of homotopy theoretic properties of fiber bundles when I was a graduate students at the University of Rochester. There was a good atmosphere of discussing various topics in topology among graduate students. My viewpoint on mathematics was formed during the exchanges with the algebraic topologists at the University of Rochester including graduate students. I would like to thank them and professor Akira Kono at Kyoto University who helped me to go to Rochester.

I became familiar with fibrations and quasifibration, together with classifying spaces, during my struggle to understand the paper [Rothenberg and Steenrod (1965)] and May's book [May (1972)] when I was a student at Kyoto University with lots of help from Professor Kouyemon Iriye, Professor Goro Nishida, and Professor Hiroshi Toda. I also learned practical uses of fibrations from Professor Fred Cohen at the University of Rochester. I am deeply grateful to them.

Professor Takao Sato carefully read through a Japanese version and pointed out lots of typos and mistakes. I am also grateful to him.

Contents

Preface   v

Acknowledgments   vii

List of Figures   xi

1. How to Bundle Fibers   1

    1.1   Simple Examples   1
    1.2   Tangent Bundles   4

2. Covering Spaces as a Toy Model   7

    2.1   Covering Spaces   7
    2.2   Paths and Their Lifts   10
    2.3   The Fundamental Group   18
    2.4   Universal Covering   27

3. Basic Properties of Fiber Bundles   33

    3.1   Defining Fiber Bundles   33
    3.2   Fiber Bundles with Structure Groups   38
    3.3   Topological Groups   44
    3.4   Compact-Open Topology   50
    3.5   Fiber Bundles and Group Action   57
    3.6   Quotient Spaces by Group Actions   67
    3.7   Principal Bundles   86

4. Classification of Fiber Bundles   99

    4.1   Maps between Fiber Bundles   99

4.2   Pullbacks   108
4.3   Fiber Bundles and Homotopy   114
4.4   Classification of Fiber Bundles: Simple Cases   126
4.5   Classifying Fiber Bundles over CW Complexes   132
4.6   CW Complexes and Homotopy   141
4.7   The First Half of the Proof of the Classification Theorem   155
4.8   Fiber Bundles and Homotopy Groups   160
4.9   Construction of Universal Bundles: Steenrod's Approach .   168
4.10 Construction of Universal Bundles: The Bar Construction   177
4.11 Covering Spaces Revisited   191

5. Fibrations   199

5.1   Why Further Generalizations of Fiber Bundles?   199
5.2   Serre Fibrations and Hurewicz Fibrations   200
5.3   Loop Spaces   205
5.4   Comparing Fiber Bundles and Fibrations   213
5.5   Deforming Continuous Maps into Fibrations   224
5.6   Homotopy Fiber Sequences   230
5.7   Iterated Loop Spaces   236
5.8   Fibrations and Homotopy Groups   243
5.9   Cofibrations   255
5.10 Duality between Fibrations and Cofibrations   259
5.11 Quasifibrations   273

6. Postscript   281

6.1   What is Homotopy Theory?   281
6.2   Many Kinds of Homotopies   284
6.3   Framework of Homotopy Theory   285

Appendix A   Related Topics   291

A.1   The Meaning of Compact-Open Topology   291
A.2   Vector Bundles   295
A.3   Simplicial Techniques   304

Bibliography   313

Index   321

List of Figures

1.1   cylinder E 2
1.2   annulus A 2
1.3   turned annulus 3
1.4   torus T^2 3
1.5   Möbius band 4
1.6   vector fields on S^2 5

2.1   A local structure of covering space 8
2.2   an open covering of S^1 9
2.3   a lift of a path 11
2.4   a homotopy ℓ_t from ℓ_0 to ℓ_1 15
2.5   a homotopy that shifts partitions 23

3.1   an open covering of S^1 35
3.2   local patching 39
3.3   local trivializations of the Möbius band 43
3.4   coordinate transformations of the Möbius band 43
3.5   regular space 54
3.6   reflection with respect to a line 60
3.7   dihedral group 61
3.8   the action of C_2 on S^1 69
3.9   rotation by angle θ 73
3.10 twist and fold 87
3.11 the boundary of the Möbius band 88
3.12 attaching segments to Figure 3.11 88
3.13 splitting of a circle 95
3.14 splitting of an annulus 96

3.15 Möbius bundle as an associated bundle 96
3.16 the cross section to the center of the Möbius band 97

4.1 M_2 105
4.2 cutting M_2 and M_0 106
4.3 the graph of the map $u : X \to [0, 1]$ 118
4.4 a partition of $X \times [0, 1]$ 121
4.5 contractible space 127
4.6 cone and hemisphere 131
4.7 a cell decomposition of S^1 134
4.8 a cell decomposition of S^2 134
4.9 S^1 with a vertical line 135
4.10 the minimal cell decomposition of S^n 136
4.11 attaching a cell 137
4.12 cone and disk are homeomorphic 144
4.13 wedge sum 146
4.14 collapsing the equator 148
4.15 folding map 148
4.16 $f + g$ 151
4.17 $g + f$ 152
4.18 a homotopy between $f + g$ and $g + f$ 152
4.19 a lift of loop 167
4.20 join 178

5.1 a fibration which is not a fiber bundle 201
5.2 concatenation of paths 203
5.3 difference of fibers 205
5.4 continuity of loop product 207
5.5 the fibration in Example 5.2.6 216
5.6 a homeomorphism of pairs 221
5.7 $S^1 \wedge S^1$ 237
5.8 torus from square 238
5.9 concatenation of n-fold loops 242
5.10 mapping cylinder 263
5.11 mapping cone 265
5.12 base points and i_f and j_f 265
5.13 a quasifibration which is not a fibration 275
5.14 quasifibration cannot be restricted 276

Chapter 1

How to Bundle Fibers

The concept of fiber bundles is one of the most important tools in modern geometry. Especially, in differential geometry, geometric structures on manifolds are usually described in terms of fiber bundles. For example, vector fields, differential forms, connections, and curvatures are defined as sections[1] of certain fiber bundles.

Unfortunately, however, the definition of fiber bundles looks complicated at first. It will not be easy to understand its geometric meanings for novices. Let us begin with simple examples to build geometric intuitions behind fiber bundles.

1.1 Simple Examples

Literally a fiber bundle is a bundle of fibers. The reader might imagine a collection of thin strings, but we do not care the length and thickness of strings in topology. Roughly speaking, (the *total space* of) a fiber bundle is a space obtained by bundling a certain space F (*fiber*) along another space B (*base space*). It would make our life complicated, however, if we allow sloppy bindings. We want them neatly packed into a bundle.

Example 1.1.1. Let E be a *cylinder* as depicted in Figure 1.1. If the height is 2 and the radius of the base disk is 1, it can be described as $E = D^2 \times [-1, 1]$ by using the closed interval $[-1, 1]$ and a unit disk $D^2 = \{(x, y) \in \mathbb{R}^2 \mid x^2 + y^2 \le 1\}$.

We may decompose E as

$$E = \bigcup_{x \in D^2} \{x\} \times [-1, 1],$$

---

[1] Definition 3.6.47.

1

which means that E is obtained by putting copies of $[-1, 1]$ on D^2 vertically. We regard E as a *bundle* of $[-1, 1]$ along D^2 and the closed intervals are called *fibers*. The dotted line in Figure 1.1 is one of fibers.

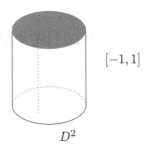

$[-1, 1]$

D^2

Fig. 1.1   cylinder E

□

Example 1.1.2. Let A be the side of the cylinder in Figure 1.1, namely an *annulus* (Figure 1.2).

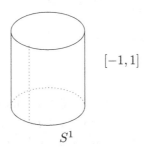

$[-1, 1]$

S^1

Fig. 1.2   annulus A

In other words, A is obtained from E by removing the inside. This annulus can be also obtained by bundling $[-1, 1]$, but in this case, along the unit circle

$$S^1 = \left\{ (x, y) \in \mathbb{R}^2 \,\middle|\, x^2 + y^2 = 1 \right\}.$$

The dotted line is one of fibers.

If we turn it, A can be regarded as a bundle of S^1 along $[-1, 1]$.

The dotted circle in Figure 1.3 is one of fibers. □

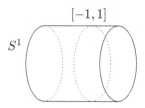

Fig. 1.3   turned annulus

Example 1.1.3. Let T^2 a 2-dimensional *torus*. Intuitively it is the surface of a doughnut as is shown in Figure 1.4.

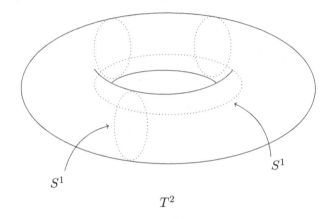

Fig. 1.4   torus T^2

More precisely, it is a subspace of $\mathbb{R}^3$ given by

$$T^2 = \{((2 + \cos\theta)\cos\varphi, (2 + \cos\theta)\sin\varphi, \sin\theta) \mid 0 \leq \theta \leq 2\pi, 0 \leq \varphi \leq 2\pi\}.$$

Any "vertical cut" of T^2 give us a circle S^1 as a section, as is indicated by the three vertical circles in Figure 1.4. The reader can also find an S^1 sitting "horizontally" in T^2. Thus T^2 can be obtained by bundling vertical circles S^1 along a horizontal circle S^1. In fact we have a homeomorphism

$$T^2 \cong S^1 \times S^1.$$

$\square$

Exercise 1.1.4. Find an explicit homeomorphism $T^2 \cong S^1 \times S^1$.

Example 1.1.5. The *Möbius band* M is the space obtained by gluing the left edge and the right edge of a rectangle after a twist of 180 degree. More precisely, it is a subspace of $\mathbb{R}^2$ given by

$$M = \left\{ \left((2 + t\cos\tfrac{\varphi}{2})\cos\varphi, (2 + t\cos\tfrac{\varphi}{2})\sin\varphi, t\sin\tfrac{\varphi}{2} \right) \,\middle|\, -\tfrac{1}{2} \le t \le \tfrac{1}{2}, 0 \le \varphi \le 2\pi \right\}.$$

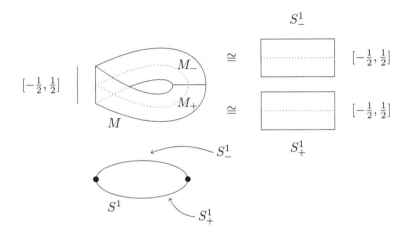

Fig. 1.5 Möbius band

A careful look tells us that, M is obtained by bundling the interval $[-\tfrac{1}{2}, \tfrac{1}{2}]$ along the unit circle S^1. However, it is not the direct product of two spaces because of a twist. Now cut S^1 into halves S^1_+ and S^1_- as is shown in Figure 1.5 and cut the Möbius band M accordingly into two parts M_+ and M_-. Then M_+ and M_- are homeomorphic to $S^1_+ \times [-\tfrac{1}{2}, \tfrac{1}{2}]$ and $S^1_- \times [-\tfrac{1}{2}, \tfrac{1}{2}]$, respectively. In other words, the bundling in M becomes a direct product if it is restricted to appropriate subspaces. $\square$

1.2 Tangent Bundles

As is advertised at the beginning of this chapter, fiber bundles help us to describe geometric structures in sophisticated ways. As one of examples, let us take a look at vector fields on the 2-dimensional sphere

$$S^2 = \left\{ (x_0, x_1, x_2) \in \mathbb{R}^3 \,\middle|\, x_0^2 + x_1^2 + x_2^2 = 1 \right\}.$$

A vector field v on S^2 is a rule which assigns each point $x \in S^2$ a tangent vector $v(x)$ at x

$$x \longmapsto v(x) \in T_x S^2. \qquad (1.1)$$

Here $T_x S^2$ is the *tangent space* of S^2 at x, which is defined by

$$T_x S^2 = \left\{ \xi \in \mathbb{R}^3 \,\middle|\, x \perp \xi \right\}.$$

In this definition, $T_x S^2$ does not look tangent to S^2, but the parallel transport of it by x is tangent to S^2 at x. This definition of $T_x S^2$ is more convenient than the actual tangent space since it is a vector space.

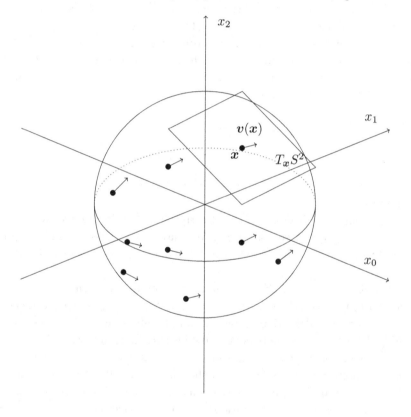

Fig. 1.6   vector fields on S^2

In Figure 1.6, the parallel transport of $T_x S^2$ by x is labelled by $T_x S^2$. Then a vector field on S^2 can be depicted by drawing a vector lying in this tangent plane at each point of S^2.

In order to use the language of set theory, which is the foundation of modern mathematics, the description of a vector field as an assignment (1.1) is not appropriate. The range of v varies when x changes. This can be resolved by collecting all tangent spaces into a single space

$$TS^2 = \coprod_{x \in S^2} T_x S^2, \tag{1.2}$$

where $\coprod$ is the disjoint union. It is topologized as a subspace of $\mathbb{R}^3 \times \mathbb{R}^3$ under the identification

$$TS^2 = \left\{ (x, \xi) \in \mathbb{R}^3 \times \mathbb{R}^3 \mid x \in S^2, x \perp \xi \right\}.$$

With this space TS^2, a vector field on S^2 can be defined as a map $v : S^2 \to TS^2$ satisfying the condition that $v(x) \in T_x S^2$ for each $x \in S^2$. This condition can be simplified by introducing a map $p : TS^2 \to S^2$ which maps $T_x S^2$ to a single point $\{x\}$. Then the condition $v(x) \in T_x S^2$ is equivalent to $p(v(x)) = x$. In other words, a vector field on S^2 is a map $v : S^2 \to TS^2$ making the diagram

commutative.

An important advantage of this description is that it allows us to import properties of maps to vector spaces. We may now discuss continuities and smoothness of vector fields, which was difficult in the previous definition (1.1).

By definition (1.2), TS^2 is obtained by "bundling" tangent spaces $T_x S^2$. The pair of this space TS^2 and the map $p : TS^2 \to S^2$ is called the *tangent bundle* of S^2. The definition of tangent bundle can be generalized to smooth manifolds and plays a fundamental role in modern geometry.

As is mentioned above, each tangent space is a vector space which defines a structure of *vector bundle* on the tangent bundle of a smooth manifold. Vector bundles can be regarded as a generalization of vector spaces in the sense that many of operations on vectors spaces can be extended to vector bundles. For example, the *cotangent bundle* is defined by taking the dual vector space of each fiber. The tensor product and the exterior power can be also defined, with which important concepts in differential geometry such as differential forms, connections, and curvatures are defined without using local coordinates.

Chapter 2

Covering Spaces as a Toy Model

In Chapter 1, the idea of bundling fibers is described by using examples. Starting with very simple examples in §1.1, a more practical example of tangent bundles is discussed in §1.2, although we have not given a general definition of tangent bundles.

Although fiber bundles are defined as "bundles of fibers", this is not a primary feature of fiber bundles. As we will see in Chapter 3, it is the relation with the notion of homotopy which makes fiber bundles useful. Furthermore examples in Chapter 1 did not give us an idea of how well fibers should be bundled.

Fortunately covering spaces provides us with a special class of fiber bundles having all important structures in simpler forms. In this chapter, we review the theory of covering spaces as a toy model of fiber bundles.

2.1 Covering Spaces

Let us begin with the definition of covering spaces.

Definition 2.1.1. Let B be a connected topological space. A *covering space* over B is a continuous map $p : E \to B$ such that each point $x \in B$ and each point $y \in p^{-1}(x)$ are equipped with open neighborhoods U_x and $\widetilde{U}_y$, respectively, satisfying the following conditions:

(1) if $y \neq y'$, $\widetilde{U}_y \cap \widetilde{U}_{y'} = \emptyset$, and
(2) $p(\widetilde{U}_y) \subset U_x$ and the restriction $p|_{\widetilde{U}_x} : \widetilde{U}_y \to U_x$ is a homeomorphism.

The spaces B and E are called the *base space* and the *total space*, respectively. The map p is called the *projection*, and $p^{-1}(x)$ is called the *fiber* over x.

Remark 2.1.2. It is common to assume that the total space E is connected in the theory of covering spaces. This condition is, however, not essential from the viewpoint of fiber bundles.

With the notion of disjoint union, we may write

$$p^{-1}(U_x) = \coprod_{y \in p^{-1}(x)} \widetilde{U}_y.$$

The condition (2) allows us to identify each $\widetilde{U}_y$ with $U_x \cong U_x \times \{y\}$ and we obtain an identification

$$p^{-1}(U_x) = \coprod_{y \in p^{-1}(x)} \widetilde{U}_y \cong \coprod_{y \in p^{-1}(x)} U_x \times \{y\} = U_x \times p^{-1}(x).$$

Thus $p^{-1}(U_x)$ consists of copies of U_x stacked as dishes sitting over U_x as is shown in Figure 2.1 when $p^{-1}(x) = \{y_1, y_2, y_3, y_4\}$.

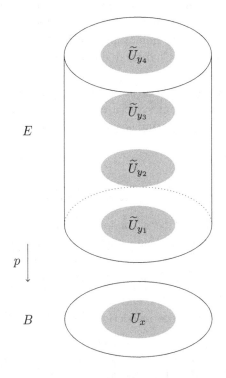

Fig. 2.1   A local structure of covering space

On the other hand, from the viewpoint of Chapter 1, we may regard $p^{-1}(U_x)$ as a bundle of $p^{-1}(x)$ along U_x. Thus the simplest example of covering spaces is the following.

Example 2.1.3. Equip a set F with a discrete topology and regard it as a topological space. Given a topological space B, the projection onto the first factor $\mathrm{pr}_1 : B \times F \to B$ is a covering space over B with fiber F. This is called a *trivial covering space*. □

One of the simplest nontrivial covering spaces is the following.

Example 2.1.4. In §1.1, the unit circle S^1 was defined as a subspace of $\mathbb{R}^2$. Here we regard it as a subspace of $\mathbb{C}$ by $S^1 = \{z \in \mathbb{C} \,|\, |z| = 1\}$. Define a map $p_2 : S^1 \to S^1$ by $p_2(z) = z^2$. Let us verify that this is a covering space.

For $w = w_1 + w_2 i \in S^1$, define

$$
U_w = \begin{cases}
\{x + yi \in S^1 \,|\, y > 0\}, & w_2 > 0 \\
\{x + yi \in S^1 \,|\, y < 0\}, & w_2 < 0 \\
\{x + yi \in S^1 \,|\, x > 0\}, & w = 1 \\
\{x + yi \in S^1 \,|\, x < 0\}, & w = -1.
\end{cases}
$$

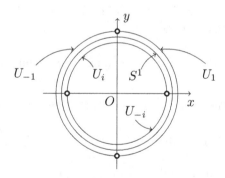

Fig. 2.2   an open covering of S^1

Let us consider the case of $w_2 > 0$. Other cases are left to the reader. By the definition of p_2,

$$
\begin{aligned}
p_2^{-1}(U_w) &= \{z \in S^1 \,|\, z^2 \in U_w\} \\
&= \{e^{i\theta} \,|\, 0 \le \theta < 2\pi, \ \sin(2\theta) > 0\} \\
&= \{e^{i\theta} \,|\, 0 < \theta < \tfrac{\pi}{2}\} \cup \{e^{i\theta} \,|\, \tfrac{3\pi}{2} < \theta < 2\pi\}.
\end{aligned}
$$

Let $w = e^{i\xi}$, then

$$p_2^{-1}(w) = \left\{ e^{\frac{i\xi}{2}}, e^{\frac{i(\xi+2\pi)}{2}} \right\} = \left\{ e^{\frac{i\xi}{2}}, -e^{\frac{i\xi}{2}} \right\}.$$

Define

$$\widetilde{U}_{w, e^{\frac{i\xi}{2}}} = \left\{ e^{i\theta} \mid 0 < \theta < \frac{\pi}{2} \right\}$$

$$\widetilde{U}_{w, -e^{\frac{i\xi}{2}}} = \left\{ e^{i\theta} \mid \frac{3\pi}{2} < \theta < 2\pi \right\},$$

then each of them is an open neighborhood of $e^{\frac{i\xi}{2}}$ and $-e^{\frac{i\xi}{2}}$, respectively. And we see that the conditions for covering space are satisfied. □

This example suggests that we may reduce the number of open neighborhoods U_x. For example, when $x' \in U_x$, we may take $U_{x'} = U_x$.

Proposition 2.1.5. *Let B be a connected space. A continuous map $p : E \to B$ is a covering space over B if and only if there exist an open covering $\mathcal{U} = \{U_\alpha\}_{\alpha \in A}$ of B, and an open neighborhood $\widetilde{U}_{\alpha,y}$ of $y \in p^{-1}(x)$ for each $\alpha \in A$ and $x \in U_\alpha$, satisfying the following conditions:*

(1) $\widetilde{U}_{\alpha,y} \cap \widetilde{U}_{\alpha,y'} = \emptyset$ if $y \neq y'$, and
(2) $p(\widetilde{U}_{\alpha,y}) \subset U_\alpha$ and the restriction $p|_{\widetilde{U}_\alpha} : \widetilde{U}_{\alpha,y} \to U_\alpha$ is a homeomorphism.

Exercise 2.1.6. The number 2 of p_2 in Example 2.1.4 can be replaced by any positive integer n. Define $p_n : S^1 \to S^1$ by $p_n(z) = z^n$. Show that this is a covering space.

Furthermore show that the map $\exp : \mathbb{R} \to S^1$ defined by $\exp(t) = e^{2\pi i t}$ is a covering space.

Note that this map $\exp$ can be regarded as the "limit" of p_n as $n \to \infty$.

2.2 Paths and Their Lifts

In §2.1, we have seen examples of covering spaces over S^1, first in Example 2.1.4 and then in Exercise 2.1.6. Are there any other covering spaces over S^1? How can we classify covering spaces over a given space?

Surprisingly, paths on the base space play an important role to answer these questions.

Definition 2.2.1. A *path* on a topological space X is a continuous map $\ell : [a, b] \to X$. The points $\ell(a)$ and $\ell(b)$ are called the *initial point* and the *end point*, respectively. Given a point x_0 in X, a path whose initial and end points are x_0 is called a *loop* on X with *base point* x_0.

Remark 2.2.2. The domain of a path is usually taken to be $[0,1]$, in particular, in the definition of the fundamental group defined in §2.3. It is more convenient, however, to allow arbitrary closed intervals as domains when we concatenate and cut paths.

One of the most important properties of covering spaces is the existence of a lift of a path as is depicted in Figure 2.3. A precise statement is as follows.

Theorem 2.2.3 (Existence of Lift of Path). *Let $p : E \to B$ be a covering space. For a path $\ell : [a, b] \to B$ on B, take a point $e \in E$ with $p(e) = \ell(a)$. Then there exists a unique path $\tilde{\ell}$ on E such that $\tilde{\ell}(a) = e$ and $p \circ \tilde{\ell} = \ell$.*

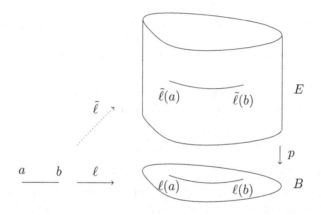

Fig. 2.3   a lift of a path

Remark 2.2.4. The existence of a lift of a path can be stated by using a diagram

Theorem 2.2.3 says that, when the outer square is commutative, there exists a unique map $\tilde{\ell}$ which makes the two triangles commutative, where e is the map which send a to e.

This can be regarded as a special case of the covering homotopy theorem for fiber bundles (Theorem 4.3.16), as we will see in §4.3.

Definition 2.2.5. The path $\tilde{\ell}$ in Theorem 2.2.3 is called a *lift* of ℓ.

The uniqueness in Theorem 2.2.3 hold for continuous maps in general.

Proposition 2.2.6. *Let $p : E \to B$ be a covering space and $f : X \to B$ a continuous map. Suppose that there exist maps $\tilde{f}_1, \tilde{f}_2 : X \to E$ satisfying $p \circ \tilde{f}_1 = p \circ \tilde{f}_2$. If X is connected and there exists a point $x_0 \in X$ with $\tilde{f}_1(x_0) = \tilde{f}_2(x_0)$, then $\tilde{f}_1 = \tilde{f}_2$.*

Proof. Define

$$X_0 = \left\{ x \in X \mid \tilde{f}_1(x) = \tilde{f}_2(x) \right\}.$$

By assumption $X_0 \neq \emptyset$. If X_0 is shown to be open and closed, the connectivity of X implies that $X_0 = X$ and we are done.

In order to show that X_0 is open, take a point $x \in X_0$. We need to find an open set V of X such that $x \subset V \subset X_0$.

Let $\{U_\alpha\}_{\alpha \in A}$ be an open covering of B satisfying the requirements of covering space. Suppose $f(x_0)$ is contained in U_α. We have a decomposition

$$p^{-1}(U_\alpha) = \coprod_{e \in p^{-1}(x_0)} \tilde{U}_{\alpha,e}$$

and the restriction $p|_{\tilde{U}_{\alpha,e}} : \tilde{U}_{\alpha,e} \to U_\alpha$ is a homeomorphism for each e.

Let $e_0 = \tilde{f}_1(x_0) = \tilde{f}_2(x_0)$. By the continuity of $\tilde{f}_1$ and $\tilde{f}_2$, there exists an open neighborhood V of x_0 such that $\tilde{f}_1(V) \subset \tilde{U}_{\alpha,e_0}$ and $\tilde{f}_2(V) \subset \tilde{U}_{\alpha,e_0}$. Since $p \circ \tilde{f}_1 = p \circ \tilde{f}_2$, we have $(p|_{\tilde{U}_{\alpha,e_0}}) \circ (\tilde{f}_1|_V) = (p|_{\tilde{U}_{\alpha,e}}) \circ (\tilde{f}_2|_V)$, from which we obtain $\tilde{f}_1|_V = \tilde{f}_2|_V$, since $p|_{\tilde{U}_{\alpha,e_0}}$ is a homeomorphism. This implies that $V \subset X_0$ and thus X_0 is open.

Let us show that X_0 is also closed. We show that the complement $X \setminus X_0$ is open. For a point $x \in X \setminus X_0$, suppose $f(x) \in U_\beta$ for $\beta \in A$. Then $p^{-1}(U_\beta)$ can be written as

$$p^{-1}(U_\beta) = \coprod_{e \in p^{-1}(x)} \tilde{U}_{\beta,e}$$

and the restriction $p|_{\tilde{U}_{\beta,e}} : \tilde{U}_{\beta,e} \to U_\beta$ is a homeomorphism. Since $x \notin X_0$, we have $\tilde{f}_1(x) \neq \tilde{f}_2(x)$ and $\tilde{U}_{\beta,\tilde{f}_1(x)} \cap \tilde{U}_{\beta,\tilde{f}_2(x)} = \emptyset$. By the continuity of these maps, there exists an open neighborhood W of x such that $\tilde{f}_1(W) \subset \tilde{U}_{\beta,\tilde{f}_1(x)}$ and $\tilde{f}_2(W) \subset \tilde{U}_{\beta,\tilde{f}_2(x)}$. These facts imply that $\tilde{f}_1(W) \cap \tilde{f}_2(W) = \emptyset$ and we obtain $W \subset X \setminus X_0$, which means that $X \setminus X_0$ is open. $\square$

In order to prove Theorem 2.2.3, an obvious idea is to construct a lift. In fact, this is what we are going to do. We cut a given path into small pieces and then construct lifts of small paths from the initial point. The problem is how fine should we cut the path. The following property of compact metric spaces is useful for this purpose.

Definition 2.2.7. Let A be a subset of a metric space (X, d). If $\sup \{d(a, a') \mid a, a' \in A\}$ exists, the value is denoted by $d(A)$ and is called the *diameter* of A.

Lemma 2.2.8 (Lebesgue's Lemma). *For an open covering $\{U_\lambda\}_{\lambda \in \Lambda}$ of a compact metric space (X, d), there exists a positive number $\sigma > 0$ with the following property: for a subset $A \subset X$ with $d(A) < \sigma$, there exists $\lambda \in \Lambda$ such that $A \subset U_\lambda$.*

The number σ is called the *Lebesgue number* of $\{U_\lambda\}_{\lambda \in \Lambda}$. The reader should be able to find a proof in standard textbooks on metric spaces.

Corollary 2.2.9. *For any open covering $\{V_\alpha\}_{\alpha \in A}$ of a closed interval $[a, b]$, there exist two decompositions of $[a, b]$*

$$a = s_0 < s_1 < \cdots < s_{m-1} < s_m < b$$

$$a < t_0 < t_1 < \cdots < t_{m-1} < t_m = b$$

satisfying the following conditions:

(1) For each $1 \leq i \leq m - 1$, $s_i < t_{i-1} < s_{i+1} < t_i$.

(2) Let $W_0 = [s_0, t_0)$, $W_1 = (s_1, t_1), \ldots, W_{m-1} = (s_{m-1}, t_{m-1})$, and $W_m = (s_m, t_m]$. Then, for each i, there exists $\alpha \in A$ such that $W_i \subset V_\alpha$.

Proof. Since $[a, b]$ is a compact metric space and $\{V_\alpha\}_{\alpha \in A}$ is its open covering, Lemma 2.2.8 applies. Let $\sigma > 0$ be a Lebesgue number. Let $m \in \mathbb{N}$ be an integer satisfying $\frac{3(b-a)}{2m+1} < \sigma$. For each $0 < i \leq m$ and $0 \leq j < m$, define $s_i = a + \frac{2i-1}{2m+1}(b-a)$ and $t_j = a + \frac{2j+2}{2m+1}(b-a)$. The diameter $d(W_i)$ of each interval W_i is either $\frac{2(b-a)}{2m+1}$ or $\frac{3(b-a)}{2m+1}$. By the definition of Lebesgue number there exists $\alpha \in A$ with $W_i \subset V_\alpha$. $\square$

Remark 2.2.10. The above conditions for partitioning is designed to make the small intervals to have the following properties;

(1) $\bigcup_i W_i = [0, 1]$.

(2) $W_i \cap W_{i+1} \neq \emptyset$.

(3) $W_i \cap W_{i+2} = \emptyset$.

Proof of Theorem 2.2.3. In order define a lift $\tilde{\ell}$, we decompose the interval $[a, b]$ into small pieces. Let $\{U_\alpha\}_{\alpha \in A}$ be an open covering of B satisfying the conditions of covering space. By the continuity of the path ℓ, the covering $[a, b] = \bigcup_{\alpha \in A} \ell^{-1}(U_\alpha)$ is an open covering of $[a, b]$. Let $W_0, \dots, W_m$ be an open covering of $[a, b]$ obtained by Corollary 2.2.9. For each i, choose $\alpha_i \in A$ satisfying $W_i \subset U_{\alpha_i}$. We first construct a lift on W_0 and then extended it to $W_0 \cup \cdots \cup W_i$ by induction on i.

By the definition of covering space, we obtain

$$p^{-1}(U_{\alpha_0}) = \coprod_{y \in p^{-1}(\ell(a))} \tilde{U}_{\alpha_0, y}.$$

Since $e \in p^{-1}(\ell(a))$, the restriction of p $p|_{\tilde{U}_{\alpha_0, e}} : \tilde{U}_{\alpha_0, e} \to U_{\alpha_0}$ is a homeomorphism. Define $\tilde{\ell}_0 = (p|_{\tilde{U}_{\alpha_1, e}})^{-1} \circ (\ell|_{\overline{W_0}})$.

Suppose that we have constructed a lift $\tilde{\ell}_i$ of $\ell|_{W_i}$ such that $\tilde{\ell}_i|_{W_{i-1} \cap W_i} = \tilde{\ell}_{i-1}|_{W_{i-1} \cap W_i}$. Let y be the end-point of $\tilde{\ell}_i$. Since $p(y) \in \ell(W_i \cap W_{i+1})$, there exists a lift $\tilde{\ell}_{i+1}$ of $\ell|_{\overline{W_{i+1}}}$ starting from y.

We obtain a path $\tilde{\ell}$ we desired by concatenating the paths $\tilde{\ell}_0, \dots, \tilde{\ell}_m$. $\square$

Example 2.2.11. Let us denote the constant path at $x_0 \in B$ by c_{x_0}. In other words, $\ell(t) = x_0$ for $t \in [a, b]$. Choose $y_0 \in E$ with $p(y_0) = x_0$. Then c_{y_0} is a lift of c_{x_0} starting from y_0. By the uniqueness of lift, any lift of c_{x_0} is a constant path. $\square$

Example 2.2.12. Let $\ell_1 : [a, b] \to B$ and $\ell_2 : [b, c] \to B$ be paths with $\ell_1(b) = \ell_2(b)$. Define a path $\ell_1 \cup \ell_2 : [a, c] \to B$ by

$$(\ell_1 \cup \ell_2)(t) = \begin{cases} \ell_1(t), & a \leq t \leq b \\ \ell_2(t), & b \leq t \leq c. \end{cases}$$

Choose a point $e_1 \in E$ in the fiber $\ell_1(a)$. Let $\tilde{\ell}_1$ be a lift of ℓ_1 with e_1 the initial point. There exists a lift $\tilde{\ell}_2$ of ℓ_2 with $\tilde{\ell}_1(b)$ the initial point, since $\tilde{\ell}_1(b) \in p^{-1}(\ell_1(b)) = p^{-1}(\ell_2(b))$.

Then $\tilde{\ell}_1 \cup \tilde{\ell}_2 : [a, c] \to E$ is a lift of $\ell_1 \cup \ell_2$ starting from e_1. $\square$

We often need to perturb paths continuously.

Definition 2.2.13. Let $\ell_0, \ell_1 : [a, b] \to X$ be paths in a topological space X with $\ell_0(a) = \ell_1(a)$ and $\ell_0(b) = \ell_1(b)$. A *homotopy* from ℓ_0 to ℓ_1 is a continuous map

$$H : [a, b] \times [0, 1] \longrightarrow X$$

satisfying the following conditions:

(1) $H(s,0) = \ell_0(s)$ for all $s \in [a,b]$,
(2) $H(s,1) = \ell_1(s)$ for all $s \in [a,b]$,
(3) $H(a,t) = \ell_0(a) = \ell_1(a)$ for all $t \in [0,1]$, and
(4) $H(b,t) = \ell_0(b) = \ell_1(b)$ for all $t \in [0,1]$.

We say ℓ_0 and ℓ_1 are *homotopic* and denote $\ell_0 \simeq \ell_1$ when such a homotopy exists.

Remark 2.2.14. The latter two conditions in the above definition imply that end-points are fixed. In order to understand the meaning of the first two conditions, for $0 \le t \le 1$, define an intermediate path by $\ell_t(s) = H(s,t)$. When $t = 0$ and $t = 1$, it agrees with ℓ_0 and ℓ_1, respectively. See Figure 2.2.14. The continuity of $H(s,t)$ in t means that the path ℓ_0 is continuously deformed into the other path ℓ_1 by H.

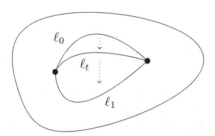

Fig. 2.4   a homotopy ℓ_t from ℓ_0 to ℓ_1

Definition 2.2.15. An arcwise connected space X is called *simply connected* if for any two points $x_0, x_1 \in X$ and two paths $\ell_0, \ell_1 : [0,1] \to X$ connecting these points, we always have $\ell_0 \simeq \ell_1$.

Remark 2.2.16. In order to define the homotopy relation among paths, they have to share the same domain. We should fix the domain of paths. A popular choice is $[0,1]$ as we have done above.

Example 2.2.17. The 2-dimensional disk D^2 is simply connected. In fact, for any paths $\ell_0, \ell_1 : [0,1] \to D^2$ with the same initial and end points, a homotopy from ℓ_0 to ℓ_1 can be defined by

$$H(s,t) = (1-t)\ell_0(s) + t\ell_1(s).$$

The same formula proves that $\mathbb{R}$ and $\mathbb{R}^n$ are simply connected. More generally any convex set in $\mathbb{R}^n$ is simply connected. $\qquad\square$

Example 2.2.18. The unit circle S^1 is arcwise connected but is not simply connected. For example, the paths

$$\ell_0, \ell_1 : [0, \pi] \longrightarrow S^1$$

defined by $\ell_0(t) = e^{it}$ and $\ell_1(t) = e^{-it}$ are not homotopic. This fact will be proved in Example 2.4.10. $\qquad\square$

Theorem 2.2.19. *Suppose that a covering space $p : E \to B$ over an arcwise connected space B, continuous paths $\ell_0, \ell_1 : [a, b] \to B$, and a homotopy $H : [a, b] \times [0, 1] \to B$ from ℓ_0 to ℓ_1 are given. Choose a lift $\tilde{\ell}_0 : [a, b] \to E$ of ℓ_0. Then there exists a unique homotopy $\tilde{H} : [a, b] \times [0, 1] \to E$ satisfying the following conditions:*

(1) $\tilde{H}(t, 0) = \tilde{\ell}_0(t)$
(2) $p \circ \tilde{H} = H$.

Furthermore, if $\ell_0(a) = \ell_1(a)$ and $\ell_0(b) = \ell_1(b)$ and if H is a homotopy which fixes initial and end points, $\tilde{H}(a, t)$ and $\tilde{H}(b, t)$ are independent of t.

Proof. The idea is basically the same as that of Theorem 2.2.3. Choose an open covering $\mathcal{U}$ which trivializes B. Then we obtain an open covering $\{H^{-1}(U)\}_{U \in \mathcal{U}}$ of $[a, b] \times [0, 1]$.

We would like to cut $[a, b] \times [0, 1]$ into small rectangles by cutting $[a, b]$ and $[0, 1]$ into small interval. Since $[a, b] \times [0, 1]$ is a compact metric space, Lebesgue's Lemma applies. Let σ be a Lebesgue number of this open covering. We cut $[a, b]$ and $[0, 1]$ so that the diagonals of small rectangles are less than σ.

A lift can be constructed by using local trivializations starting from $\tilde{H}|_{[a,b] \times \{0\}} = \tilde{\ell}_0$. Details are omitted. The uniqueness follows from Proposition 2.2.6.

When H fixes the initial and end points, if we regard $\tilde{H}(a, s)$ as a path, it is a lift of a constant path. By the uniqueness of lift, we see that $\tilde{H}(a, s)$ is a constant paths. The same argument applies to $\tilde{H}(b, s)$. $\qquad\square$

Lifts of paths and homotopies can be used to prove that, over an arcwise connected base space, all fibers of a covering space have the same cardinality.

The correspondences between fibers are given as follows.

Definition 2.2.20. Let $p : E \to B$ be a covering space. For a path ℓ connecting $x_0, x_1 \in B$, define a map $\varphi_\ell : p^{-1}(x_0) \to p^{-1}(x_1)$ as follows: for $y \in p^{-1}(x_0)$, let $\tilde{\ell}_y$ be a lift of ℓ with y the initial point. Since $\tilde{\ell}_y(1) \in p^{-1}(x_1)$, we may define $\varphi_\ell(y) = \tilde{\ell}_y(1)$.

This map has the following property.

Lemma 2.2.21. *Let $p : E \to B$ be a covering space. The map defined in Definition 2.2.20 has the following properties:*

(1) Let $\ell_0, \ell_1 : [a, b] \to B$ be two paths connecting $x_0, x_1 \in B$. If $\ell_0 \simeq \ell_1$, then $\varphi_{\ell_0} = \varphi_{\ell_1}$.

(2) If $\ell_1 : [a, b] \to B$ and $\ell_2 : [b, c] \to B$ are paths on B with $\ell_1(b) = \ell_2(b)$. Then we have $\varphi_{\ell_1 \cup \ell_2} = \varphi_{\ell_2} \circ \varphi_{\ell_1}$.

(3) We have $\varphi_{c_{x_0}} = 1_{p^{-1}(x_0)}$, where c_{x_0} is the constant path to $x_0 \in B$.

Proof. (1) Let H be a homotopy for $\ell_0 \simeq \ell_1$. For $y \in p^{-1}(x_0)$, let $\tilde{\ell}_0$ be a lift of ℓ_0 with y the initial point. Then by Theorem 2.2.19, there exists a unique lift $\tilde{H}$ of H satisfying $\tilde{H}|_{[0,1] \times \{0\}} = \tilde{\ell}_0$. Since $\tilde{H}(b, t)$ is independent of t, $\varphi_{\ell_0}(y) = \varphi_{\ell_1}(y)$.

(2) By Example 2.2.12.

(3) By Example 2.2.11. $\qquad\qquad\qquad\qquad\qquad\qquad\qquad\qquad\qquad\qquad$ □

When the domain of paths are fixed to $[0, 1]$, concatenations of paths are defined as follows.

Definition 2.2.22. Let $\ell_1, \ell_2 : [0, 1] \to X$ be paths on a topological space X with $\ell_1(1) = \ell_2(0)$. Define the *concatenation* $\ell_1 * \ell_2$ by

$$(\ell_1 * \ell_2)(t) = \begin{cases} \ell_1(2t), & 0 \le t \le \frac{1}{2} \\ \ell_2(2t - 1), & \frac{1}{2} \le t \le 1. \end{cases}$$

Corollary 2.2.23. *Let $p : E \to B$ be a covering space. For paths $\ell_1, \ell_2 : [0, 1] \to B$ on B with $\ell_1(1) = \ell_2(0)$, we have $\varphi_{\ell_1 * \ell_2} = \varphi_{\ell_2} \circ \varphi_{\ell_1}$.*

Proposition 2.2.24. *Let $p : E \to B$ be a covering space over an arcwise connected space B. Then for any $x_0, x_1 \in X$, there exists a bijection between $p^{-1}(x_0)$ and $p^{-1}(x_1)$.*

Proof. Since B is arcwise connected, there exists a path $\ell : [0, 1] \to B$ from x_0 to x_1, which gives us a map $\varphi_\ell : p^{-1}(x_0) \to p^{-1}(x_1)$. On the other hand, the path $\nu(\ell)$ defined by $\nu(\ell)(t) = \ell(1 - t)$ is a path from x_1 to x_0. And we have a map $\varphi_{\nu(\ell)} : p^{-1}(x_1) \to p^{-1}(x_0)$. It suffices to show that these maps φ_ℓ are $\varphi_{\nu(\ell)}$ inverse to each other.

By Corollary 2.2.23, we have $\varphi_{\nu(\ell)} \circ \varphi_\ell = \varphi_{\ell * \nu(\ell)}$. Define

$$H(s,t) = \begin{cases} \ell(2st), & 0 \le s \le \frac{1}{2} \\ \ell((2 - 2s)t), & \frac{1}{2} \le s \le 1. \end{cases}$$

Then this is a homotopy between $\ell * \nu(\ell)$ and the constant path $c_{\ell(0)}$ to $\ell(0)$. By the homotopy invariance in Lemma 2.2.21, we have $\varphi_{\ell * \nu(\ell)} = \varphi_{c_{\ell(0)}}$. Lemma 2.2.21 also tells us that $\varphi_{c_{\ell(0)}}$ is the identity. Similarly $\varphi_\ell \circ \varphi_{\nu(\ell)}$ can be shown to be the identity map. □

2.3   The Fundamental Group

We have seen in Proposition 2.2.24 that, when the base space is arcwise connected, a bijection between fibers can be constructed by using lifts of paths. Recall from §2.2 that a path $\ell : [0,1] \to X$ with $\ell(0) = \ell(1) = x_0$ is called a *loop* based at x_0. Given a loop on the base space, the map in Proposition 2.2.24 is a permutation of the fiber over the base point of the loop. Such permutations are closely related to the structure of the covering space.

Definition 2.3.1. Fix a point x_0 in a topological space X and define

$$\Omega(X, x_0) = \{\ell : [0,1] \to X \,|\, \text{continuous}, \ell(0) = \ell(1) = x_0\}.$$

This is called the *loop space* of X. When the base point is obvious from the context, it is simply denoted by ΩX.

Remark 2.3.2. The reader might wonder why $\Omega(X, x_0)$ is called "space". We will define a topology on this set and make it into a topological space in §5.2.

Definition 2.3.3. If a topological X has a particular point x_0 designated, the pair (X, x_0) is called a *based space* or *pointed space*. The point x_0 is called the *base point* of X.

Let ℓ be a loop on the base space B of a covering space $p : E \to B$ based at x_0. By Proposition 2.2.24, it induces a bijection $\varphi_\ell : p^{-1}(x_0) \to p^{-1}(x_0)$. Let us take a look at a couple of simple examples.

Example 2.3.4. Recall the covering space $p_2 : S^1 \to S^1$ in Example 2.1.4. Let $\ell(t) = e^{2\pi t i}$. This is a loop based at $x_0 = 1$. We have $p_2^{-1}(x_0) = \{1, -1\}$.

In order to find the map $\varphi_\ell : p_2^{-1}(x_0) \to p_2^{-1}(x_0)$, we need to understand lifts of ℓ. The lifts with initial points 1 and -1 are given by $\tilde{\ell}_1(t) = e^{\pi t i}$ and $\tilde{\ell}_{-1}(t) = e^{\pi(t+1)i}$, respectively. And we have

$$\varphi_\ell(x) = \begin{cases} e^{\pi i} = -1, & x = 1 \\ e^{2\pi i} = 1, & x = -1. \end{cases}$$

In other words, the map φ_ℓ switches 1 and -1.

What about the covering space $p_n : S^1 \to S^1$ in Exercise 2.1.6. The fiber over $x_0 = 1$ is $p_n^{-1}(1) = \{1, \zeta_n, \zeta_n^2, \ldots, \zeta_n^{n-1}\}$, where $\zeta_n = e^{\frac{2\pi i}{n}}$. Let ℓ be the loop used in p_2. The lift $\tilde{\ell}_k$ whose initial point is ζ_n^k is given by $\tilde{\ell}_k(t) = e^{\frac{2\pi(t+k)i}{n}}$. And we have $\varphi_\ell(\zeta_n^k) = \zeta_n^{k+1}$. In other words, the map φ_ℓ is a shift of the sequence $1, \zeta_n, \zeta_n^2, \ldots, \zeta_n^{n-1}$ by one.

On the other hand, in the case of a trivial covering space $S^1 \times F \to S^1$, for any loop ℓ on the base space, φ_ℓ is the identity map. □

These examples suggest that we may tell if a covering space is the same as the trivial covering space by investigating the maps φ_ℓ. To be precise, we need to make the meaning of "same" clear.

Definition 2.3.5. Let $p : E \to B$ and $p' : E' \to B$ be covering spaces. A *map of covering spaces* is a continuous map $f : E \to E'$ which makes the diagram

$$\begin{array}{ccc} E & \xrightarrow{f} & E' \\ {\scriptstyle p}\downarrow & & \downarrow{\scriptstyle p'} \\ B & = & B \end{array}$$

commutative.

If the restriction $f|_{p^{-1}(x)} : p^{-1}(x) \to (p')^{-1}(x)$ is a bijection for all $x \in B$, f is said to be an *isomorphism* of covering spaces. When such a map f exists, these two covering spaces are said to be *isomorphic*. When $E = E'$, an isomorphism from E to E is called a *covering transformation*.

Remark 2.3.6. This relation "isomorphic" is an equivalence relation, which allows us to use this relation to classify covering spaces. The fact that this is an equivalence relation will be proved in a more general setting of fiber bundles as Proposition 4.1.8.

We have the following sufficient condition for a covering space to be trivial.

Theorem 2.3.7. *If B is simply connected and locally arcwise connected, then any covering space over B is isomorphic to the trivial covering space.*

Here we need a technical assumption "locally arcwise connected". Such local conditions often appear in the study of covering spaces. For the convenience of the reader, we summarize the definitions here.

Definition 2.3.8. Suppose that a topological space X has an open basis of the following kind:

(1) connected
(2) arcwise connected
(3) simply connected
(4) contractible.

We say that X is (*locally connected*) in the case of (1), *locally arcwise connected* in the case of (2), *locally simply connected* in the case of (3), and *locally contractible* in the case of (4).

Remark 2.3.9. See Definition 4.4.1 for contractibility.

Proof of Theorem 2.3.7. Suppose B is simply connected and $p : E \to B$ be a covering space. Choose and fix $x_0 \in B$ and define $F = p^{-1}(x_0)$.

We first need to define a map of covering spaces $f : E \to B \times F$. Since B is arcwise connected, for each $e \in E$, there exists a path $\ell : [0,1] \to B$ from $p(e)$ to x_0. By the correspondence in Definition 2.2.20, we obtain a map $\varphi_\ell : p^{-1}(p(e)) \to p^{-1}(x_0) = F$. Now define $f(e) = (p(e), \varphi_\ell(e))$. Since B is simply connected, this is independent of the choice of ℓ and f is well defined.

Let us show that this map is continuous. Note that F has the discrete topology as a subspace of E. It suffices to show that $f^{-1}(U \times \{y\})$ is open for any open subset U of B and $y \in F$. For $e \in f^{-1}(U \times \{y\})$, $p(e) \in U$. Since U is open, there exists an open neighborhood $V \subset U$ of $p(e)$ such that p is trivialized. Namely, we have a decomposition

$$p^{-1}(V) = \coprod_{z \in p^{-1}(p(e))} \widetilde{V}_z.$$

By the assumption of local arcwise-connectivity, we may assume that V is arcwise connected. Let us show that $f(\widetilde{V}_e) \subset V \times \{y\}$. For $z \in \widetilde{V}_e$, choose a path ℓ' which connects $p(z)$ and $p(e)$ in V. The path $\ell' * \ell$ is a path connecting $p(z)$ and x_0. Its lift $\widetilde{\ell' * \ell}$ with z the initial point is given by $\widetilde{\ell'} * \widetilde{\ell}$, where $\widetilde{\ell'}$ is the lift of ℓ' with z the initial point.

The end point of $\widetilde{\ell' * \ell}$ is y, and we have $f(z) \in V \times \{y\}$. Thus we see that $f(\widetilde{V}_e) \subset V \times \{y\} \subset U \times \{y\}$, and we have shown that $f^{-1}(U \times \{y\})$ is open.

The commutativity of the diagram

$$
\begin{array}{ccc}
E & \xrightarrow{\ f\ } & B \times F \\
\big\downarrow{\scriptstyle p} & & \big\downarrow{\scriptstyle \mathrm{pr}_1} \\
B & =\!=\!= & B
\end{array}
$$

follows from the definition. By Proposition 2.2.24, f is a bijection on each fiber and thus f is an isomorphism of covering spaces. $\qquad\square$

In order to classify covering spaces in general, we need to understand deeper structures of the loop space $\Omega(X, x_0)$. One of the most significant properties of loop spaces is the existence of products.

Definition 2.3.10. For loops $\ell_1, \ell_2 \in \Omega(X, x_0)$, the concatenation $\ell_1 * \ell_2$ again belongs to $\Omega(X, x_0)$ and we have a map

$$
\mu : \Omega(X, x_0) \times \Omega(X, x_0) \longrightarrow \Omega(X, x_0).
$$

This map is called the *loop product*.

The loop product does not make the loop space into a group, but something close to a group. Let us recall the definition of group.

Definition 2.3.11. A *group* consists of a set G, a special element $e \in G$, and maps

$$
\mu : G \times G \longrightarrow G
$$
$$
\nu : G \longrightarrow G
$$

satisfying the following conditions

(1) For any $g \in G$, we have $\mu(e, g) = \mu(g, e) = g$. In other words, the diagram

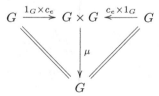

$$
G \xrightarrow{\ 1_G \times c_e\ } G \times G \xleftarrow{\ c_e \times 1_G\ } G
$$

is commutative. Here $c_e : G \to G$ is the constant map to e.

(2) For any triple $g, h, k \in G$, we have $\mu(\mu(g, h), k) = \mu(g, \mu(h, k))$. In other words, the diagram

$$
\begin{array}{ccc}
G \times G \times G & \xrightarrow{\ \mu \times 1_G\ } & G \times G \\
{\scriptstyle 1_G \times \mu}\big\downarrow & & \big\downarrow{\scriptstyle \mu} \\
G \times G & \xrightarrow{\quad \mu \quad} & G
\end{array}
$$

is commutative.

(3) For any $g \in G$, $\mu(\nu(g), g) = \mu(g, \nu(g)) = e$. In other words, the diagram

$$
\begin{array}{ccccc}
G \times G & \xleftarrow{\ 1_G \times \nu\ } & G \times G & \xrightarrow{\ \nu \times 1_G\ } & G \times G \\
{\scriptstyle \mu}\big\downarrow & & {\scriptstyle \Delta}\big\uparrow & & \big\downarrow{\scriptstyle \mu} \\
G & \xleftarrow{\quad c_e \quad} & G & \xrightarrow{\quad c_e \quad} & G
\end{array}
$$

is commutative, where Δ is the diagonal map.

The element e is called the *unit*, $\mu(g, h) = g \cdot h$ is called the *product* of g and h, and $\nu(g) = g^{-1}$ is called the *inverse* to g. Maps μ and ν are called the *structure map* of G.

The reader might not be familiar with the commutative diagrams in this definition but these are convenient ways to describe properties without taking elements. This point will be made clear when we define topological groups in Definition 3.3.1.

Unfortunately, the loop product in ΩX satisfies none of these three requirements in the definition of group. Let us take a look at the associativity.

For $\ell_1, \ell_2, \ell_3 \in \Omega X$, we have

$$
((\ell_1 * \ell_2) * \ell_3)(t) = \begin{cases} (\ell_1 * \ell_2)(2t) & 0 \le t \le \frac{1}{2} \\ \ell_3(2t - 1) & \frac{1}{2} \le t \le 1 \end{cases}
$$

$$
= \begin{cases} \ell_1(4t) & 0 \le t \le \frac{1}{4} \\ \ell_2(4t - 1) & \frac{1}{4} \le t \le \frac{1}{2} \\ \ell_3(2t - 1) & \frac{1}{2} \le t \le 1 \end{cases}
$$

$$
(\ell_1 * (\ell_2 * \ell_3))(t) = \begin{cases} \ell_1(2t) & 0 \le t \le \frac{1}{2} \\ (\ell_2 * \ell_3)(2t - 1) & \frac{1}{2} \le t \le 1 \end{cases}
$$

$$
= \begin{cases} \ell_1(2t) & 0 \le t \le \frac{1}{2} \\ \ell_2(4t - 2) & \frac{1}{2} \le t \le \frac{3}{4} \\ \ell_3(4t - 3) & \frac{3}{4} \le t \le 1 \end{cases}.
$$

And we see that $((\ell_1 * \ell_2) * \ell_3)$ and $(\ell_1 * (\ell_2 * \ell_3))$ are different. But the difference is the way $[0,1]$ is partitioned. It does not seem to be an essential difference. In fact, we can adjust the difference by using a homotopy.

Proposition 2.3.12. *For loops ℓ_1, ℓ_2, ℓ_3 in a based space (X, x_0), we have*

$$((\ell_1 * \ell_2) * \ell_3) \simeq (\ell_1 * (\ell_2 * \ell_3)).$$

Proof. A homotopy $H : [0,1] \times [0,1] \rightarrow X$ between $((\ell_1 * \ell_2) * \ell_3)$ and $(\ell_1 * (\ell_2 * \ell_3))$ can be defined as follows.

The idea is to deform the partition of $[0,1]$ with respect to $((\ell_1 * \ell_2) * \ell_3)$ to the partition of $[0,1]$ with respect to $(\ell_1 * (\ell_2 * \ell_3))$ as is drawn in 2.3.

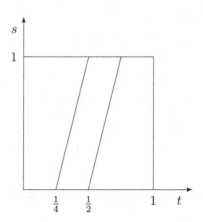

Fig. 2.5   a homotopy that shifts partitions

The parameter s is for homotopy and t is for loops. The left line is $s = 4t - 1$. The right line is obtained by shifting it by $\frac{1}{4}$ along the t-axis. A homotopy can be given by

$$H(t, s) = \begin{cases} \ell_1(\frac{4}{s+1}t), & 0 \leq t \leq \frac{s+1}{4} \\ \ell_2(4t - s - 1), & \frac{s+1}{4} \leq t \leq \frac{s+2}{4} \\ \ell_3(\frac{4}{2-s}(t - \frac{s+2}{4})), & \frac{s+2}{4} \leq t \leq 1 \end{cases}.$$

We need to verify that $H(-, s)$ is an element of ΩX for each s or $H(\partial[0,1], s) = x_0$. But it can be easily verified by the definition of H. $\square$

Let us consider the unit next. An obvious candidate for the unit is the constant path to x_0, i.e. c_{x_0}. However,

$$(\ell * c_{x_0})(t) = \begin{cases} \ell(2t), & 0 \le t \le \frac{1}{2} \\ x_0, & \frac{1}{2} \le t \le 1 \end{cases}$$

$$(c_{x_0} * \ell)(t) = \begin{cases} x_0, & 0 \le t \le \frac{1}{2} \\ \ell(2t-1), & \frac{1}{2} \le t \le 1 \end{cases}$$

and they are different from ℓ. But these are homotopic to ℓ.

Proposition 2.3.13. *For a loop in a based space* (X, x_0), *we have*
$$\ell * c_{x_0} \simeq \ell \simeq c_{x_0} * \ell.$$

Proof. Let us verify the left hand side. The path $\ell * c_{x_0}$ is obtained by squeezing ℓ into the first half of $[0, 1]$ and by using c_{x_0} in the latter half. We obtain a homotopy by changing the time interval for the constant path c_{x_0}. The definition of homotopy is easier then the case of Proposition 2.3.12 and is left to the reader as an exercise.   □

Exercise 2.3.14. Complete the proof of the above proposition by explicitly constructing a homotopy between $\ell * c_{x_0}$ and ℓ.

Finally let us consider inverses.

Definition 2.3.15. For $\ell \in \Omega(X, x_0)$, define $\nu(\ell) \in \Omega(X, x_0)$ by $(\nu(\ell))(t) = \ell(1-t)$.

This is a candidate for an inverse to ℓ. Again $\ell * \nu(\ell)$ does not agree with c_{x_0} but they are homotopic.

Proposition 2.3.16. *For* $\ell \in \Omega(X, x_0)$, *we have*
$$\ell * \nu(\ell) \simeq c_{x_0} \simeq \nu(\ell) * \ell.$$

Proof. We only prove the left hand side. For $\ell \in \Omega(X, x_0)$, we have
$$(\ell * \nu(\ell))(t) = \begin{cases} \ell(2t) & 0 \le t \le \frac{1}{2} \\ \ell(2-2t) & \frac{1}{2} \le t \le 1 \end{cases}.$$
This is a path in which a particle moves along ℓ in the first half of its time and then go back to the initial point along the same path in the reverse direction in the second half. In order to construct a homotopy between such a path and c_{x_0}, an obvious idea is to start going back in an early time.

A concrete definition is given by
$$H(t, s) = \begin{cases} \ell(2ts) & 0 \le t \le \frac{1}{2} \\ \ell((2-2t)s) & \frac{1}{2} \le t \le 1 \end{cases}.$$

This is a homotopy between $\ell * \nu(\ell)$ and c_{x_0}. Note that this homotopy is already used in the proof of Proposition 2.2.24. $\quad\square$

These three facts imply that, in a loop space $\Omega(X)$, the defining properties of groups hold if we replace $=$ by $\simeq$. In other words, if we can make $\simeq$ into $=$, then we obtain a group. This can be done by taking the quotient space by the relation.

Definition 2.3.17. When an equivalence relation $\sim$ on a set S is given, the set of equivalence classes is denoted by $S/\sim$ and is called the *quotient set* of S by $\sim$.

Proposition 2.3.18. *The relation $\simeq$ is an equivalence relation on $\Omega(X, x_0)$.*

Proof. (1) For $\ell \in \Omega(X, x_0)$, define $H(t, s) = \ell(t)$. Then this is a homotopy for $\ell \simeq \ell$.

(2) Suppose that $\ell, \ell' \in \Omega(X, x_0)$ are homotopic $\ell \simeq \ell'$. Let H be the homotopy. Define $G(t, s) = H(t, 1 - s)$. Then G is a homotopy from ℓ' to ℓ.

(3) Suppose that $\ell \simeq \ell'$, $\ell' \simeq \ell''$. Let H and H' be their homotopies. Define

$$G(t, s) = \begin{cases} H(t, 2s), & 0 \le s \le \frac{1}{2} \\ H'(t, 2s - 1), & \frac{1}{2} \le t \le 1. \end{cases}$$

Then this is a homotopy from ℓ to ℓ'' and we have $\ell \simeq \ell''$.

$\quad\square$

Definition 2.3.19. The quotient set of $\Omega(X, x_0)$ by $\simeq$ is denoted by $\pi_1(X, x_0)$ and is called the *fundamental group* of X based at x_0. The equivalence class of $\ell \in \Omega(X, x_0)$ is denoted by $[\ell]$ and is called the *homotopy class* of ℓ.

Corollary 2.3.20. *For a based space (X, x_0), the loop product makes $\pi_1(X, x_0)$ into a group.*

Remark 2.3.21. The reader might wonder why the number 1 is put in the fundamental group. This is because loops are 1-dimensional. More generally, for any positive integer n, a group $\pi_n(X, x_0)$ call the n-th homotopy group is defined. This is a subject of §4.8.

The simply-connectivity can be defined by using the fundamental group.

Lemma 2.3.22. *An arcwise connected space X is simply connected if and only if $\pi_1(X, x_0)$ is the trivial group, i.e. the group having only the unit element, for all $x_0 \in X$.*

A proof will be given in §4.6.

Let us go back to covering spaces. By Lemma 2.2.21, the map φ_ℓ is independent of the homotopy class of ℓ. Thus we obtain a map

$$\varphi : p^{-1}(x_0) \times \pi_1(X, x_0) \longrightarrow p^{-1}(x_0) \qquad (2.1)$$

by assigning $\varphi_\ell(y)$ to $[\ell] \in \pi_1(X, x_0)$ and $y \in p^{-1}(x_0)$. This is an action of the fundamental group on $p^{-1}(x_0)$.

Definition 2.3.23. Let G be a group and S a set. An *action* of G on S (from the right) is a map $\mu : S \times G \to S$ satisfying the following conditions:

(1) Let e be the unit of G. Then for any $x \in S$, we have $\mu(x, e) = x$. In other words, the diagram

$$
\begin{array}{ccc}
S & \xrightarrow{\;1_s \times c_e\;} & S \times G \\
 & \searrow{\scriptstyle =} & \downarrow{\scriptstyle \mu} \\
 & & S
\end{array}
$$

is commutative, where $c_e : S \to G$ is the constant map to e.

(2) For $g, h \in G$ and $x \in S$, we have $\mu(\mu(x, g), h) = \mu(x, gh)$. In other words, the diagram

$$
\begin{array}{ccc}
S \times G \times G & \xrightarrow{\;\mu \times 1_G\;} & S \times G \\
\downarrow{\scriptstyle 1_S \times \mu_G} & & \downarrow{\scriptstyle \mu} \\
S \times G & \xrightarrow{\quad \mu \quad} & S
\end{array}
$$

is commutative, where $\mu_G : G \times G \to G$ is the product of G.

We often denote $\mu(x, g) = x \cdot g$ for simplicity.

Left actions $G \times S \to S$ are defined analogously by switching S and G.

Definition 2.3.24. The projection onto the first coordinate $\mathrm{pr}_1 : S \times G \to S$ is an action of G on S. This is called the *trivial action*.

Corollary 2.3.25. *The map (2.1) is an action of the fundamental group $\pi_1(X, x_0)$ on the fiber $p^{-1}(x_0)$.*

Proof. The conditions for group action have essentially been proved as Lemma 2.2.21, where domains of paths are allowed be any closed interval and concatenations of paths are defined without adjusting parameters. It is not difficult to modify proofs of these proposition. □

Definition 2.3.26. This action of the fundamental group $\pi_1(X, x_0)$ on the fiber $p^{-1}(x_0)$ is called the *monodromy*.

Example 2.3.27. When $p : E \to B$ is a trivial covering space, the action (2.1) is trivial. □

Group actions are studied in §3.5.

2.4 Universal Covering

We have seen in §2.3 that, given a covering space $p : E \to B$, the fundamental group $\pi_1(B, x_0)$ acts on the fiber $p^{-1}(x_0)$ from the right. We may rewrite this action into a left action. The proof is left to the reader.

Lemma 2.4.1. *Given a right action μ of a group G on a set S, define $\mu^{\mathrm{op}}(g, x) = \mu(x, g^{-1})$. Then μ^{op} is a left action of G on S.*

Left actions are more convenient since they have the following description.

Proposition 2.4.2. *Given a set S, let $\mathrm{Aut}(S)$ be the group of bijections from S to itself. The product is given by the composition of maps. Then there exists a one-to-one correspondence between left actions of G on S and group homomorphisms $G \to \mathrm{Aut}(S)$.*

Proof. Let $\mu : G \times S \to S$ be a left action. For $g \in G$, define $\mathrm{ad}(\mu)(g) \in \mathrm{Aut}(S)$ by $\mathrm{ad}(\mu)(g)(x) = \mu(g, x)$. For simplicity, we denote $gx = \mu(g, x)$ for $g \in G, x \in X$.

We first need to verify that $\mathrm{ad}(\mu)(g) : S \to S$ takes values in $\mathrm{Aut}(S)$ for any $g \in G$. This can be done by showing that $\mathrm{ad}(\mu)(g^{-1})$ is the inverse.

In fact, for any $x \in S$,

$$
\begin{aligned}
\text{ad}(\mu)(g^{-1}) \circ \text{ad}(\mu)(g)(x) &= \text{ad}(\mu)(g^{-1})(\mu(g,x)) \\
&= \text{ad}(\mu)(g^{-1})(gx) \\
&= \mu(g^{-1}, gx) \\
&= g^{-1}(gx) \\
&= (g^{-1}g)x \\
&= ex \\
&= x
\end{aligned}
$$

and we have $\text{ad}(\mu)(g^{-1}) \circ \text{ad}(\mu)(g) = 1_S$. A similar calculation shows that $\text{ad}(\mu)(g) \circ \text{ad}(\mu)(g^{-1}) = 1_S$. And $\text{ad}(\mu)(g)$ belongs to $\text{Aut}(S)$.

We next show that $\text{ad}(\mu)$ is a homomorphism, i.e. for $g, h \in G$, $\text{ad}(\mu)(gh) = \text{ad}(\mu)(g) \circ \text{ad}(\mu)(h)$. Note that for $x \in S$, we have $\text{ad}(\mu)(gh)(x) = \mu(gh, x) = (gh)x$ and

$$
\begin{aligned}
\text{ad}(\mu)(g) \circ \text{ad}(\mu)(h)(x) &= \text{ad}(\mu)(g)(\mu(h,x)) \\
&= \text{ad}(\mu)(g)(hx) \\
&= \mu(g, hx) \\
&= g(hx) \\
&= (gh)x.
\end{aligned}
$$

Thus we have $\text{ad}(\mu)(gh) = \text{ad}(\mu)(g) \circ \text{ad}(\mu)(h)$.

Conversely, when a homomorphism $f : G \to \text{Aut}(S)$ is given, an action $\mu_f : G \times S \to S$ can be defined by $\mu_f(g, x) = f(g)(x)$. We leave it to the reader to verify that this is in fact an action and that $f \mapsto \mu_f$ is an inverse correspondence to $\mu \to \text{ad}(\mu)$. $\qquad\square$

Let $p : E \to B$ be a covering space. By making the right action of the fundamental group $\pi_1(B, x_0)$ on the fiber $F_{x_0} = p^{-1}(x_0)$ over x_0 into a left action, we obtain a group homomorphism

$$
\text{ad}(\varphi^{\text{op}}) : \pi_1(B, x_0) \longrightarrow \text{Aut}(F_{x_0}).
$$

Suppose we have another covering space $p' : E' \to B$ with monodromy action φ'. We have a group homomorphism

$$
\text{ad}((\varphi')^{\text{op}}) : \pi_1(B, x_0) \longrightarrow \text{Aut}(F'_{x_0}),
$$

where $F'_{x_0} = (p')^{-1}(x_0)$.

If there exists an isomorphism of covering spaces $f : E \to E'$, we have a bijection of each fibers $f|_{F_{x_0}} : F_{x_0} \to F'_{x_0}$, which induces an isomorphism of groups

$$f_* : \mathrm{Aut}(F_{x_0}) \longrightarrow \mathrm{Aut}(F'_{x_0})$$

by $f_*(\sigma) = (f|_{F_{x_0}})^{-1} \circ \sigma \circ (f|_{F_{x_0}})$. A relation among these maps can be described by a commutative diagram.

Proposition 2.4.3. *The following diagram is commutative*

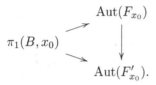

The proof is left to the reader.

Definition 2.4.4. Let μ and μ' be left actions of a group on sets S and S', respectively. We say μ and μ' are *equivalent* if there exists a bijection $f : S \to S'$ making the following diagram commutative

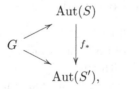

where f_* is a homomorphism defined by $f_*(\sigma) = f \circ \sigma \circ f^{-1}$.

With this terminology, Proposition 2.4.3 says that actions of the fundamental group on fibers obtained from isomorphic covering spaces are equivalent. A natural question is a converse to this fact.

Problem 2.4.5. Given a group homomorphism $f : \pi_1(B, x_0) \to \mathrm{Aut}(F)$ or an action of $\pi_1(B, x_0)$ on F, can we construct a covering space over B with fiber F, in such a way that if actions are equivalent, obtained covering spaces are equivalent?

If we could answer this question affirmatively, we would be able to classify covering spaces by actions of the fundamental group of the base space on fibers. To this end, we need universal coverings.

Definition 2.4.6. A covering space $p : E \to B$ is called a *universal covering* if E is simply connected. We also say that E is a universal covering of B.

Example 2.4.7. The covering space $\exp : \mathbb{R} \to S^1$ in Exercise 2.1.6 is a universal covering. □

We need a couple of conditions for the existence of universal coverings.

Definition 2.4.8. A topological space X is called *semilocally simply connected* if each $x \in X$ has a neighborhood U satisfying the following conditions: any loop $\ell \in \Omega(U, x)$ is homotopic to the constant loop c_x at x as elements of $\Omega(X, x)$.

Theorem 2.4.9. *If B is locally arcwise connected, semilocally simply connected, and arcwise connected, then there exists a universal covering over B.*

A proof of this fact will be given in §4.11. Our discussion on Problem 2.4.5 by using this fact is also postponed to §4.11, since it is more natural and easier to understand the problem after learning basic properties of fiber bundles.

Instead we end this section and hence this introductory chapter by the following example in order to advertise the usefulness of universal coverings.

Example 2.4.10. Recall the universal covering in Example 2.4.7. Take $x_0 = (1, 0)$ as a base point of S^1. An element $[\ell]$ in $\pi_1(S^1, x_0)$ is represented by a path $\ell : [0, 1] \to S^1$ with $\ell(0) = \ell(1) = x_0$.

Since $\exp(0) = x_0$, there exists a unique lift $\tilde{\ell} : [0, 1] \to \mathbb{R}$ of ℓ with 0 the initial point. Following Definition 2.2.20, define $\varphi([\ell]) = \tilde{\ell}(1)$. And we obtain a map

$$\varphi : \pi_1(S^1, x_0) \longrightarrow \mathbb{Z}.$$

By Theorem 2.2.19, this is independent of the choice of a representative. It turns out that this map is an isomorphism of groups.

First of all, Corollary 2.2.23 implies that the φ is a homomorphism. It is surjective since, for $n \in \mathbb{Z}$, the path defined by $\ell_n(t) = e^{2\pi i n t}$ satisfies $\varphi([\ell_n]) = n$.

In order to show the injectivity, it suffices to show that $\text{Ker}\,\varphi = \{[c_{x_0}]\}$. If $[\ell] \in \text{Ker}\,\varphi$, the lift $\tilde{\ell}$ of ℓ with 0 the initial point should satisfy $\tilde{\ell}(1) = 0$. In other words, $\tilde{\ell}$ is a loop in $\mathbb{R}$ with 0 the base point. As we have seen in Example 2.2.17, $\mathbb{R}$ is simply connected, which means that there exists a homotopy H between $\tilde{\ell}$ and the constant loop c_0 at 0. Then $\exp \circ H$ is a homotopy between ℓ and c_{x_0}, and we have $[\ell] = [c_{x_0}]$. $\qquad \square$

Chapter 3

Basic Properties of Fiber Bundles

We have seen examples of spaces constructed by bundling a space along another space in Chapter 1. We investigated covering spaces in Chapter 2 as a toy model of the theory of fiber bundles.

3.1 Defining Fiber Bundles

One of the most simply-minded definitions of a fiber bundle is the following.

Definition 3.1.1. A *fiber map* with the *total space* E and the *base space* B is a surjective continuous map

$$p : E \longrightarrow B.$$

For each $x \in B$, the inverse image $p^{-1}(x)$ is called the *fiber* over x.

Some authors call this a fiber bundle. See, for example, [Baum *et al.* (2007)]. The definition is, of course, too crude to be of any use. It does refer to fibers but it does not say anything about how fibers are bundled. Let us give a definition of fiber bundles. The reader should check that it is a generalization of the examples in Chapter 1 and covering spaces.

Definition 3.1.2. A *fiber bundle* consists of

(1) three spaces B, E, and F, called the *total space*, the *base space*, and the *fiber*, respectively,
(2) a continuous map $p : E \to B$ called the *projection*,
(3) for each $x \in B$, an open neighborhood U_x of x and a homeomorphism

$$\varphi_x : p^{-1}(U_x) \xrightarrow{\cong} U_x \times F$$

making the following diagram commutative

$$p^{-1}(U_x) \xrightarrow{\ \varphi_x\ } U_x \times F$$
$$p \searrow \qquad \swarrow \mathrm{pr}_1$$
$$U_x,$$

where pr_1 is the projection onto the first factor.

Each map φ_x is called a *local trivialization*. For simplicity, we often denote a fiber bundle by a map

$$p : E \longrightarrow B$$

or a triple (B, E, F) or a sequence of maps

$$F \longrightarrow E \xrightarrow{\ p\ } B.$$

We also say that E is a fiber bundle over B with fiber F.

Example 3.1.3. When $p : E \to B$ is a covering space, each point $x \in B$ has a neighborhood U with a homeomorphism $p^{-1}(U) \cong U \times p^{-1}(x)$ which is compatible with projections. By Proposition 2.2.24, when B is arcwise connected, there is a bijection between any two fibers. Since each fiber has the discrete topology, all fibers are homeomorphic and we can regard $p : E \to B$ as a fiber bundle. $\square$

As is the case of covering spaces, we may take $U_{x'} = U_x$ and $\varphi_{x'} = \varphi_x$ when $x' \in U_x$.

Proposition 3.1.4. *Given three topological spaces B, E, and F, a continuous map $p : E \to B$ is a fiber bundle with fiber F if and only if B has an open covering $B = \bigcup_{\alpha \in A} U_\alpha$ and there exists a homeomorphism $\varphi_\alpha : p^{-1}(U_\alpha) \xrightarrow{\cong} U_\alpha \times F$ for each α such that the diagram*

$$p^{-1}(U_\alpha) \xrightarrow{\ \varphi_\alpha\ } U_\alpha \times F$$
$$p \searrow \qquad \swarrow \mathrm{pr}_1$$
$$U_\alpha$$

is commutative.

Remark 3.1.5. The maps φ_α in this proposition are also called local trivializations.

The first three examples in §1.1, i.e. cylinder, annulus and torus, should be regarded as simplest type of fiber bundles. Such fiber bundles are called trivial.

Definition 3.1.6. When $E = B \times F$ and $p : E \to B$ is the projection onto the first factor, this is called a *trivial bundle*.

Example 3.1.7. By Theorem 2.3.7, a covering space whose base space is simply connected and locally arcwise connected is a trivial bundle. □

Example 3.1.8. Let us show that the Möbius band is a fiber bundle over S^1 with fiber $[-\frac{1}{2}, \frac{1}{2}]$.

Define a projection $p : M \to S^1$ by $p(x, y, z) = (\cos\varphi, \sin\varphi)$ for

$$(x, y, z) = \left((2 + t\cos\tfrac{\varphi}{2})\cos\varphi, (2 + t\cos\tfrac{\varphi}{2})\sin\varphi, t\sin\tfrac{\varphi}{2}\right) \in M.$$

In Example 1.1.5, we split S^1 into the upper and lower halves to "trivialize" the Möbius band. By the above definition, however, we need to decompose S^1 into a union of open subsets.

Define

$$U_1 = \{(\cos\theta, \sin\theta) \mid 0 < \theta < 2\pi\}$$
$$= S^1 \setminus \{(1, 0)\}$$
$$U_2 = \{(\cos\theta, \sin\theta) \mid \theta \neq \pi\}$$
$$= \{(\cos\theta, \sin\theta) \mid \pi < \theta < 3\pi\}$$
$$= S^1 \setminus \{(-1, 0)\}$$

as is shown in Figure 3.1.

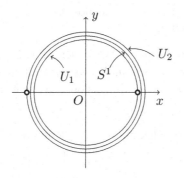

Fig. 3.1   an open covering of S^1

Since U_1 and U_2 are complements of single points, they are open in S^1 and we have $S^1 = U_1 \cup U_2$. By definition, we have

$$p^{-1}(U_1) = \left\{ ((2 + t\cos\tfrac{\varphi}{2})\cos\varphi, (2 + t\cos\tfrac{\varphi}{2})\sin\varphi, t\sin\tfrac{\varphi}{2}) \,\middle|\, -\tfrac{1}{2} \leq t \leq \tfrac{1}{2}, 0 < \varphi < 2\pi \right\}$$

$$p^{-1}(U_2) = \left\{ ((2 + t\cos\tfrac{\varphi}{2})\cos\varphi, (2 + t\cos\tfrac{\varphi}{2})\sin\varphi, t\sin\tfrac{\varphi}{2}) \,\middle|\, -\tfrac{1}{2} \leq t \leq \tfrac{1}{2}, \pi < \varphi < 3\pi \right\}.$$

In order to construct a homeomorphism $p^{-1}(U_1) \cong U_1 \times [-\tfrac{1}{2}, \tfrac{1}{2}]$, define maps

$$f_1 : (0, 2\pi) \times \left[-\tfrac{1}{2}, \tfrac{1}{2}\right] \longrightarrow p^{-1}(U_1)$$
$$g_1 : (0, 2\pi) \longrightarrow U_1$$

by

$$f_1(\varphi, t) = \left((2 + t\cos\tfrac{\varphi}{2})\cos\varphi, (2 + t\cos\tfrac{\varphi}{2})\sin\varphi, t\sin\tfrac{\varphi}{2} \right)$$
$$g_1(\varphi) = (\cos\varphi, \sin\varphi).$$

Then f_1 and g_1 are homeomorphisms. Now define

$$\varphi_1 = (g_1 \times 1_{[-\frac{1}{2}, \frac{1}{2}]}) \circ f_1^{-1} : p^{-1}(U_1) \longrightarrow U_1 \times \left[-\tfrac{1}{2}, \tfrac{1}{2}\right].$$

Then φ_1 is a homeomorphism. Similarly a homeomorphism $\varphi_2 : p^{-1}(U_2) \to U_2 \times \left[-\tfrac{1}{2}, \tfrac{1}{2}\right]$ is defined and the map $p : M \to S^1$ is a fiber bundle with fiber $\left[-\tfrac{1}{2}, \tfrac{1}{2}\right]$.   □

Exercise 3.1.9. Why did we use an open covering in the definition of fiber bundles? What's wrong with a closed covering used in 4 for the Möbius band?

Exercise 3.1.10. Prove that the maps p, f_1, and g_1 in Example 3.1.8 are homeomorphisms. And find a homeomorphism $\varphi_2 : p^{-1}(U_2) \to U_2 \times [-\tfrac{1}{2}, \tfrac{1}{2}]$.

The next fiber bundle is one of the most famous and important examples of nontrivial fiber bundles.

Example 3.1.11 (Hopf Bundle). We present the unit 3-dimensional sphere S^3 as the total space of a fiber bundle over the unit sphere S^2 with fiber the unit circle S^1. In general the n-dimensional unit sphere S^n is defined by

$$S^n = \left\{ (x_0, \ldots, x_n) \in \mathbb{R}^{n+1} \,\middle|\, x_0^2 + \cdots + x_n^2 = 1 \right\}.$$

Here we make use of complex numbers to define S^1, S^2, and S^3 as follows

$$S^1 = \left\{ z \in \mathbb{C} \,\middle|\, |z|^2 = 1 \right\} = \left\{ z \in \mathbb{C} \,\middle|\, |z| = 1 \right\}$$
$$S^2 = \left\{ (x, z) \in \mathbb{R} \times \mathbb{C} \,\middle|\, x^2 + |z|^2 = 1 \right\}$$
$$S^3 = \left\{ (z_1, z_2) \in \mathbb{C}^2 \,\middle|\, |z_1|^2 + |z_2|^2 = 1 \right\}.$$

Define a continuous map $p : S^3 \to S^2$ by $p(z_1, z_2) = \left(2|z_1|^2 - 1, 2z_1\overline{z_2}\right)$. It is left to the reader to verify that p is well-defined, namely if $(z_1, z_2) \in S^3$, then $p(z_1, z_2) \in S^2$.

For local trivializations, we use the open covering $S^2 = U_+ \cup U_-$ of S^2 given by

$$U_+ = S^2 \setminus \{(1, 0)\}$$
$$U_- = S^2 \setminus \{(-1, 0)\}.$$

We would like to find homeomorphisms

$$\varphi_+ : p^{-1}(U_+) \longrightarrow U_+ \times S^1$$
$$\varphi_- : p^{-1}(U_-) \longrightarrow U_- \times S^1.$$

For $(z_1, z_2) \in p^{-1}(U_+)$, define

$$\varphi_+(z_1, z_2) = \left(2|z_1|^2 - 1, 2z_1\overline{z_2}, \tfrac{z_2}{|z_2|}\right).$$

Since $\left(2|z_1|^2 - 1, 2z_1\overline{z_2}\right) \in U_+$, $\left(2|z_1|^2 - 1, 2z_1\overline{z_2}\right) \neq (1, 0)$ and thus $z_2 \neq 0$. This map makes the triangle in the definition of local trivialization commutative, since the first two coordinates agrees with $p(z_1, z_2)$. It remains to show that φ_+ is a homeomorphism. Define a map $\psi_+ : U_+ \times S^1 \to p^{-1}(U_+)$ by

$$\psi_+(x, z, w) = \left(\frac{zw}{2\sqrt{\frac{1-x}{2}}}, w\sqrt{\frac{1-x}{2}}\right).$$

This is an inverse to φ_+.

We leave it to the reader to define $\varphi_- : p^{-1}(U_-) \to U_- \times S^1$ and its inverse as Exercise 3.1.12.

This fiber bundle $p : S^3 \to S^2$ with fiber S^1 is called the *Hopf bundle*. $\square$

Exercise 3.1.12. Define maps

$$\varphi_- : p^{-1}(U_-) \longrightarrow U_- \times S^1$$
$$\psi_- : U_- \times S^1 \longrightarrow p^{-1}(U_-)$$

by mimicking the constructions of maps φ_+ and ψ_+ in Example 3.1.11 and verify that they are inverse to each other.

Even though the Möbius band is one of the simplest examples of non-trivial fiber bundles, it was not straightforward to find local trivializations let alone the Hopf bundle.

In order to show that a given map $p : E \to B$ is a fiber bundle, it will not be a good idea to try to find explicit local trivializations. It will be quite useful if we know ways to construct fiber bundles from known one. If we want to show that p is not a fiber bundle, we need to find a property that is satisfied by fiber bundles but not by p. In this chapter, we study general properties fiber bundles. In Chapter 4, we show that any fiber bundle can be obtained by "pulling back a universal bundle".

Exercise 3.1.13. Prove that the tangent bundle $p : TS^2 \to S^2$ of S^2 appeared in §1.2 is a fiber bundle with fiber $\mathbb{R}^2$.

3.2   Fiber Bundles with Structure Groups

By Proposition 3.1.4, a map $p : E \to B$ is a fiber bundle with fiber F, if and only if there exists an open covering $B = \bigcup_{\alpha \in A} U_\alpha$ of B and a homeomorphism (local trivialization) $\varphi_\alpha : p^{-1}(U_\alpha) \xrightarrow{\cong} U_\alpha \times F$ for each λ making the diagram

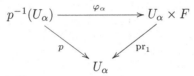

commutative. On the other hand, this description has too simple to be used in geometric applications. For example, it is an important fact that fibers of the tangent bundle of a smooth manifold (section 1.2) have structures of vector bundles. There is no room in the above definition of fiber bundles to incorporate this fact. We need to impose conditions on fiber bundles. Where in the definition of fiber bundles should we impose such conditions? Since twistings of fibers are controlled by local trivializations, we should add conditions to local trivializations. Let us consider the meaning of local trivializations.

Discussion 3.2.1. Notice that the decomposition $E = \bigcup_{\alpha \in A} p^{-1}(U_\alpha)$ and homeomorphisms $p^{-1}(U_\alpha) \cong U_\alpha \times F$ suggest us to write the total space E as

$$E \cong \bigcup_{\alpha \in A} U_\alpha \times F. \tag{3.1}$$

Of course, this is not a rigorous description, but we can see that the total space of a fiber bundle is obtained by gluing trivial fiber bundles. Then what is a precise meaning of gluing?

Let us glue two trivial bundles, say $U_\alpha \times F$ and $U_\beta \times F$ to see what is going on. According to (3.1), when we glue $U_\alpha \times F$ and $U_\beta \times F$, we use homeomorphisms (local trivializations)

$$U_\alpha \times F \cong p^{-1}(U_\alpha)$$
$$U_\beta \times F \cong p^{-1}(U_\beta)$$

and the the union of $p^{-1}(U_\alpha)$ and $p^{-1}(U_\beta)$ in E.
Let us name the local trivialization as

$$\varphi_\alpha : p^{-1}(U_\alpha) \longrightarrow U_\alpha \times F$$
$$\varphi_\beta : p^{-1}(U_\beta) \longrightarrow U_\beta \times F$$

to keep track of points. Then, for $(x, y) \in U_\alpha \times F$ and $(x', y') \in U_\beta \times F$, (x, y) are (x', y') are identified if and only if $\varphi_\alpha^{-1}(x, y) = \varphi_\beta^{-1}(x', y')$. See Figure 3.2.1.

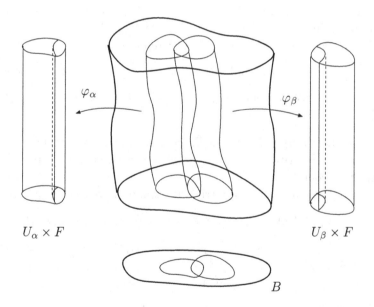

$U_\alpha \times F$ $\qquad\qquad\qquad\qquad\qquad U_\beta \times F$

B

Fig. 3.2 local patching

Let us make the above discussion precise. Define a relation $\sim$ on the union $U_\alpha \times F \amalg U_\beta \times F$ as follows

$$(x, y) \sim (x', y') \Longleftrightarrow \varphi_\alpha^{-1}(x, y) = \varphi_\beta^{-1}(x', y'). \tag{3.2}$$

The quotient space[1] by the equivalence relation generated by $\sim$

$$(U_\alpha \times F \amalg U_\beta \times F)/_\sim$$

is the space obtained by gluing $U_\alpha \times F$ and $U_\beta \times F$.

If we would like to impose certain conditions on the way trivial bundles are glued together, we should do it to this equivalence relation. Note that the above relation (3.2) can be restated as

$$(x, y) \sim (x', y') \iff \varphi_\beta \circ \varphi_\alpha^{-1}(x, y) = (x', y').$$

This is more convenient, since we only need to deal with a single map $\varphi_\beta \circ \varphi_\alpha^{-1}$ instead of two maps φ_α and φ_β. In the diagram

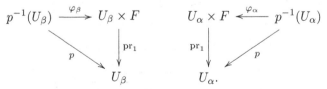

when we want to know if (x, y) and (x', y') are equivalent, we can just go horizontally instead of going down to $p^{-1}(U_\alpha \cap U_\beta)$ and then going up. We should regard that the map $\varphi_\beta \circ \varphi_\alpha^{-1}$ represents the gluing.

In order to understand this map $\varphi_\beta \circ \varphi_\alpha^{-1}$, the first step is the following.

Lemma 3.2.2. *For any* $(x, y) \in (U_\alpha \cap U_\beta) \times F$, *we have* $(\varphi_\beta \circ \varphi_\alpha^{-1})(x, y) \in \{x\} \times F$.

Proof. Let $(x, y) \in (U_\alpha \cap U_\beta) \times F$. In order to show $\varphi_\beta \circ \varphi_\alpha^{-1}(x, y) \in \{x\} \times F$, it suffices to verify $\mathrm{pr}_1 \circ \varphi_\beta \circ \varphi_\alpha^{-1}(x, y) = x$, where pr_1 is the projection onto the first coordinate.

By the definition of local trivialization, we have a commutative diagram

$$
\begin{array}{ccc}
p^{-1}(U_\beta) \xrightarrow{\ \varphi_\beta\ } U_\beta \times F & \qquad & U_\alpha \times F \xleftarrow{\ \varphi_\alpha\ } p^{-1}(U_\alpha) \\
\searrow_p \quad \downarrow^{\mathrm{pr}_1} & & {}^{\mathrm{pr}_1}\downarrow \quad \swarrow_p \\
U_\beta & & U_\alpha.
\end{array}
$$

And we have

$$\mathrm{pr}_1 \circ \varphi_\beta \circ \varphi_\alpha^{-1}(x, y) = p \circ \varphi_\alpha^{-1}(x, y)$$
$$= \mathrm{pr}_1(x, y)$$
$$= x.$$

$\qquad\square$

Definition 3.2.3. Under the assumption of above lemma, write

$$(\varphi_\beta \circ \varphi_\alpha^{-1})(x, y) = \left(x, \Phi^{\alpha\beta}(x)(y)\right).$$

We regard $\Phi^{\alpha\beta}(x)$ as a map which assigns $\Phi^{\alpha\beta}(x)(y) \in F$ to $y \in F$. In other words,

$$\Phi^{\alpha\beta}(x)(y) = \mathrm{pr}_2 \circ \varphi_\beta \circ \varphi_\alpha^{-1}(x, y),$$

where pr_2 is the projection onto the second factor.

Lemma 3.2.4. *For any $x \in U_\alpha \cap U_\beta$, the map $\Phi^{\alpha\beta}(x) : F \to F$ is a homeomorphism.*

Proof. Since φ_α are φ_β homeomorphisms, their restrictions $\varphi_\beta|_{p^{-1}(x)}$ and $\varphi_\alpha^{-1}|_{\{x\} \times F}$ are homeomorphisms. Since

$$\Phi^{\alpha\beta}(x) = \mathrm{pr}_2 \circ \left(\varphi_\beta|_{p^{-1}(x)}\right) \circ \left(\varphi_\alpha^{-1}|_{\{x\} \times F}\right) \circ i_x,$$

$\Phi^{\alpha\beta}(x)$ is a homeomorphism. Here $i_x : F \to \{x\} \times F$ is the map defined by $i_x(y) = (x, y)$. $\square$

Thus we obtain a correspondence from a point $x \in U_\alpha \cap U_\beta$ to a homeomorphism $\Phi^{\alpha\beta}(x)$. In order to regard this correspondence as a map, we need to name the range.

Definition 3.2.5. For a topological space X, define

$$\mathrm{Homeo}(X) = \{f : X \to X \mid \text{homeomorphism}\}.$$

We leave it to the reader to verify the following fact.

Proposition 3.2.6. *The set $\mathrm{Homeo}(X)$ can be made into a group by compositions of maps.*

Now we see that, under the notations in Definition 3.2.3, local gluings in a fiber bundle are given by maps

$$\Phi^{\alpha\beta} : U_\alpha \cap U_\beta \longrightarrow \mathrm{Homeo}(F).$$

Definition 3.2.7. These maps $\{\Phi^{\alpha\beta}\}$ are called *coordinate transformations*.

With no condition, coordinate transformations can take values in any homeomorphisms. If we want to restrict the way fibers are glued together, we should require that the coordinate transformations take values in appropriate subsets.

Definition 3.2.8 (Temporary Definition of Structure Group). Let G be a subgroup of Homeo(F). A fiber bundle (B, E, F) has G as a *structure group* if the coordinate transformations $\Phi^{\alpha\beta}$ takes values in G for any α, β. In other words, there exists a map $\overline{\Phi}^{\alpha\beta}$ making the following diagram commutative

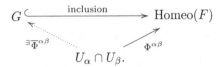

As is labelled as "Temporary Definition," there are some rooms for improvements. A couple of questions might have already came up to the reader's mind.

(1) Why does G have to be a subgroup, instead of being just a subset?
(2) If H is a subgroup of Homeo(F) containing G, i.e. $G \subset H \subset$ Homeo(F), a fiber bundle having G as a structure group also has H as its structure group. There seems to be an ambiguity in the definition of structure group.

Exercise 3.2.9. Find your own justification for (1).

One of answers to the question (2) is the notion of reduction, which will be introduced in Definition 3.5.26.

Another reason why Definition 3.2.8 is temporary is that we did not say anything on the topology of Homeo(X) and continuities of maps $G \hookrightarrow$ Homeo(X) and $\overline{\Phi}^{\alpha\beta}$. A precise definition of structure group will be given in section 3.5.

This definition contains, on the other hand, essential features of structure groups so that we can play with typical examples to get intuitions.

Example 3.2.10. Let $E \cong B \times F \to B$ be a trivial bundle. It has a unique local trivialization, i.e. the identity map $E = B \times F$. In this case, the coordinate transformation is define by the identity map from $B \times F$ to $B \times F$. In other words, it is the identity map from F to F. Thus any trivial bundle has the trivial group (the group consisting only of the unit) as structure group. $\square$

A more interesting example is the Möbius band.

Example 3.2.11. Let $M \to S^1$ be the Möbius band in Example 3.1.8. Recall that an open covering $S^1 = U_1 \cup U_2$ of the base space S^1 is given by

$$U_1 = \{(\cos\theta, \sin\theta) \,|\, 0 < \theta < 2\pi\} = S^1 \setminus \{(1,0)\}$$
$$U_2 = \{(\cos\theta, \sin\theta) \,|\, \pi < \theta < 3\pi\} = S^1 \setminus \{(-1,0)\}$$

as is described in Figure 3.1.

The local trivializations

$$\varphi_1 : p^{-1}(U_1) \longrightarrow U_1 \times [-\tfrac{1}{2}, \tfrac{1}{2}]$$
$$\varphi_2 : p^{-1}(U_2) \longrightarrow U_2 \times [-\tfrac{1}{2}, \tfrac{1}{2}]$$

are described in Figure 3.3.

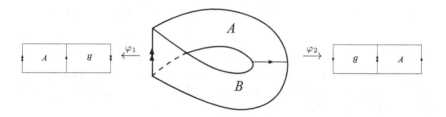

Fig. 3.3   local trivializations of the Möbius band

The intersection $U_1 \cap U_2$ consists of two connected components. Let us denote the upper and lower components in Figure 3.3 by A and B, respectively.

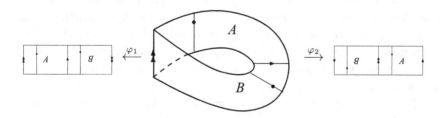

Fig. 3.4   coordinate transformations of the Möbius band

The coordinate transformation $\Phi^{12} : U_1 \cap U_2 \to \mathrm{Homeo}([-\tfrac{1}{2}, \tfrac{1}{2}])$ is determined by the composition

$$(U_1 \cap U_2) \times [-\tfrac{1}{2}, \tfrac{1}{2}] \xrightarrow{\varphi_1^{-1}} p^{-1}(U_1 \cap U_2) \xrightarrow{\varphi_2} (U_1 \cap U_2) \times [-\tfrac{1}{2}, \tfrac{1}{2}].$$

By Figure 3.4, we see that

$$\Phi^{12}(x)(t) = \begin{cases} t, & x \in A, \ t \in [-\tfrac{1}{2}, \tfrac{1}{2}] \\ -t, & x \in B, \ t \in [-\tfrac{1}{2}, \tfrac{1}{2}] \end{cases}.$$

Let us denote $g(t) = -t$. Then $g \in \mathrm{Homeo}([-\tfrac{1}{2}, \tfrac{1}{2}])$ and Φ^{12} takes values in $\left\{ 1_{[-\frac{1}{2}, \frac{1}{2}]}, g \right\}$. And we have a map

$$\Phi^{12} : U_1 \cap U_2 \longrightarrow \left\{ 1_{[-\frac{1}{2}, \frac{1}{2}]}, g \right\}.$$

Since $g \circ g(t) = g(-t) = -(-t) = t$, the group $\left\{ 1_{[-\frac{1}{2}, \frac{1}{2}]}, g \right\}$ is the cyclic group C_2 of order 2. Thus the Möbius band has C_2 as a structure group. $\square$

Both the annulus A and the Möbius band M are fiber bundles with base space S^1 and fiber $[\tfrac{1}{2}, -\tfrac{1}{2}]$. By Example 3.2.10 and Example 3.2.11, we see that the former has the trivial group and the latter has C_2 as trivial groups. These examples show that the twisting in the Möbius band is encoded in its structure group C_2.

3.3　Topological Groups

As we have seen in the previous section, structure groups have important information on fiber bundles.

Recall that our temporary definition of structure groups are given in terms of coordinate transformations, i.e. maps $\Phi^{\alpha\beta} : U_\alpha \cap U_\beta \to \mathrm{Homeo}(F)$. The domain $U_\alpha \cap U_\beta$ of the map $\Phi^{\alpha\beta}$ is a topological space and the map $\Phi^{\alpha\beta}$ is defined by continuous maps φ_α and φ_β. These facts suggest to put topology on $\mathrm{Homeo}(F)$ and regard it as a topological space. Furthermore, since we introduced structure groups in order to impose restrictions on coordinate transformations, it is reasonable to require certain continuity on coordinate transformations. For this purpose, we need the notion of topological group.

The reader might have wondered why we used commutative diagrams in our definition of group in Definition 2.3.11. Here is a reason.

Definition 3.3.1. When a group G is a topological space and the structure maps μ and ν are continuous, it is called a *topological group*.

Let us begin with simple examples.

Example 3.3.2. The additive group of real numbers $\mathbb{R}$ is a topological group. □

Example 3.3.3. The multiplicative group $\mathbb{R}^\times$ of nonzero elements in $\mathbb{R}$ is a topological group.
The subspace $\mathbb{R}_{>0} = \{x \in \mathbb{R} \mid x > 0\}$ is also a topological group. □

Example 3.3.4. The multiplicative group $\mathbb{C}^\times = \{z \in \mathbb{C} \mid z \neq 0\}$ of nonzero complex numbers is a topological group. The unit circle $S^1 = \{z \in \mathbb{C} \mid |z| = 1\}$ is, as a subspace of $\mathbb{C}^\times$, a topological group. □

Topological groups defined by matrices are quite important.

Example 3.3.5. Let us denote the set of real $n \times n$ matrices by $M_n(\mathbb{R})$. Under the identification $M_n(\mathbb{R}) = \mathbb{R}^{n^2}$, we regard $M_n(\mathbb{R})$ as a topological space. The addition of matrices is continuous and makes $M_n(\mathbb{R})$ into a topological group. Example 3.3.2 is a special case of $n = 1$. □

Example 3.3.6. The set of invertible $n \times n$ real matrices $\mathrm{GL}_n(\mathbb{R}) = \{A \in M_n(\mathbb{R}) \mid \det A \neq 0\}$ is a subspace of $M_n(\mathbb{R})$. The matrix multiplication makes $\mathrm{GL}_n(\mathbb{R})$ into a topological group. This is called the *general linear group*. Example 3.3.3 is a special case of $n = 1$. □

Exercise 3.3.7. Prove that μ and ν in $\mathrm{GL}_n(\mathbb{R})$ are continuous.

Example 3.3.8. Define $\mathrm{SL}_n(\mathbb{R}) = \{A \in M_n(\mathbb{R}) \mid \det A = 1\}$. This is a subspace of $\mathrm{GL}_n(\mathbb{R})$. The structure maps μ, ν of $\mathrm{GL}_n(\mathbb{R})$ in Example 3.3.6 satisfy

$$\mu(\mathrm{SL}_n(\mathbb{R}) \times \mathrm{SL}_n(\mathbb{R}))) \subset \mathrm{SL}_n(\mathbb{R})$$
$$\nu(\mathrm{SL}_n(\mathbb{R})) \subset \mathrm{SL}_n(\mathbb{R}).$$

Thus restrictions of μ and ν make $\mathrm{SL}_n(\mathbb{R})$ into a topological group. This is called the *special linear group*. □

Example 3.3.9. Let $\mathrm{O}(n) = \{A \in M_n(\mathbb{R}) \mid {}^t\!AA = E_n\}$. This is also a subspace of $\mathrm{GL}_n(\mathbb{R})$. Since $\mathrm{O}(n)$ is closed under the matrix multiplication, it becomes a topological group. This is called the *orthogonal group*. □

Example 3.3.10. The subset $\mathrm{SO}(n) = \{A \in \mathrm{O}(n) \mid \det A = 1\} = \mathrm{O}(n) \cap \mathrm{SL}_n(\mathbb{R})$ of $\mathrm{O}(n)$ is closed under the matrix multiplication and becomes a topological group. This is called the *special orthogonal group*. □

Exercise 3.3.11. Prove that $O(n)$ and $SO(n)$ are topological groups by the matrix multiplication.

Example 3.3.12. There are complex versions of the above five examples (Examples from 3.3.5 to 3.3.10):

$$M_n(\mathbb{C}) = \{\text{complex } n \times n \text{ matrices}\}$$
$$GL_n(\mathbb{C}) = \{A \in M_n(\mathbb{C}) \mid \det A \neq 0\}$$
$$SL_n(\mathbb{C}) = \{A \in M_n(\mathbb{C}) \mid \det A = 1\}$$
$$U(n) = \{A \in M_n(\mathbb{C}) \mid {}^t\overline{A}A = E_n\}$$
$$SU(n) = \{A \in U(n) \mid \det A = 1\}$$

$M_n(\mathbb{C})$ becomes a topological group by the addition of matrices. Other spaces become topological groups by the matrix multiplication.

The topological groups $GL_n(\mathbb{C})$, $SL_n(\mathbb{C})$, $U(n)$, $SU(n)$ are called the *complex general linear group*, the *complex special linear group*, the *unitary group*, and the *special unitary group*, respectively.   □

Exercise 3.3.13. Prove that $U(n)$ is a topological group.

Example 3.3.14. Let $\mathbb{H}$ be Hamilton's *quaternions*. It is a vector space over $\mathbb{R}$ having a basis $\{1, i, j, k\}$ on which multiplication is defined by the following table:

		i	j	k
i		-1	k	$-j$
j		$-k$	-1	i
k		j	$-i$	-1

Here we define 1 to be the unit. The multiplication of $\mathbb{H}$ satisfies the associativity, although it is not commutative.

For $w = a + bi + cj + dk \in \mathbb{H}$, define its *conjugate* by $\overline{w} = a - bi - cj - dk$. It is straight forward to verify that

$$w\overline{w} = \overline{w}w = a^2 + b^2 + c^2 + d^2.$$

Thus if $w \neq 0$, it has an inverse $w^{-1} = \frac{1}{a^2+b^2+c^2+d^2}\overline{w}$ and $\mathbb{H}^\times = \mathbb{H} \setminus \{0\}$ becomes a topological group.   □

Example 3.3.15. Since the multiplication in $\mathbb{H}$ is not commutative, the definition of the determinant cannot be extended to elements of $M_n(\mathbb{H})$. On the other hand $GL_n(\mathbb{H})$ and analogues of orthogonal and unitary groups can be defined.

$$GL_n(\mathbb{H}) = \{A \in M_n(\mathbb{H}) \mid A \text{ has an inverse matrix}\}$$
$$Sp(n) = \{A \in M_n(\mathbb{H}) \mid {}^t\overline{A}A = A{}^t\overline{A} = E_n\}.$$

$\square$

Exercise 3.3.16. Prove that $GL_n(\mathbb{H})$ and $Sp(n)$ are topological groups.

Remark 3.3.17. The quaternions $\mathbb{H}$ is obtained by adding an element j with $j^2 = -1$ to $\mathbb{C}$. This is an analogue of the process we obtain $\mathbb{C}$ from $\mathbb{R}$ by adding an element i. It is no wonder if somebody has already tried to add an element ℓ with $\ell^2 = -1$ to $\mathbb{H}$ to extend it to an 8-dimensional real vector space with multiplication.

In fact, inspired by Hamilton's discovery of quaternions, John Graves defined the set of *octonions* $\mathbb{O}$ in 1843. Octonions are often called Cayley numbers, since they were independently discovered by Author Cayley.

Unfortunately, the multiplication of octonions is not associative. The set of nonzero elements $\mathbb{O}^\times = \mathbb{O} \setminus \{0\}$ is not a group, let alone $GL_n(\mathbb{O})$. On the other hand, important topological groups (Lie groups) are constructed from octonions. Details can be found, for example, in Yokota's book [Yokota (2009)].

In the above examples, $SL_n(\mathbb{R})$ is contained in $GL_n(\mathbb{R})$ and the structure maps are restrictions of those of $GL_n(\mathbb{R})$. We use the following terminology to state such a situation.

Definition 3.3.18. Let G be a topological group and H a subspace. If the group operation of G can be restricted to make H into a topological group, we say that H is a *subgroup* of G. If H is a closed subset of G, H is called a *closed subgroup)* of G.

Remark 3.3.19. Let μ and ν be structure maps of G. Then a subspace H of G is a subgroup if and only if $\mu(H \times H) \subset H$ and $\nu(H) \subset H$.

Example 3.3.20. The following are inclusions of closed subgroups:

$$\mathbb{R}_{>0} \subset \mathbb{R}^\times$$
$$S^1 \subset \mathbb{C}^\times$$
$$\mathrm{SL}_n(\mathbb{R}) \subset \mathrm{GL}_n(\mathbb{R})$$
$$\mathrm{SO}(n) \subset \mathrm{O}(n) \subset \mathrm{GL}_n(\mathbb{R})$$
$$\mathrm{SL}_n(\mathbb{C}) \subset \mathrm{GL}_n(\mathbb{C})$$
$$\mathrm{U}(n) \subset \mathrm{GL}_n(\mathbb{C})$$
$$\mathrm{SU}(n) \subset U(n)$$
$$\mathrm{Sp}(n) \subset \mathrm{GL}_n(\mathbb{H})$$
$$\mathrm{GL}_n(\mathbb{R}) \subset \mathrm{GL}_n(\mathbb{C}) \subset \mathrm{GL}_n(\mathbb{H})$$
$$\mathrm{O}(n) \subset \mathrm{U}(n) \subset \mathrm{Sp}(n),$$

where the last two inclusions are induced by the inclusions $\mathbb{R} \subset \mathbb{C} \subset \mathbb{H}$. □

Any group can be made into a topological group.

Proposition 3.3.21. *For any group G, the discrete topology makes G into a topological group.*

In order to compare two groups, we need the following.

Definition 3.3.22. For groups G, H, a *homomorphism* from G to H is a map which makes the diagram

$$
\begin{array}{ccc}
G \times G & \xrightarrow{\mu_G} & G \\
{\scriptstyle f \times f}\downarrow & & \downarrow{\scriptstyle f} \\
H \times H & \xrightarrow{\mu_H} & H
\end{array}
$$

commutative, where μ_G and μ_H are multiplications of G and H, respectively.

When G and H are topological groups, we require that a homomorphism $f : G \to H$ to be continuous.

If there exists another homomorphism $g : H \to G$ with $f \circ g = 1_H$ and $g \circ f = 1_G$, f is said to be an isomorphism. When there exists a homomorphism $f : G \to H$, we say G and H are *isomorphic* and denote $G \cong H$.

Remark 3.3.23. In many textbooks of group theory, a map $f : G \to H$ is defined to be an isomorphism if f is a homomorphism and a bijection. When G and H are topological groups, this is not equivalent to our definition.

For example, let $\mathbb{R}_\delta$ be $\mathbb{R}$ with discrete topology. Then the identity map $f : \mathbb{R}_\delta \to \mathbb{R}$ is a bijective homomorphisms of topological groups. However, the inverse of f is not continuous and f cannot be an isomorphism of topological groups.

Example 3.3.24. Define $f : \mathbb{R} \to \mathbb{R}_{>0}$ by $f(x) = e^x$. Then f is a homomorphism of topological groups. In fact, f is an isomorphism. $\square$

Example 3.3.25. The map $\exp : \mathbb{R} \to S^1$ in Exercise 2.1.6 is a homomorphism of topological groups, but is not an isomorphism. $\square$

Example 3.3.26. By linear algebra, the map $\det : \mathrm{GL}_n(\mathbb{R}) \to \mathbb{R}^\times$ defined by taking the determinant is a homomorphism of topological groups. $\square$

Exercise 3.3.27. Define a map $r : S^1 \to \mathrm{SO}(2)$ by

$$r(e^{i\theta}) = \begin{pmatrix} \cos\theta & -\sin\theta \\ \sin\theta & \cos\theta \end{pmatrix}.$$

Show that r is an isomorphism of topological groups.

Example 3.3.28. The map r in the previous exercise can be generalized as follows. Any $A \in M_n(\mathbb{C})$ can be written as $A = A_0 + A_1 i$ by real matrices A_0, A_1. With these matrices, define a map $r : M_n(\mathbb{C}) \to M_{2n}(\mathbb{R})$ by

$$r(A) = \begin{pmatrix} A_0 & -A_1 \\ A_1 & A_0 \end{pmatrix}.$$

The restriction of r to the general linear group induces a homomorphism $r : \mathrm{GL}_n(\mathbb{C}) \to \mathrm{GL}_{2n}(\mathbb{R})$ which is an embedding of topological spaces. This map can be used to regard $\mathrm{GL}_n(\mathbb{C})$ as a closed subgroup of $\mathrm{GL}_{2n}(\mathbb{R})$. Analogous maps define inclusions of closed subgroups

$$\mathrm{GL}_n(\mathbb{C}) \subset \mathrm{GL}_{2n}(\mathbb{R})$$
$$\mathrm{GL}_n(\mathbb{H}) \subset \mathrm{GL}_{2n}(\mathbb{C})$$
$$\mathrm{U}(n) \subset \mathrm{O}(2n)$$
$$\mathrm{Sp}(n) \subset \mathrm{U}(2n).$$

$\square$

The following is a "complexification" of Exercise 3.3.27.

Exercise 3.3.29. Show that there exists an isomorphism of topological groups $\mathrm{SU}(2) \cong \mathrm{Sp}(1)$ by showing that

$$\mathrm{SU}(2) = \left\{ \begin{pmatrix} a & -b \\ \bar{b} & \bar{a} \end{pmatrix} \,\middle|\, a, b \in \mathbb{C}, |a|^2 + |b|^2 = 1 \right\}.$$

Exercise 3.3.30. Prove that SU(2) is homeomorphic to S^3 as topological spaces.

Exercise 3.3.31. Fined a surjective homomorphism from SU(2) to SO(3) and show that the kernel of the map is a group of order 2.

3.4   Compact-Open Topology

Let us go back to the discussion of fiber bundles. Let $p : E \to B$ be a fiber bundle with fiber F. Denote the coordinate transformations by $\Phi^{\alpha\beta} : U_\alpha \cap U_\beta \to \mathrm{Homeo}(F)$. According to the definition in section 3.2, a subgroup G of $\mathrm{Homeo}(F)$ with $\mathrm{Im}\,\Phi^{\alpha\beta} \subset G$ is called a structure group of this fiber bundle.

Since these maps $\Phi^{\alpha\beta}$ are defined by using continuous maps, it is natural to expect that we can make them continuous by introducing an appropriate topology on $\mathrm{Homeo}(F)$. Furthermore, by Proposition 3.2.6, $\mathrm{Homeo}(F)$ is a group under the composition. It would be useful if $\mathrm{Homeo}(F)$ becomes a topological group with this topology. In other words, we would like to define a topology on $\mathrm{Homeo}(F)$ satisfying the following requirements:

(1) The maps $\Phi^{\alpha\beta}$ are continuous.
(2) $\mathrm{Homeo}(F)$ becomes a topological group.

Let us first consider the condition (1). The map $\Phi^{\alpha\beta}(x) : F \to F$ is defined by $\Phi^{\alpha\beta}(x)(y) = \mathrm{pr}_2 \circ \varphi_\beta \circ \varphi_\alpha^{-1}(x, y)$. It is defined by fixing the first entry of the domain of the composition

$$(U_\alpha \cap U_\beta) \times F \xrightarrow{\varphi_\alpha^{-1}} p^{-1}(U_\alpha \cap U_\beta) \xrightarrow{\varphi_\beta} (U_\alpha \cap U_\beta) \times F \xrightarrow{\mathrm{pr}_2} F$$

and then by regarding the second entry as a variable.

More generally, we have the following construction.

Definition 3.4.1. Given a map $\varphi : X \times Y \to Z$, fix $x \in X$ and define a map $\mathrm{ad}(\varphi)(x) : Y \to Z$ by $\mathrm{ad}(\varphi)(x)(y) = \varphi(x, y)$.

Lemma 3.4.2. *If $\varphi : X \times Y \to Z$ is continuous, then, for any $x \in X$, $\mathrm{ad}(\varphi)(x) : Y \to Z$ is continuous.*

Proof. We show that $\mathrm{ad}(\varphi)(x)$ is continuous at each point $y \in Y$. In other words, for each open neighborhood U of $\mathrm{ad}(\varphi)(x)(y)$, we are going to find an open neighborhood V of y such that $\mathrm{ad}(\varphi)(x)(V) \subset U$.

Since $\mathrm{ad}(\varphi)(x)(y) = \varphi(x, y)$, $\varphi(x, y) \in U$, and φ is continuous, there exists an open neighborhood $W \times V$ of (x, y) in $X \times Y$ such that $\varphi(W \times V) \subset U$.

Then $\mathrm{ad}(\varphi)(x)(V) = \varphi(\{x\} \times V) \subset \varphi(W \times V) \subset U$, and $\mathrm{ad}(\varphi)(x)$ is continuous. $\qquad\square$

Definition 3.4.3. For topological spaces X and Y, define

$$\mathrm{Map}(X, Y) = \{f : X \to Y \,|\, \text{continuous}\}.$$

By Lemma 3.4.2, a continuous map $\varphi : X \times Y \to Z$ gives rise to an element $\mathrm{ad}(\varphi)(x) \in \mathrm{Map}(Y, Z)$ for $x \in X$. Thus we obtain a map

$$\mathrm{ad}(\varphi) : X \longrightarrow \mathrm{Map}(Y, Z).$$

This is called the *adjoint*[2] map to φ.

With this notation and terminology, coordinate transformations $\{\Phi^{\alpha\beta}\}$ of a fiber bundle are given as adjoints

$$\Phi^{\alpha\beta} = \mathrm{ad}(\mathrm{pr}_2 \circ \varphi_\beta \circ \varphi_\alpha^{-1}) : U_\alpha \cap U_\beta \longrightarrow \mathrm{Map}(F, F)$$

to the composition

$$(U_\alpha \cap U_\beta) \times F \xrightarrow{\varphi_\alpha^{-1}} p^{-1}(U_\alpha \cap U_\beta) \xrightarrow{\varphi_\beta} (U_\alpha \cap U_\beta) \times F \xrightarrow{\mathrm{pr}_2} F.$$

From this viewpoint, the requirement (1) reduces to the problem of defining a topology on $\mathrm{Map}(Y, Z)$ in such a way that, for any continuous map $\varphi : X \times Y \to Z$, the adjoint $\mathrm{ad}(\varphi) : X \to \mathrm{Map}(Y, Z)$ is continuous. This problem was studied by Fox in [Fox (1945)], in which of compact-open topology was introduced.

Definition 3.4.4. Let X and Y be topological spaces. For subspaces $K \subset X$ and $U \subset Y$, define

$$W(K, U) = \{f : X \to Y \,|\, \text{continuous}, f(K) \subset U\}.$$

Define a topology on $\mathrm{Map}(X, Y)$ by using

$$\mathcal{B} = \{W(K, U) \,|\, K \subset X \text{ compact}, U \subset Y \text{ open}\}$$

as a subbasis. In other words, it is the topology defined by declaring subsets obtained by taking

(1) finite intersections, and then

---

[2]This is a terminology in category theory. See textbooks of category theory, such as [Mac Lane (1998); Riehl (2016)], for details.

(2) arbitrary unions

of $W(K, U)$ for compact set $K \subset X$ and open set $U \subset Y$.
This topology is called the *compact-open topology* on $\mathrm{Map}(X, Y)$.

In the rest of this book, we always regard the set of continuous maps $\mathrm{Map}(X, Y)$ as a topological space by the compact-open topology.

Remark 3.4.5. The reader might wonder why we need to use compact sets in X and open sets in Y. One of interpretations of compact-open topology is described in the appendix §A.1.

The following is one of the most important properties of compact-open topology.

Lemma 3.4.6. *For a continuous map* $\varphi : X \times Y \to Z$, *its adjoint* $\mathrm{ad}(\varphi) :$ $X \to \mathrm{Map}(Y, Z)$ *is continuous.*

Proof. We show that $\mathrm{ad}(\varphi)$ is continuous at x for any $x \in X$. It suffices to find an open neighborhood V of x with $\mathrm{ad}(\varphi)(V) \subset W$ for any open neighborhood W of $\mathrm{ad}(\varphi)(x)$. By the definition of compact-open topology, we may assume that $W = W(C, U)$ for a compact set $C \subset Y$ and an open set $U \subset Z$. Thus it suffices to fined an open neighborhood V of x with $\mathrm{ad}(\varphi)(V) \subset W(C, U)$, when $\mathrm{ad}(\varphi)(x) \in W(C, U)$.

By the assumption $\mathrm{ad}(\varphi)(x) \in W(C, U)$, we have $\mathrm{ad}(\varphi)(x)(C) \subset U$ or $\varphi(\{x\} \times C) \subset U$, which implies that, for any $y \in C$, $\varphi(x, y) \in U$. By the continuity of φ, there exist open neighborhoods V_y of x and W_y of y such that $\varphi(V_y \times W_y) \subset U$. Since $C \subset \bigcup_{y \in C} W_y$ and C is compact, we may choose $W_{y_1}, \ldots, W_{y_n}$ with

$$C \subset W_{y_1} \cup \cdots \cup W_{y_n}.$$

Define $V = \bigcap_{i=1}^{n} V_{y_i}$. For any $x \in V, y \in C$, choose i with $y \in W_{y_i}$. Then $(x, y) \in V \times W_{y_i} \subset V_{y_i} \times W_{y_i}$ and we have

$$\mathrm{ad}(\varphi)(x)(y) = \varphi(x, y) \in \varphi(V_{y_i} \times W_{y_i}) \subset U.$$

$\square$

Corollary 3.4.7. *Let* $p : E \to B$ *be a fiber bundle with fiber* F. *If we regard* $\mathrm{Homeo}(F)$ *as a subspace of* $\mathrm{Map}(F, F)$, *the coordinate transformations* $\Phi^{\alpha\beta} : U_\alpha \cap U_\beta \to \mathrm{Homeo}(F)$ *are continuous.*

By this corollary, the compact-open topology has the property (1) that we need. The next problem is if the other requirement (2) is satisfied, namely if Homeo(F) becomes a topological space by the compact-open topology. The problem is the continuity of the maps

$$\mu : \text{Homeo}(F) \times \text{Homeo}(F) \longrightarrow \text{Homeo}(F)$$
$$\nu : \text{Homeo}(F) \longrightarrow \text{Homeo}(F).$$

Unfortunately, we need to impose a couple of conditions on F.

Definition 3.4.8. We say a topological space X is *locally compact*, if for any $x \in X$ and its one neighborhood V, there exists an open neighborhood U of x such that the closure $\overline{U}$ is compact and $\overline{U} \subset V$.

Theorem 3.4.9. *Let X, Y, and Z be topological spaces. If Y is locally compact Hausdorff, the map*

$$\mu : \text{Map}(Y, Z) \times \text{Map}(X, Y) \longrightarrow \text{Map}(X, Z)$$

defined by the composition of maps is continuous.

Corollary 3.4.10. *If X is locally compact Hausdorff, the composition*

$$\mu : \text{Homeo}(X) \times \text{Homeo}(X) \longrightarrow \text{Homeo}(X)$$

is continuous.

We need the following fact to prove Theorem 3.4.9, whose proof can be found in any one of textbooks on point-set topology.

Proposition 3.4.11. *Any locally compact Hausdorff space is regular.*

For the convenience of the reader, we recall the definition of regular spaces.

Definition 3.4.12. Let X be a Hausdorff space. X is called *regular* if, for any $x \in X$ and a closed set $A \subset X$ with $x \notin A$, there exist open sets U, V of X such that $x \in U$, $A \subset V$, and $U \cap V = \emptyset$. (See Figure 3.5.)

Proof of Theorem 3.4.9. It suffices to show that, for any $(f, g) \in \text{Map}(Y, Z) \times \text{Map}(X, Y)$ and an open neighborhood W of $\mu(f, g)$, there exists an open neighborhood W' of (f, g) satisfying $\mu(W') \subset W$.

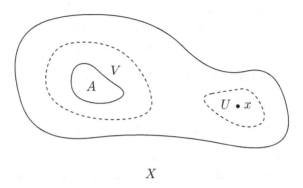

$$X$$

Fig. 3.5   regular space

We may assume that $W = W(C, U)$ by a compact set $C \subset X$ and an open set $U \subset Z$. Then we have

$$\mu(f, g) \in W \iff \mu(f, g) \in W(C, U)$$
$$\iff f \circ g \in W(C, U)$$
$$\iff f(g(C)) \subset U$$
$$\iff g(C) \subset f^{-1}(U)$$
$$\iff \text{for any } y \in g(C), \, y \notin Y \setminus f^{-1}(U).$$

Since Y is locally compact Hausdorff, Y is regular by Proposition 3.4.11, which means that, for any $y \in g(C)$, there exists an open neighborhood V_y of y and an open set W_y containing $Y \setminus f^{-1}(U)$ such that $V_y \cap W_y = \emptyset$. Then we have $V_y \subset Y \setminus W_y \subset f^{-1}(U)$. Since $Y \setminus W_y$ is closed, we have

$$\overline{V}_y \subset Y \setminus W_y \subset f^{-1}(U).$$

We may also assume that $\overline{V}_x$ is compact, since Y is locally compact.

In summary, we have shown that, for any $y \in g(C)$, there exists an open neighborhood V_y of y such that $\overline{V}_y$ is compact and $\overline{V}_y \subset f^{-1}(U)$.

By taking the union for all $y \in g(C)$, we obtain

$$g(C) \subset \bigcup_{y \in g(C)} V_y \subset \bigcup_{y \in g(C)} \overline{V}_y \subset f^{-1}(U).$$

Since C is compact, so is $g(C)$. And we may choose $y_1, \ldots, y_n \in g(C)$ so that $g(C) \subset \bigcup_{i=1}^{n} V_{y_i}$. Now define $V = \bigcup_{i=1}^{n} V_{y_i}$ and we have $g \in W(C, V)$ and

$f \in W(\overline{V}, U)$. Furthermore, define $W' = W(\overline{V}, U) \times W(C, V)$. Then W' is an open neighborhood of (f, g) satisfying $(f, g) \in W' \subset \mu^{-1}(W(C, U))$, since

$$\mu(W(\overline{V}, U) \times W(C, V)) \subset W(C, U).$$

This completes the proof of the continuity of μ. □

We need to impose a stronger condition for the continuity of the map $\nu : \mathrm{Homeo}(X) \to \mathrm{Homeo}(X)$ which assigns the inverse. The problem is the defining condition of the compact-open topology is not symmetric. An obvious solution is to used the following topology.

Definition 3.4.13. For topological spaces X and Y, define

$$\mathcal{B}^S = \left\{ W(K, U) \,\middle|\, \begin{array}{l} K \subset X \text{ is closed, } U \subset Y \text{ is open,} \\ \text{and either } K \text{ or } Y \backslash U \text{ is compact.} \end{array} \right\}$$

The topology of $\mathrm{Map}(X, Y)$ generated by $\mathcal{B}^S$ as a subbasis is called the *symmetrized compact-open topology*.

Remark 3.4.14. The symmetrized compact-open topology is obtained from the usual compact-open topology by adding open sets to make the defining conditions symmetric. This is much less popular than the compact-open topology but still useful. For example, it is used in a paper by Atiyah and Segal [Atiyah and Segal (2004)] on twisted K-theory.

Proposition 3.4.15. *For any topological space X, the map $\nu :$ $\mathrm{Homeo}(X) \to \mathrm{Homeo}(X)$ is continuous is $\mathrm{Homeo}(X)$ is topologized by the symmetrized compact-open topology.*

Proof. It suffices to show that, for any $W(K, U) \in \mathcal{B}^S$, $\nu^{-1}(W(K, U))$ is open. For $f \in \mathrm{Homeo}(X)$, we have

$$f \in \nu^{-1}(W(K, U)) \Longleftrightarrow f^{-1}(K) \subset U$$
$$\Longleftrightarrow K \subset f(U)$$
$$\Longleftrightarrow X \backslash K \supset f(X \backslash U)$$
$$\Longleftrightarrow f^{-1}(X \backslash K) \supset X \backslash U.$$

In other words, $\nu^{-1}(W(K, U)) = W(X \backslash U, X \backslash K)$, which implies that ν is continuous. □

If X is compact Hausdorff, ν becomes continuous even if $\mathrm{Homeo}(X)$ is equipped with the compact-open topology.

Exercise 3.4.16. Prove that ν is continuous if X is compact Hausdorff and $\mathrm{Homeo}(X)$ is topologized by the compact-open topology.

Corollary 3.4.17. *Under the assumption of Proposition 3.4.15, ν : $\mathrm{Homeo}(X) \to \mathrm{Homeo}(X)$ is a homeomorphism.*

Thus we have found conditions under which the requirement (2) stated at the beginning of this section holds.

Corollary 3.4.18. *If X is locally compact Hausdorff and $\mathrm{Homeo}(X)$ has the symmetric compact-open topology or if X is compact Hausdorff and $\mathrm{Homeo}(X)$ has the compact-open topology, $\mathrm{Homeo}(X)$ is a topological group.*

These two conditions for $\mathrm{Homeo}(X)$ being a topological group are not very satisfactory. For example, when $X = \mathbb{R}^n$, we need to use the symmetrized compact-open topology, since it is not compact. It turns out, however, this is too much to require for practical purposes. For small subgroups of $\mathrm{Homeo}(\mathbb{R}^n)$, the compact-open topology suffices.

Example 3.4.19. Regard $\mathrm{GL}_n(\mathbb{R})$ as a subgroup of $\mathrm{Homeo}(\mathbb{R}^n)$. As we will see in Exercise A.1.13, the subspace topology of the compact-open topology of $\mathrm{Homeo}(\mathbb{R}^n)$ and the topology of $\mathrm{GL}_n(\mathbb{R})$ in Example 3.3.6 agree. Thus $\mathrm{GL}_n(\mathbb{R})$ becomes a topological group by the compact-open topology. $\square$

Furthermore we often do not need to use the map ν : $\mathrm{Homeo}(X) \to$ $\mathrm{Homeo}(X)$. For example, when we define actions of topological groups, which is the subject of the next section, we do not need inverses. Hence the compact-open topology suffices. When we remove the existence of inverse from the definition of topological groups, we obtain the following algebraic structure.

Definition 3.4.20. A *topological monoid* consists of

- a topological space M,
- an element $e \in M$, and
- a continuous map $\mu : M \times M \to M$

making the following diagrams commutative

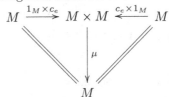

$$M \times M \times M \xrightarrow{\mu \times 1_M} M \times M$$

$$\Big\downarrow{\scriptstyle 1_M \times \mu} \qquad\qquad \Big\downarrow{\scriptstyle \mu}$$

$$M \times M \xrightarrow{\qquad \mu \qquad} M$$

Example 3.4.21. For a locally compact Hausdorff space X, $\mathrm{Map}(X, X)$ becomes a topological monoid by the compact-open topology and the composition of maps. Thus $\mathrm{Homeo}(X)$ is also a topological monoid under these conditions. □

In the rest of this book, we regard $\mathrm{Homeo}(X)$ as a topological monoid by the compact-open topology, instead of making it into a topological group by the symmetrized compact-open topology. In many cases, we only need to deal with small subgroups of $\mathrm{Homeo}(X)$ which happen to be topological groups, as is the case of Example 3.4.19.

3.5 Fiber Bundles and Group Action

In the definition of structure groups in §3.2, the description of coordinate transformations by the homeomorphism group $\mathrm{Homeo}(F)$ plays a central role. The homeomorphism group $\mathrm{Homeo}(F)$ is topologized by the compact-open topology in §3.4. In order to make the definition of structure groups more precise, we consider actions of $\mathrm{Homeo}(F)$ on the fiber F.

Although we have already defined group action in §2.3, groups are not topological group and actions are actions on sets. We need to extend the definition of actions of groups to actions of topological groups on topological spaces. An obvious condition to require is the continuity of actions.

Definition 3.5.1. Let G be a topological group and X a topological space. A *right action* of G on X is a continuous map $\mu : X \times G \to X$ satisfying the conditions in Definition 2.3.23. A left action $G \times X \to X$ is define analogously.

Remark 3.5.2. We do not need inverses in the definition of group actions. Thus Definition 2.3.23 and Definition 3.5.1 can be used to define actions of monoids without a change.

The commutative diagrams used in the definition of group actions look similar to the diagram appeared in the definition of groups in Definition 2.3.11.

Example 3.5.3. Let G a topological group with multiplication given by $\mu : G \times G \to G$. Regard the right factor of the domain of μ as a topological group, the left factor as a topological space, and the range of μ as the same topological spaces as the left fact of the domain. Then μ is a right action of G on G itself. We may, of course, regard μ as a left action of G on itself.

In particular, for a subgroup H of G, the restriction of multiplication of G

$$\mu|_{H \times G} : H \times G \longrightarrow G$$

is a left action of H on G. □

Another important example of an action of a group on itself is given by the *conjugation*.

Example 3.5.4. For a topological group G, define $\mu^c : G \times G \to G$ by $\mu^c(g, h) = ghg^{-1}$ for $(g, h) \in G \times G$. Then this is a left action of G on itself. □

Let us take a look at a couple of more examples of group actions.

Example 3.5.5. Let $G = \mathrm{GL}_n(\mathbb{R})$ and $X = \mathbb{R}^n$. Then, for $A \in G$ and $v \in \mathbb{R}^n$, the matrix multiplication $\mu(A, v) = Av$ defines a left action of G on X. □

Example 3.5.6. Let X be a topological space. Denote the symmetric group of n letters by Σ_n. Define $\mu : X^n \times \Sigma_n \to X^n$ by $\mu((x_1, \cdots, x_n), \sigma) = (x_{\sigma(1)}, \cdots, x_{\sigma(n)})$ for $(x_1, \cdots, x_n) \in X^n$ and $\sigma \in \Sigma_n$. Then this is a right action of Σ_n on X^n. Here we regard Σ_n as a topological group by the discrete topology. □

Remark 3.5.7. In the rest of this book, we regard finite groups as topological groups by the discrete topology.

We obtain the following example by restricting the action in Example 3.5.5.

Example 3.5.8. Let $G = \mathrm{O}(n)$ and $X = S^{n-1}$. Restrict the action of $\mathrm{GL}_n(\mathbb{R})$ on $\mathbb{R}^n$ by inclusions $\mathrm{O}(n) \subset \mathrm{GL}_n(\mathbb{R})$ and $S^{n-1} \subset \mathbb{R}^n$. Then we obtain an action of $\mathrm{O}(n)$ on S^{n-1}. We need to verify that

$$A \in \mathrm{O}(n), \ v \in S^{n-1} \Longrightarrow Av \in S^{n-1}. \tag{3.3}$$

Note that

$$v \in S^{n-1} \Longleftrightarrow |v|^2 = 1 \Longleftrightarrow \langle v, v \rangle = 1 \Longleftrightarrow {}^t v v = 1.$$

Here we regard v as a column vector, i.e. $n \times 1$ matrix. Thus it suffices to show $^t(Av)(Av) = 1$, which follows from

$$^t(Av)(Av) = {}^t v {}^t A A v = {}^t v v = 1,$$

where we used $^t A A = E_n$ by the definition of $O(n)$. And (3.3) is verified.

$\square$

We used the following fact in the previous example and Example 3.5.3.

Lemma 3.5.9. *Let G a topological group action on a topological space X from the left by $\mu : G \times X \to X$. For a subgroup H of G and a subspace A of X satisfying $\mu(H \times A) \subset A$, the restriction $\mu|_{H \times A}$ is an action of H on A.*

Example 3.5.10. Consider the case $n = 2$ in the previous example, which is an action of $O(2)$ on S^1.

Since $A \in O(2)$, $\det A = \pm 1$, which means that

$$\det : O(2) \longrightarrow \{\pm 1\}$$

is a surjective homomorphism. By the fundamental theorem on homomorphism, we obtain an isomorphism $O(2)/\operatorname{Ker}(\det) \cong \{\pm 1\}$. By definition, $\operatorname{Ker}(\det) = SO(2)$. Thus $O(2)$ has a coset decomposition

$$O(2) = SO(2) \cup T \cdot SO(2)$$

by an element $T \in O(2) \setminus SO(2)$. In other words, if $A \in O(2)$, either $A \in SO(2)$ or $A = TB$ for $B \in SO(2)$.

By Exercise 3.3.27, we have

$$SO(2) = \left\{ \begin{pmatrix} \cos\theta & -\sin\theta \\ \sin\theta & \cos\theta \end{pmatrix} \,\middle|\, 0 \le \theta < 2\pi \right\}.$$

Choose $T = \begin{pmatrix} -1 & 0 \\ 0 & 1 \end{pmatrix}$. This is a reflection with respect to the y-axis. Thus the action of $O(2)$ on S^1 is given by rotations, the reflection with respect to the y-axis, or composition of them. $\square$

Remark 3.5.11. The group $SO(n)$ is sometimes called the *rotation group*. In fact, elements of $SO(n)$ can be written as compositions of rotations around lines in $\mathbb{R}^n$ which go through the origin.

The group $O(2)$ contains the following important elements.

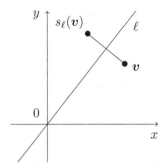

Fig. 3.6   reflection with respect to a line

Exercise 3.5.12. Let ℓ be a line in $\mathbb{R}^2$ which go through the origin and s_ℓ be the *reflection* with respect to ℓ. (See Figure 3.6.)

Find a matrix expression of the linear map s_ℓ and show that it belongs to O(2). Find also an expression of s_ℓ a product of T and an element in SO(2).

Exercise 3.5.13. For $0 \leq \theta < 2\pi$, define

$$R_\theta = \begin{pmatrix} \cos\theta & -\sin\theta \\ \sin\theta & \cos\theta \end{pmatrix}.$$

Let D_{2n} be the subgroup of O(2) generated by $R_{\frac{2\pi}{n}}$ and T for a positive integer $n \geq 3$. (See Figure 3.7.) This group D_{2n} is called the *dihedral group* of the regular n-gon.

(1) Verify that the action of O(2) on $\mathbb{R}^2$ can be restricted to an action of D_{2n} on the regular n-gon in $\mathbb{R}^2$ centered at the origin.

(2) Show that D_{2n} is the group of order $2n$ and find a presentation of D_{2n} by generators and relations.

Let us go back to the discussion of group actions in general. Since an action of G on X is a map

$$\mu : G \times X \longrightarrow X,$$

we may take its adjoint

$$\mathrm{ad}(\mu) : G \longrightarrow \mathrm{Map}(X, X).$$

Recall that we have seen in Proposition 2.4.2 that left actions of a group G on a set X are in one-to-one correspondence to homomorphisms

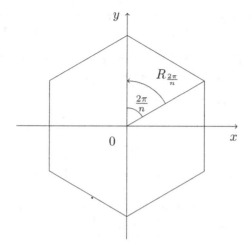

Fig. 3.7   dihedral group

$\mathrm{ad}(\mu) : G \to \mathrm{Aut}(X)$. When G and X have topologies, we would like to replace $\mathrm{Aut}(X)$ by $\mathrm{Homeo}(X)$.

Lemma 3.5.14. *Let G be a topological group action on a topological group X from the left by $\mu : G \times X \to X$. Then the image of the adjoint $\mathrm{ad}(\mu)$: $G \to \mathrm{Map}(X, X)$ is contained in $\mathrm{Homeo}(X)$ and thus we have a map $\mathrm{ad}(\mu)$: $G \to \mathrm{Homeo}(X)$. And this is a continuous homomorphism.*

Proof. We have seen in the proof of Proposition 2.4.2 that $\mathrm{ad}(\mu)(g)$ and $\mathrm{ad}(\mu)(g^{-1})$ are inverse to each other. Thus $\mathrm{ad}(\mu)(g)$ is a homeomorphism for each $g \in G$.

The continuity of $\mathrm{ad}(\mu)$ follows from Proposition 3.4.6. It is already shown in the proof of Proposition 2.4.2 that $\mathrm{ad}(\mu)$ is a homomorphism. $\square$

We have seen that an action $G \times X \to X$ defines a continuous homomorphism $G \to \mathrm{Homeo}(X)$. Conversely can we obtain an action of G on X from such a homomorphism? When we forget the continuity, the correspondence is stated at the end of the proof of Proposition 2.4.2. In order to study the continuity, we use the following notation.

Definition 3.5.15. For a map $\varphi : X \to \mathrm{Map}(Y, Z)$, define $\mathrm{ad}^{-1}(\varphi)$: $X \times Y \to Z$ by $\mathrm{ad}^{-1}(\varphi)(x, y) = \varphi(x)(y)$.

Remark 3.5.16. The notation ad^{-1} is used since it is inverse to the map ad, which is $\mathrm{ad} : \mathrm{Map}(X \times Y, Z) \to \mathrm{Map}(X, \mathrm{Map}(Y, Z))$ by Lemma 3.4.6.

Lemma 3.5.17. *Suppose that Y is locally compact Hausdorff. If $\varphi : X \to \mathrm{Map}(Y, Z)$ is continuous, then so is $\mathrm{ad}^{-1}(\varphi) : X \times Y \to Z$.*

Proof. For an open set $U \subset Z$,

$$\left(\mathrm{ad}^{-1}(\varphi)\right)^{-1}(U) = \{(x, y) \in X \times Y \mid \varphi(x)(y) \in U\}.$$

For $(x_0, y_0) \in \left(\mathrm{ad}^{-1}(\varphi)\right)^{-1}(U)$, $\varphi(x_0)^{-1}(U)$ is open since $\varphi(x_0) : Y \to Z$ is continuous. $Y \setminus \varphi(x_0)^{-1}(U)$ is a closed set of Y which does not contain y_0.

By the assumption and Proposition 3.4.11, Y is regular. There exists an open neighborhood V of y_0 such that $\overline{V} \cap (Y \setminus \varphi(x_0)^{-1}(U)) = \emptyset$ or $\varphi(x_0)(\overline{V}) \subset U$.

Since Y is locally compact, we may assume that $\overline{V}$ is compact. Then $\varphi^{-1}(W(\overline{V}, U)) \times V$ is an open neighborhood of (x_0, y_0) with $\varphi^{-1}(W(\overline{V}, U)) \times V \subset \left(\mathrm{ad}^{-1}(\varphi)\right)^{-1}(U)$. Thus $\mathrm{ad}^{-1}(\varphi)$ is continuous. $\quad\square$

Corollary 3.5.18. *Let G be a topological group and X a locally compact Hausdorff space. If $\varphi : G \to \mathrm{Homeo}(X)$ is a continuous homomorphism, then $\mathrm{ad}^{-1}(\varphi) : G \times X \to X$ is an action of G on X.*

Proof. By Lemma 3.5.17, $\mathrm{ad}^{-1}(\varphi)$ is continuous. The fact that $\mathrm{ad}^{-1}(\varphi)$ satisfies the conditions of action can be verified immediately. $\quad\square$

Since the identity map

$$1_{\mathrm{Homeo}(X)} : \mathrm{Homeo}(X) \longrightarrow \mathrm{Homeo}(X)$$

is a continuous homomorphism, we obtain a left action of $\mathrm{Homeo}(X)$ on X.

Corollary 3.5.19. *If X is locally compact Hausdorff, then the topological monoid $\mathrm{Homeo}(X)$ acts on X continuously.*

Thanks to Corollary 3.5.18, we may identify actions of a topological group G on X and continuous homomorphisms $G \to \mathrm{Homeo}(X)$, when X is locally compact Hausdorff.

Now we are ready to discuss structure groups of fiber bundles. In §3.2, we proposed a temporary definition of a structure group G of a fiber bundle with fiber F as an inclusion $G \hookrightarrow \mathrm{Homeo}(F)$ of a subgroup. Since it is a homomorphism, if F is locally compact Hausdorff, it defines an action of G on F, if it is continuous. We should also use more general homomorphisms.

Definition 3.5.20 (Precise Definition of Structure Group). Let G be a topological group acting on a locally compact Hausdorff space F by a continuous homomorphism $G \to \mathrm{Homeo}(F)$.

A fiber bundle (B, E, F) is said to have G as a *structure group* if, for any α and β, there exists a continuous map $\overline{\Phi}^{\alpha\beta}$ making the diagram

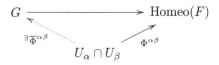

is commutative, where $\{\Phi^{\alpha\beta}\}$ are coordinate transformations of the fiber bundle.

Remark 3.5.21. In the rest of this book, the term "structure group" is used in this sense. We also assume that the fiber of a fiber bundle is always locally compact Hausdorff.

Remark 3.5.22. It should be noted that when we use a fiber bundle with a structure group, we should regard local trivializations, coordinate transformations, and the action of the structure group on the fiber as parts of defining data of the fiber bundle.

Example 3.5.23. Let us verify that the Hopf bundle $S^3 \xrightarrow{p} S^2$ in Example 3.1.11 has S^1 as a structure group. Recall that the local trivializations

$$\varphi_+ : p^{-1}(U_+) \longrightarrow U_+ \times S^1$$
$$\varphi_- : p^{-1}(U_-) \longrightarrow U_- \times S^1$$
$$\varphi_+^{-1} : U_+ \times S^1 \longrightarrow p^{-1}(U_+)$$

are given by

$$\varphi_+(z_1, z_2) = \left(2|z_1|^2 - 1, 2z_1\bar{z}_2, \frac{z_2}{|z_2|} \right)$$

$$\varphi_-(z_1, z_2) = \left(2|z_1|^2 - 1, 2z_1\bar{z}_2, \frac{z_1}{|z_1|} \right)$$

$$\varphi_+^{-1}(x, z, w) = \left(\frac{zw}{2\sqrt{\frac{1-x}{2}}}, w\sqrt{\frac{1-x}{2}} \right),$$

where

$$U_+ = S^2 \setminus \{(1, 0)\}$$
$$U_- = S^2 \setminus \{(-1, 0)\}.$$

Thus

$$\varphi_- \circ \varphi_+^{-1} : (U_+ \cap U_-) \times S^1 \longrightarrow (U_+ \cap U_-) \times S^1$$

is given by

$$\varphi_- \circ \varphi_+^{-1}(x, z, w) = \varphi_- \left(\frac{zw}{2\sqrt{\frac{1-x}{2}}}, w\sqrt{\frac{1-x}{2}} \right)$$

$$= \left(2\frac{|zw|^2}{4(\frac{1-x}{2})} - 1, 2\frac{zw}{2\sqrt{\frac{1-x}{2}}}\overline{w}\sqrt{\frac{1-x}{2}}, \frac{\frac{zw}{2\sqrt{\frac{1-x}{2}}}}{\left|\frac{zw}{2\sqrt{\frac{1-x}{2}}}\right|} \right)$$

$$= \left(\frac{|z|^2}{1-x} - 1, z, \frac{z}{|z|}w \right)$$

$$= \left(x, z, \frac{z}{|z|}w \right).$$

In other words, the coordinate transformation $\Phi^{+-} : U_+ \cap U_- \to \mathrm{Homeo}(S^1)$ is given by $\Phi^{+-}(x,z)(w) = \frac{z}{|z|}w$. And $\frac{z}{|z|} \in S^1$. Define a map $\overline{\Phi}^{+-}$: $U_+ \cap U_- \to S^1$ by $\overline{\Phi}^{+-}(x,z) = \frac{z}{|z|}$. Then the diagram

$$
\begin{array}{ccc}
S^1 & \xrightarrow{\mathrm{ad}(\mu)} & \mathrm{Homeo}(S^1) \\
& \overline{\Phi}^{+-} \nwarrow \quad \nearrow \Phi^{+-} & \\
& U_+ \cap U_- &
\end{array}
$$

is commutative. Here $\mathrm{ad}(\mu) : S^1 \to \mathrm{Homeo}(S^1)$ is the homomorphism defined by the action of S^1 on itself given by the multiplications in S^1. Thus the structure group of the Hopf bundle can be made as small as S^1. □

Recall from §1.2 that the fiber of the tangent bundle $TS^2 \to S^2$ is $\mathbb{R}^2$. It turns out that the coordinate transformations are linear maps. Such fiber bundles form an important class of fiber bundles with structure groups.

Definition 3.5.24. A fiber bundle is called a (real) *vector bundle* of *rank n*, if its fiber is a vector space V of dimension n over $\mathbb{R}$ and the structure group is $\mathrm{GL}(V)$. When its fiber is a complex vector space V and the structure group is $\mathrm{GL}(V)$, it is called a *complex vector bundle*.

Example 3.5.25. The tangent bundle of S^2 in §1.2 can be generalized to smooth manifolds. When the manifold M is of dimension n, the tangent bundle $TM \to M$ becomes a vector bundle of rank n. A proof can be found in any textbook on differential topology or differential geometry. □

As we will see in §A.2, the structure group of a rank n real vector bundle can be "shrinked" from $\mathrm{GL}_n(\mathbb{R})$ to the orthogonal group $O(n)$ if the base space is paracompact Hausdorff. In order to make a precise statement, we need the notion of reduction of structure groups.

Definition 3.5.26. Let $\xi = (B, E, F)$ be a fiber bundle with structure group G. If the coordinate transformations $\Phi^{\alpha\beta} : U_\alpha \cap U_\beta \to G$ of this fiber bundle takes values in a subgroup H of G, we say that the structure group of ξ can be *reduced* to H.

Example 3.5.27. By Example 3.5.23, the structure group of the Hopf bundle can be reduced to S^1. When a point (x, z) moves around in $U_+ \cap U_-$, the point $\frac{z}{|z|}$ covers all over S^1. Thus the structure group can not be reduced any more.   □

The operations on vector spaces such as direct sum, dual vector space, tensor product, and exterior product can be extended to vector bundles. For the reader's convenience, they are briefly described in §A.2. These operations allow us to define important vector bundles such as the cotangent bundle T^*M and its exterior powers $\wedge^n T^*M$ from the tangent bundle TM. For general fiber bundles, the following operation is fundamental.

Proposition 3.5.28. *Let $\xi = (p : E \to B)$ and $\xi' = (p' : E' \to B')$ be fiber bundles with fibers F and F', structure groups G and G', respectively. The the product*

$$p \times p' : E \times E' \longrightarrow B \times B'$$

is a fiber bundle with fiber $F \times F'$ and structure group $G \times G'$.

Proof. Let $\{\varphi_\alpha : p^{-1}(U_\alpha) \to U_\alpha \times F\}_{\alpha \in A}$ and $\{\psi_\beta : {p'}^{-1}(V_\beta) \to V_\beta \times F\}_{\beta \in B}$ be local trivializations of ξ and ξ', respectively. Then $\{U_\alpha \times V_\beta\}_{(\alpha,\beta) \in A \times B}$ is an open covering of $B \times B'$. Under the identification

$$(U_\alpha \times V_\beta) \cap (U_{\alpha'} \times V_{\beta'}) = (U_\alpha \cap U_{\alpha'}) \times (V_\beta \cap V_{\beta'}),$$

we obtain maps

$$\Phi^{\alpha\alpha'} \times \Psi^{\beta\beta'} : (U_\alpha \cap U_{\alpha'}) \times (V_\beta \cap V_{\beta'}) \longrightarrow G \times G'$$

which form coordinate transformations of the product bundle.   □

Corollary 3.5.29. *Let $p : E \to B$ be a fiber bundle with fiber F and structure group G, then, for any topological space X*

$$p \times 1_X : E \times X \longrightarrow B \times X$$

is a fiber bundle with fiber F and structure group G.

We have begun §3.2 by saying that fiber bundles are obtained by gluing trivial bundles and that the gluing information is described as coordinate transformations. Before we end this section, let us give a precise statement of this observation as a theorem.

We first need to define "quotient spaces" which is a topological version of quotient sets in Definition 2.3.17.

Definition 3.5.30. For a topological space X and an equivalence relation $\sim$ on X, define a topology $\mathcal{O}$ on $X/_{\sim}$ by

$$\mathcal{O} = \left\{ U \subset X/_{\sim} \,\middle|\, \pi^{-1}(U) \text{ is open in } X \right\},$$

where $\pi : X \to X/_{\sim}$ is the projection, i.e. $p(x) = [x]$. This map is called the *quotient map*.

When $X/_{\sim}$ is topologized by this topology, it is called the *quotient space* of X by $\sim$. The topology $\mathcal{O}$ is called the *quotient topology*.

Suppose an open covering $\{U_\alpha\}_{\alpha \in A}$ of B and a family of maps $\{\varphi^{\alpha\alpha'} : U_\alpha \cap U_{\alpha'} \to G\}_{\alpha,\alpha' \in A}$ are given. Then a precise way of constructing a fiber bundle over B by gluing trivial bundles $U_\alpha \times F$ can be stated as follows.

Theorem 3.5.31. *Let G be a topological group acting on a topological space F from the left. Suppose an open covering $\{U_\alpha\}_{\alpha \in A}$ of B and a family of maps $\{\Phi^{\alpha\alpha'} : U_\alpha \cap U_{\alpha'} \to G\}_{\alpha,\alpha' \in A}$ satisfying the following conditions is given:*

(1) $\Phi^{\alpha\alpha}$ is the constant map to the identity.
(2) For any $x \in U_\alpha \cap U_{\alpha'}$, $\Phi^{\alpha'\alpha}(x) = \Phi^{\alpha\alpha'}(x)^{-1}$.
(3) For any $x \in U_{\alpha_1} \cap U_{\alpha_2} \cap U_{\alpha_3}$, $\Phi^{\alpha_2\alpha_3}(x)\Phi^{\alpha_1\alpha_2}(x) = \Phi^{\alpha_1\alpha_3}(x)$.

Define an equivalence relation $\sim$ on $\coprod_\alpha U_\alpha \times F$ by

$$(x,y) \sim (x',y') \iff x = x' \text{ and } \Phi^{\alpha'\alpha}(x)y = y'$$

for $(x,y) \in U_\alpha \times F$ and $(x',y') \in U_{\alpha'} \times F$. Let us denote the quotient space by

$$E = \left(\coprod_{\alpha \in A} U_\alpha \times F \right)\bigg/_{\sim}$$

and define a map $p : E \to B$ by $p([x,y]) = x$. Then this is a fiber bundle with fiber F, structure group G, and coordinate transformations $\{\Phi^{\alpha\alpha'}\}_{\alpha,\alpha' \in A}$.

Remark 3.5.32. Note that the three conditions on coordinate transformations guarantee that the relation $\sim$ is an equivalence relation.

Proof. The continuity of the map p follows from the property of quotient maps, whose proof is Exercise 3.5.33 below.

Let us define local trivializations. For $e \in E$, choose α with $p(e) \in U_\alpha$. By the definition of the equivalence relation, there exists a unique $y \in F$ with $e = [(p(e), y)]$. This correspondence defines a bijection

$$\varphi_\alpha : p^{-1}(U_\alpha) \longrightarrow U_\alpha \times F.$$

By the definition of the product topology, this map is continuous if both

$$\mathrm{pr}_1 \circ \varphi_\alpha : p^{-1}(U_\alpha) \longrightarrow U_\alpha$$
$$\mathrm{pr}_2 \circ \varphi_\alpha : p^{-1}(U_\alpha) \longrightarrow F$$

are continuous. The former is continuous since it coincides with a restriction of p. In other to prove the continuity of the latter, it suffices to show that $(\mathrm{pr}_2 \circ \varphi_\alpha)^{-1}(V)$ is open in $p^{-1}(U_\alpha)$ for any open $V \subset F$. By the definition of quotient topology, it is enough to show that $\pi^{-1}((\mathrm{pr}_2 \circ \varphi_\alpha)^{-1}(V))$ is open in $\coprod_{\beta \in A} U_\beta \times F$, where

$$\pi : \coprod_{\alpha \in A} U_\alpha \times F \longrightarrow E$$

is the projection. This follows from

$$\pi^{-1}((\mathrm{pr}_2 \circ \varphi_\alpha)^{-1}(V)) = \coprod_{\beta \in A} (U_\alpha \cap U_\beta) \times \Phi^{\beta\alpha}(V)$$

and the fact that $\Phi^{\beta\alpha}$ is a homeomorphism. Thus p is a fiber bundle with fiber F.

By the definition of the equivalence relation, coordinate transformations are given by $\Phi^{\alpha\alpha'}$. □

Exercise 3.5.33. Suppose an equivalence relation $\sim$ is defined on a topological space X. Let $p : X \to X/\sim$ be the quotient map. Show that, if a continuous map $f : X \to Y$ satisfies $x \sim x' \Rightarrow f(x) = f(x')$, then the map $\bar{f} : X/\sim \to Y$ defined by $\bar{f}([x]) = f(x)$ is continuous.

3.6 Quotient Spaces by Group Actions

As we have seen so far, actions of topological groups play an essential role in the study of fiber bundles with structure groups. Another important role of topological groups in the study of fiber bundles is the construction of quotient spaces.

Definition 3.6.1. Suppose a group G acts on a topological space X from the left. Define an equivalence relation $\sim_G$ on X by

$$x \sim_G x' \iff x' = gx \text{ for some } g \in G$$

for $x, x' \in X$. The quotient space $X_{\sim_G}$ by this equivalence relation is denoted by X/G and is called the *quotient space* of X by G.

The equivalence class in X/G represented by $x \in X$ is called *orbit* of x under the action of G.

Exercise 3.6.2. Verify that the relation $\sim_G$ is an equivalence relation.

Remark 3.6.3. The reader might wonder why X/G is quotiented out from the right while G acts on X from the left. In fact, it is common to write the quotient space by a left action by $G \backslash X$ and reserve X/G for the quotient space by a right action.

Since we do not need to distinguish left and right actions in this book, we use X/G for both cases.

Remark 3.6.4. When a group G acts on X from the left, the set $Gx = \{gx \,|\, g \in G\} \subset X$ coincides with the equivalence class represented by x. Namely Gx is the orbit of x. Thus we may write $X/G = \{Gx \,|\, x \in X\}$.

Example 3.6.5. As we have seen in Example 3.5.3, when H is a subgroup of G, the multiplication of G induces an action of H on G. The quotient space is, when topologies are forgotten, the set of cosets G/H. $\quad\square$

Example 3.6.6. Regard the unit 2-dimensional sphere and the unit circle as

$$S^2 = \left\{ (x, z) \in \mathbb{R} \times \mathbb{C} \,\middle|\, x^2 + |z|^2 = 1 \right\}$$
$$S^1 = \left\{ z \in \mathbb{C} \,\middle|\, |z|^2 = 1 \right\}$$

as we have done in Example 3.1.11. By Example 3.3.4, S^1 is a topological group by the multiplication of complex numbers. Define

$$z \cdot (x, w) = (x, zw)$$

for $z \in S^1$ and $(x, w) \in S^2$. Then this is an action of S^1 on S^2. This is a rotation of S^2 around the x-axis. The orbit of (x, w) is the vertical circle that contains (x, w). $\quad\square$

When we deal with quotient spaces by group actions, it is better to interpret the quotient space X/G as the space obtained from X by identifying points that correspond to each other by an action of an element of G rather than the set of equivalence classes.

Example 3.6.7. Let $X = S^1 = \{z \in \mathbb{C} \,|\, |z| = 1\}$ and $G = C_2 = 1, -1$, the cyclic group of order 2 under the multiplication. Define an action of C_2 on S^1 by

$$1 \cdot z = z$$
$$(-1) \cdot z = -z.$$

The quotient space S^1/C_2 can be obtained as follows.

The action of -1 maps $z \in S^1$ to the antipodal point $-z$.

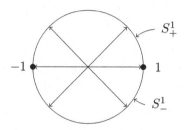

Fig. 3.8   the action of C_2 on S^1

Decompose S^1

$$S^1 = S_+^1 \cup S_-^1 \cup \{1\} \cup \{-1\}$$

as is shown in Figure 3.8. Then the action of -1 interchanges S_+^1 and S_-^1, and 1 and -1, respectively. Thus S_+^1 and S_-^1 are identified by the action C_2, which gives rise to a semicircle, and then the endpoints 1 and -1 are identified, resulting in a circle. In other words, S^1/C_2 can be identified with S^1.

In order to make this argument precise, we define a map $f : S^1/C_2 \to S^1$ by $f([z]) = z^2$. We claim that this is a homeomorphism. We need to verify the following:

(1) f is well-defined. In other words, the definition is independent of a choice of representative.
(2) f is continuous.
(3) f is a bijection.
(4) f^{-1} is continuous.

The well-definedness and the bijectivity are immediate since $z \sim_{C_2} z'$ is equivalent to $z = z'$ or $z = -z'$.

To show that f is continuous, let U be an open set of S^1. By the definition of topology of S^1/C_2, it suffices to show that $p^{-1}(f^{-1}(U))$ is open.

Note that $p^{-1}(f^{-1}(U)) = (f \circ p)^{-1}(U)$ and the composition $f \circ p : S^1 \to S^1$ is continuous since it is given by $(f \circ p)(z) = z^2$. Thus f is continuous. It remains to show that f^{-1} is continuous or f is an open map or a closed map, which follows from the next Lemma 3.6.8. $\square$

The proof of the following fact used in the above proof is left to the reader.

Lemma 3.6.8. *If X is compact and Y is Hausdorff, any continuous map $f : X \to Y$ is a closed map. Thus if f is a continuous bijection, f is a homeomorphism.*

In order to apply this fact to Example 3.6.7, we need to show that the quotient space S^1/C_2 is compact, which is guaranteed by the following lemma, which follows from the fact that the image of a compact set by a continuous map is compact.

Lemma 3.6.9. *Let $\sim$ be an equivalence relation on a topological space X. If X is compact, so is $X/\sim$.*

The argument used in Example 3.6.7 to show that f is a homeomorphism is a fundamental technique when we deal with maps from quotient spaces.

Definition 3.6.10. A map $f : X \to Y$ between topological spaces is called a *quotient map* if

(1) f is surjective, and
(2) for any $U \subset Y$, U is open, whenever $f^{-1}(U)$ is open.

The following are well-known but important cases. The proof is left to the reader.

Lemma 3.6.11. *Let $f : X \to Y$ be a surjective map. If f is open or closed, then f is a quotient map.*

Corollary 3.6.12. *If X is compact and Y is Hausdorff, then any continuous surjective map $f : X \to Y$ is a quotient map.*

The following interpretation is immediate from the definition.

Lemma 3.6.13. *When $f : X \to Y$ is a quotient map, define a relation $\sim_f$ on X by*

$$x \sim_f x' \iff f(x) = f(x').$$

Then this is an equivalence relation and we have a homeomorphism $Y \cong X/_{\sim_f}$.

Lemma 3.6.14. *Suppose we have a commutative diagram*

in which π is a quotient map. Then g is continuous if and only if f is continuous.

Proof. By Lemma 3.6.13, we may assume that $Y = X/_{\sim}$. By Exercise 3.5.33, g is continuous. Conversely if g is continuous, f is continuous as a composition of continuous maps. $\square$

Note that the map $f : S^1/C_2 \to S^1$ in Example 3.6.7 is related to the covering space in Example 2.1.4 in the sense that the diagram

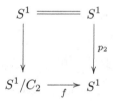

is commutative. This example can be generalized as follows.

Exercise 3.6.15. Let $X = S^1$. We regard the cyclic group C_n of order n as $C_n = \{1, \zeta_n, \zeta_n^2, \ldots, \zeta_n^{n-1}\}$ by using a primitive root of unity $\zeta_n = e^{\frac{2\pi i}{n}}$. The group operation is given by the complex multiplication. Define an action of C_n on X also by the complex multiplication. Show that we have a homeomorphism $S^1/C_n \cong S^1$ and describe a relation to the covering space p_n in Exercise 2.1.6.

The following is a higher dimensional analogue.

Example 3.6.16. Define an action of C_2 on the n-dimensional sphere $S^n = \{(x_0, \ldots, x_n) \in \mathbb{R}^{n+1} \mid x_0^2 + \cdots + x_n^2 = 1\}$ by $(-1) \cdot (x_0, \ldots, x_n) = (-x_0, \ldots, -x_n)$. The quotient space is denoted by $\mathbb{R}P^n$ and is called the *real projective space* of dimension n. $\square$

Example 3.6.17. As is the case of S^3 in Example 3.1.11, regard an odd dimensional sphere

$$S^{2n+1} = \left\{ (z_0, \ldots, z_n) \in \mathbb{C}^{n+1} \,\middle|\, |z_0|^2 + \cdots + |z_n|^2 = 1 \right\}$$

as a subspace of the complex vector space $\mathbb{C}^{n+1}$. Define an action of the cyclic group C_p on S^{2n+1} by $\zeta_p(z_0, \ldots, z_n) = (\zeta_p z_0, \ldots, \zeta_p z_n)$ for $(z_0, \ldots, z_n) \in S^{2n+1}$. The quotient space by this action is denoted by L_p^{2n+1} and is called the *lens space*. $\square$

Example 3.6.18. Regard S^1 as $S^1 = \{ z \in \mathbb{C} \mid |z| = 1 \}$ and make it into a topological group by the complex multiplication as we have done in Example 3.3.4. Define an action of S^1 on S^{2n+1} by $\omega(z_0, \ldots, z_n) = (\omega z_0, \ldots, \omega z_n)$ for $\omega \in S^1$ and $(z_0, \ldots, z_n) \in S^{2n+1}$. The quotient space of this action is denoted by $\mathbb{C}P^n$ and is called the *complex projective space*. This is a $2n$-dimensional manifold. $\square$

Exercise 3.6.19. Prove that $\mathbb{C}P^1$ is homeomorphic to S^2. Or find a homeomorphism $S^3/S^1 \cong S^2$.

Standard definitions of projective spaces $\mathbb{R}P^n$ and $\mathbb{C}P^n$ are as follows.

Exercise 3.6.20. Define actions

$$\mathbb{R}^{\times} \times (\mathbb{R}^{n+1} \setminus \{0\}) \longrightarrow \mathbb{R}^{n+1} \setminus \{0\}$$
$$\mathbb{C}^{\times} \times (\mathbb{C}^{n+1} \setminus \{0\}) \longrightarrow \mathbb{C}^{n+1} \setminus \{0\}$$

of the multiplicative groups $\mathbb{R}^{\times}$ and $\mathbb{C}^{\times}$ on $\mathbb{R}^{n+1} \setminus \{0\}$ and $\mathbb{C}^{n+1} \setminus \{0\}$ by the same formula $\omega \cdot (x_0, \ldots, x_n) = (\omega x_0, \ldots, \omega x_n)$.
Show that we have homeomorphisms

$$(\mathbb{R}^{n+1} \setminus \{0\})/\mathbb{R}^{\times} \cong \mathbb{R}P^n$$
$$(\mathbb{C}^{n+1} \setminus \{0\})/\mathbb{C}^{\times} \cong \mathbb{C}P^n.$$

Remark 3.6.21. It is known that we may also represent projective spaces as a subspace of the space of square matrices. For example, by Exercise 3.3.31, we have a homeomorphism $\mathbb{R}P^3 \cong SO(3)$.

In some cases quotient spaces become quite simple.

Example 3.6.22. Recall Example 3.5.8. We have

$$O(n) = \left\{ A \in M_n(\mathbb{R}) \,\middle|\, {}^t\!AA = 1_n \right\}$$
$$S^{n-1} = \left\{ (x_1, \ldots, x_n) \in \mathbb{R}^n \,\middle|\, x_1^2 + \cdots + x_n^2 = 1 \right\}$$

and the action $O(n) \times S^{n-1} \to S^{n-1}$ of $O(n)$ on S^{n-1} is given by the matrix multiplication. Let us consider the quotient space.

Let us begin with the case $n = 2$. By Example 3.5.10, $O(2)$ contains

$$SO(2) = \{A \in O(2) \mid \det A = 1\}$$
$$= \left\{ \begin{pmatrix} \cos\theta & -\sin\theta \\ \sin\theta & \cos\theta \end{pmatrix} \middle| 0 \le \theta < 2\pi \right\}$$

as a subgroup. Take $v, w \in S^1$ and let θ be the angle from v to w. We have $R_\theta v = w$ as is shown in Figure 3.9, where R_θ is the matrix used in Exercise 3.5.13.

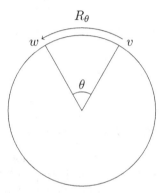

Fig. 3.9   rotation by angle θ

In other words, any two points in S^1 can be mapped to each other by the action of $O(2)$, which means that all points in S^1 are equivalent under this action and the quotient space consists of a single point $S^1/O(2) = \{*\}$. The above argument also show that we already have $S^1/SO(2) = \{*\}$.

This argument can be extended to show that $S^{n-1}/SO(n) = \{*\}$, which will be discussed in Example 3.6.42. □

This situation suggests the following definition.

Definition 3.6.23. Suppose a topological group G acts on a topological space X. We say this action is *transitive* if, for any $x, y \in X$, there exists $g \in G$ with $x = gy$.

With this terminology, what we have done in Example 3.6.22 is to show that the action of $O(2)$ or $SO(2)$ on S^1 is transitive. Thus we obtain the following.

Corollary 3.6.24. *If a topological group G acts on a topological space X transitively, then the quotient space X/G consists of a single point. In other words, we have $X = Gx$ for any $x \in X$.*

The following is a typical transitive action.

Proposition 3.6.25. *Let G be a topological group and H a subgroup. Define an action $\bar{\mu} : G \times G/H \to G/H$ of G on G/H by $\bar{\mu}(g, g'H) = \mu(g, g')H$ by using the multiplication $\mu : G \times G \to G$ of G. Then this is a transitive action.*

The proof is obvious from the definition, once the continuity of $\bar{\mu}$ is proved, and is omitted.

The continuity of $\bar{\mu}$ can be verified as follows. Consider the commutative diagram

$$
\begin{array}{ccc}
G \times G & \xrightarrow{\ \mu\ } & G \\
\downarrow{\scriptstyle 1_G \times p} & & \downarrow{\scriptstyle p} \\
G \times G/H & \xrightarrow[\ \bar{\mu}\ ]{} & G/H.
\end{array}
$$

Since $p \circ \mu$ is continuous, if $1_G \times p$ were to be a quotient map, Lemma 3.6.14 tells us that $\bar{\mu}$ is continuous. Although both p and 1_G are quotient maps, the product $1_G \times p$ may not be so in general. The following is a famous example.

Example 3.6.26. Regard $\mathbb{Q}$ as a topological space by the subspace topology of $\mathbb{R}$. Let $\sim_{\mathbb{N}}$ be the equivalence relation which identifies $\mathbb{N}$ to a single point and $p : \mathbb{Q} \to \mathbb{Q}/\sim_{\mathbb{N}}$ the quotient map. Then the product $1_{\mathbb{Q}} \times p : \mathbb{Q} \times \mathbb{Q} \to \mathbb{Q} \times (\mathbb{Q}/\sim_{\mathbb{N}})$ is not a quotient map. See p. 143 Example 7 of Munkres' book [Munkres (2000)]. $\square$

Most spaces studied in topology and geometry, such as manifolds and CW complexes, behave much more nicely than $\mathbb{Q}$. In fact, the product map $1_G \times p$ is a quotient map in this case. Hence $\bar{\mu}$ is continuous. In order to see this, the following facts can be used.

Lemma 3.6.27. *For a topological group G and its subgroup H, the projection $p : G \to G/H$ is an open map.*

Proof. For an open set U in G, we have

$$
p^{-1}(p(U)) = \bigcup_{h \in H} Uh.
$$

Since the map $G \to G$ obtained by the multiplication of h is a homeomorphism, each Uh is an open set of G and so is $p^{-1}(p(U))$. By the definition of quotient topology $p(U)$ is open in G/H. $\qquad\square$

Lemma 3.6.28. *Any open surjective map $f : X \to Y$ is a quotient map. Thus for any topological space Z, the product map $f \times 1_Z : X \times Z \to Y \times Z$ is a quotient map.*

The proof is immediate from the definition and is omitted.

For more general cases, we have the following useful fact which can be found in [Komatsu *et al.* (1967)] written in Japanese. We include a proof for the convenience of the reader.

Theorem 3.6.29. *Let $f : X \to Y$ be a quotient map. Suppose that, for an open set $V \subset X$ with $f^{-1}(f(V)) = V$ and $x \in V$, there exists an open set U of X satisfying the following conditions:*

(1) $x \in U$,
(2) $U = f^{-1}(f(U))$, and
(3) $\overline{U}$ is compact and $\overline{U} \subset V$.

Then for any quotient map $g : X' \to Y'$, $f \times g : X \times X' \to Y \times Y'$ is a quotient map.

Proof. Take $B \subset Y \times Y'$ and suppose that $(f \times g)^{-1}(B)$ is open in $X \times X'$. We show that for any $(y, y') \in B$, there exist open neighborhoods V and V' of y and y' such that $V \times V' \subset B$. Since both f and g are quotient maps, it suffices to find open subsets $U \subset X$ and $U \subset X'$ satisfying the following conditions

(1) $U = f^{-1}(f(U))$,
(2) $U' = g^{-1}(g(U'))$, and
(3) $(y, y') \in (f \times g)(U \times U') \subset B$.

Take any $x_0 \in f^{-1}(y)$ and $x'_0 \in g^{-1}(y')$. Define

$$V = \mathrm{pr}_1\left((X \times \{x'_0\}) \cap (f \times g)^{-1}(B)\right).$$

We claim that this is open in X. For $x \in V$, since $(f \times g)^{-1}(B)$ is open, there exist open neighborhoods W and W' of x and x'_0, respectively, such that

$$(x, x'_0) \in W \times W' \subset (f \times g)^{-1}(B).$$

Then $x \in W \subset V$ and x is an interior point.

We next claim that $f^{-1}(f(V)) = V$. By the commutativity of the diagram

$$X \times X' \xrightarrow{\ f \times g\ } Y \times Y'$$

$$\downarrow{\mathrm{pr}_1} \qquad\qquad \downarrow{\mathrm{pr}_1}$$

$$X \xrightarrow{\quad f \quad} Y,$$

we have

$$f(V) = f(\mathrm{pr}_1\left((X \times \{x_0'\}) \cap (f \times g)^{-1}(B)\right))$$
$$= \mathrm{pr}_1((f \times g)\left((X \times \{x_0'\}) \cap (f \times g)^{-1}(B)\right))$$
$$= \mathrm{pr}_1((f(X) \times \{g(x_0')\}) \cap B).$$

In other words, $y \in f(V)$ if and only if $(y, g(x_0')) \in B$. Thus

$$x \in f^{-1}(f(V)) \iff (f(x), g(x_0')) \in B$$
$$\iff (x, x_0') \in (f \times g)^{-1}(B)$$
$$\iff x \in V.$$

By assumption, there exists an open set U of X such that $x_0 \in U$, $U = f^{-1}(f(U))$, $\overline{U} \subset V$, and $\overline{U}$ is compact. Define

$$U' = \left\{ x' \in X' \,\middle|\, \overline{U} \times \{x'\} \subset (f \times g)^{-1}(B) \right\}$$

so that (3) holds.

Let us show that U' is open. For any $x' \in U'$, we have $\overline{U} \times \{x'\} \subset (f \times g)^{-1}(B)$ by definition. Since $(f \times g)^{-1}(B)$ is open, for each $x \in \overline{U}$, there exist open neighborhood O_x and O_x' of x and x', respectively, such that $O_x \times O_x' \subset (f \times g)^{-1}(B)$. The collection $\{O_x\}_{x \in \overline{U}}$ is an open cover of a compact set $\overline{U}$. Choose $x_1, \ldots, x_k \in \overline{U}$ such that $\overline{U} \subset \bigcup_{i=1}^k O_{x_i}$. Then $\bigcap_{i=1}^k O_{x_i}'$ is an open neighborhood of x' in U'.

For (2), it suffices to show

$$\overline{U} \times g^{-1}(g(U')) \subset (f \times g)^{-1}(B),$$

which follows from the following inclusions

$$\overline{U} \times g^{-1}(g(U')) \subset f^{-1}(f(\overline{U})) \times g^{-1}(g(U'))$$
$$= (f \times g)^{-1}((f \times g)(\overline{U} \times U'))$$
$$\subset (f \times g)^{-1}((f \times g)((f \times g)^{-1}(B)))$$
$$= (f \times g)^{-1}(B).$$

Now the proof is completed. $\qquad\qquad\qquad\qquad\qquad\qquad\quad \square$

The following two famous facts follow from this theorem.

Proposition 3.6.30. *Suppose X is locally compact Hausdorff. Then, for any quotient map $f : Y \to Z$, the product $1_X \times f : X \times Y \to X \times Z$ is a quotient map.*

Remark 3.6.31. This is a famous result of J.H.C. Whitehead [Whitehead (1948)].

Proposition 3.6.32. *Suppose X is compact and Y is Hausdorff. Then, for any surjective continuous map $f : X \to Y$ and a quotient map $g : Z \to W$, the product $f \times g : X \times Z \to Y \times W$ is a quotient map.*

Another caveat on spaces of the form G/H is that, even if G is Hausdorff, the quotient G/H may not be Hausdorff.

Exercise 3.6.33. Regard $\mathbb{R}$ as a topological group by the addition and $\mathbb{Q}$ as a subgroup. Show that the quotient $\mathbb{R}/\mathbb{Q}$ is not Hausdorff.

Exercise 3.6.34. Show that, for a topological group G and a closed subgroup H, the quotient G/H is Hausdorff.

Let us go back to the discussion on transitive actions. It is an interesting and useful fact that, in many cases the converse to Proposition 3.6.25 holds. In other words, if the action of G on X is transitive, X is often of the form G/H.

Theorem 3.6.35. *Let G be a compact topological group and X a Hausdorff space. Suppose an action $\mu : G \times X \to X$ is transitive. For $x_0 \in X$, define $H = \{g \in G \mid gx_0 = x_0\}$. Then the following hold:*

(1) H is a closed subgroup of G.
(2) The map $\overline{\varphi} : G/H \to X$ defined by $\overline{\varphi}(gH) = gx_0$ is a homeomorphism.
(3) The diagram

$$
\begin{array}{ccc}
G \times X & \xrightarrow{\ \mu\ } & X \\
{\scriptstyle 1_G \times \overline{\varphi}} \uparrow & & \uparrow {\scriptstyle \overline{\varphi}} \\
G \times G/H & \xrightarrow{\ \overline{\mu}\ } & G/H
\end{array}
$$

is commutative, where $\overline{\mu}$ is the action given by Proposition 3.6.25.

Remark 3.6.36. This theorem says that, when G acts on X transitively, we may identify X with G/H including the action of G.

Even when the action is not transitive, the action of G on the orbit Gx of a point $x \in X$ is always transitive. Thus the orbit can be written as $Gx \cong G/H$ for a closed subgroup H.

We also note that, in order to obtain the identification $X \cong G/H$ as sets with G-actions, we do not need the compactness assumption on G.

Proof of Theorem 3.6.35.   (1) We first show that H is a closed subset of G. Define a continuous map $\varphi : G \to X$ by $\varphi(g) = gx_0$. By definition $H = \varphi^{-1}(\{x_0\})$. Since X is Hausdorff, $\{x_0\}$ is closed and thus H is a closed subset of G.

For $h, h' \in H$, we have

$$(hh')x_0 = h(h'x_0) = hx_0 = x_0$$

which implies $hh' \in H$. On the other hand, $hx_0 = x_0$ implies $x_0 = h^{-1}x_0$ and which implies $h^{-1} \in H$ and H is a subgroup of G.

(2) It is easy to verify by the definition of H that $\overline{\varphi}$ is well-defined. Let us show that $\overline{\varphi}$ is bijective. By assumption, the action of G is transitive, which means that any $x \in X$ can be written as $gx_0 = x$ for some $g \in G$. Then we have $\overline{\varphi}(gH) = gx_0 = x$. Thus $\overline{\varphi}$ is surjective. In order to show that it is injective, suppose $\overline{\varphi}(gH) = \overline{\varphi}(g'H)$. Since $gx_0 = g'x_0$, we have $x_0 = g^{-1}g'x_0$ and $g^{-1}g' \in H$, which implies $gH = g'H$. And $\overline{\varphi}$ is injective.

The continuity follows from Lemma 3.6.14 by the commutativity of the diagram

the continuity of φ, and the fact that p is a quotient map.

The continuity of $\overline{\varphi}^{-1}$ can be proved by using Lemma 3.6.8 and Lemma 3.6.9.

(3) The commutativity of the diagram can be verified easily and the proof is omitted.

$\square$

Definition 3.6.37. The subgroup $H = \{g \in G \mid gx_0 = x_0\}$ appeared in Theorem 3.6.35 is called the *isotropy subgroup* of G at x_0. We denote it by $\mathrm{Iso}_{x_0}(G)$.

Remark 3.6.38. The isotropy subgroup is often denoted by G_{x_0}. It is also called the fixed point subgroup of x_0.

Definition 3.6.39. An action of a group G on X is called a *free action* if $\mathrm{Iso}_x(G)$ is a trivial group for any $x \in X$.

Remark 3.6.40. An action is free if and only if, for each $x \in X$, the orbit Gx can be identified with G including the action of G.

Let us apply Theorem 3.6.35 to Example 3.5.8. We first need the following fact.

Proposition 3.6.41. *The orthogonal group* $\mathrm{O}(n)$ *is a compact topological group.*

Proof. We already know that $\mathrm{O}(n)$ is a topological group by Exercise 3.3.11. Since $\mathrm{O}(n) \subset M_n(\mathbb{R}) \cong \mathbb{R}^{n^2}$, in order to show that $\mathrm{O}(n)$ is compact, we show that $\mathrm{O}(n)$ is a bounded closed set of $\mathbb{R}^{n^2}$.

For $A = (a_{ij}) \in \mathrm{O}(n)$, denote ${}^t A = (b_{ij})$. Since $A^t A = E_n$,

$$\sum_{j=1}^{n} a_{ij} b_{jk} = \begin{cases} 1 & i = k \\ 0 & i \neq k. \end{cases}$$

Since $b_{ij} = a_{ji}$,

$$\sum_{j=1}^{n} a_{ij} a_{kj} = \begin{cases} 1 & i = k \\ 0 & i \neq k. \end{cases}$$

In particular, we have $\displaystyle\sum_{j=1}^{n} a_{ij}^2 = 1$ for each i. Thus we have

$$\sum_{i,j=1}^{n} a_{ij}^2 = n.$$

In other words, $\mathrm{O}(n)$ is contained in the disk of radius $\sqrt{n}$ centered at the origin in $\mathbb{R}^{n^2}$.

Furthermore the map $s : M_n(\mathbb{R}) \to M_n(\mathbb{R})$ defined by $s(A) = A^t A$ is continuous and $\mathrm{O}(n) = s^{-1}(\{E_n\})$. Since $M_n(\mathbb{R}) = \mathbb{R}^{n^2}$ is Hausdorff, $\{E_n\}$ is closed and so is $\mathrm{O}(n)$. $\square$

Example 3.6.42. Recall that, in Example 3.5.8, the action of $O(n)$ on S^{n-1} is defined by

$$\mu : O(n) \times S^{n-1} \longrightarrow S^{n-1}$$
$$(A, v) \longmapsto \mu(A, v) = Av.$$

The sphere S^{n-1} is Hausdorff, since it is a subspace of a Hausdorff space $\mathbb{R}^n$. By Proposition 3.6.41, $O(n)$ is compact. In order to apply Theorem 3.6.35, it remains to verify that the action is transitive.

For $v \in S^{n-1}$, choose an orthonormal basis $\{a_1, a_2, \ldots, a_{n-1}, a_n = v\}$ of $\mathbb{R}^n$ containing v and define a matrix A by

$$A = \begin{pmatrix} {}^t a_1 \\ \vdots \\ {}^t a_n \end{pmatrix}.$$

Here we regard $a_1, \ldots, a_n$ as column vectors or $1 \times n$ matrices. Then we have

$$A^t A = \begin{pmatrix} {}^t a_1 \\ \vdots \\ {}^t a_n \end{pmatrix} (a_1, \ldots, a_n)$$

$$= \begin{pmatrix} {}^t a_1 a_1 & {}^t a_1 a_2 & \cdots & {}^t a_1 a_n \\ {}^t a_2 a_1 & \ddots & & {}^t a_2 a_n \\ \vdots & & \ddots & \vdots \\ {}^t a_n a_1 & {}^t a_n a_2 & \cdots & {}^t a_n a_n \end{pmatrix}.$$

Since $\{a_1, \ldots, a_n\}$ is an orthonormal basis and ${}^t a_i a_j$ is the inner product of a_i and a_j, we have

$$ {}^t a_i a_j = \begin{cases} 1 & i = j \\ 0 & i \neq j \end{cases}.$$

Or

$$A^t A = 1_n$$

and $A \in O(n)$. With this matrix A, we have

$$Av = \begin{pmatrix} {}^t a_1 v \\ \vdots \\ {}^t a_n v \end{pmatrix} = \begin{pmatrix} {}^t a_1 a_n \\ \vdots \\ {}^t a_n a_n \end{pmatrix} = \begin{pmatrix} 0 \\ \vdots \\ 0 \\ 1 \end{pmatrix}.$$

Denote $e_n = {}^t(0, \ldots, 0, 1)$. Then we have shown that, for any $v \in S^{n-1}$, there exists $A \in O(n)$ with $Av = e_n$. For another point $u \in S^{n-1}$, take $B \in O(n)$ with $Bu = e_n$. Then we have

$$A^{-1}Bu = A^{-1}e_n = v.$$

Since $A^{-1}B \in O(n)$, we see that the action of $O(n)$ on S^{n-1} is transitive. Now Theorem 3.6.35 gives us a homeomorphism

$$S^{n-1} \cong O(n)/\mathrm{Iso}_v(O(n)).$$

In order to obtain an explicit description of $\mathrm{Iso}_v(O(n))$, we take $v = e_n$. Then $A = (a_{ij}) \in \mathrm{Iso}_{e_n}(O(n))$ if and only of $Ae_n = e_n$, or

$$\begin{pmatrix} a_{11} & \cdots & a_{1n} \\ \vdots & \ddots & \vdots \\ a_{n-11} & \cdots & a_{n-1n} \\ a_{n1} & \cdots & a_{nn} \end{pmatrix} \begin{pmatrix} 0 \\ \vdots \\ 0 \\ 1 \end{pmatrix} = \begin{pmatrix} 0 \\ \vdots \\ 0 \\ 1 \end{pmatrix},$$

which implies

$$\begin{pmatrix} a_{1n} \\ \vdots \\ a_{n-1n} \\ a_{nn} \end{pmatrix} = \begin{pmatrix} 0 \\ \vdots \\ 0 \\ 1 \end{pmatrix}.$$

Thus

$$A = \begin{pmatrix} a_{11} & \cdots & a_{1n-1} & 0 \\ \vdots & \ddots & \vdots & \vdots \\ a_{n-11} & \cdots & a_{n-1n-1} & 0 \\ a_{n1} & \cdots & a_{nn-1} & 1 \end{pmatrix}.$$

Since $A^t A = E_n$,

$$\begin{pmatrix} a_{11} & \cdots & a_{1n-1} & 0 \\ \vdots & \ddots & \vdots & \vdots \\ a_{n-11} & \cdots & a_{n-1n-1} & 0 \\ a_{n1} & \cdots & a_{nn-1} & 1 \end{pmatrix} \begin{pmatrix} a_{11} & \cdots & a_{n-11} & a_{n1} \\ \vdots & \ddots & \vdots & \vdots \\ a_{1n-1} & \cdots & a_{n-1n-1} & a_{nn-1} \\ 0 & \cdots & 0 & 1 \end{pmatrix} =$$

$$\begin{pmatrix} 1 & 0 & \cdots & 0 \\ 0 & \ddots & & \vdots \\ \vdots & & \ddots & 0 \\ 0 & \cdots & 0 & 1 \end{pmatrix}$$

and we have $a_{n1}^2 + \cdots + a_{nn-1}^2 + 1 = 1$, which implies $a_{n1} = \cdots = a_{nn-1} = 0$. Thus A can be written as

$$A = \begin{pmatrix} & & 0 \\ A' & & \vdots \\ & & 0 \\ 0 \cdots 0\, 1 \end{pmatrix}.$$

Here we denote $A' = \begin{pmatrix} a_{11} & \cdots & a_{1n} \\ \vdots & \ddots & \vdots \\ a_{n-11} & \cdots & a_{n-1n-1} \end{pmatrix}$. The condition $A^t A = E_n$ implies $A'^t A' = E_{n-1}$. In other words, if $A \in \mathrm{Iso}_{e_n}(\mathrm{O}(n))$, then A can be written as

$$A = \begin{pmatrix} & & 0 \\ A' & & \vdots \\ & & 0 \\ 0 \cdots 0\, 1 \end{pmatrix}$$

by $A' \in \mathrm{O}(n-1)$. Conversely if $A' \in \mathrm{O}(n-1)$,

$$\begin{pmatrix} & & 0 \\ A' & & \vdots \\ & & 0 \\ 0 \cdots 0\, 1 \end{pmatrix} \in \mathrm{Iso}_{e_n}(\mathrm{O}(n)).$$

Thus if we regard $\mathrm{O}(n-1)$ as a subgroup of $\mathrm{O}(n)$ by the correspondence

$$A' \longmapsto \begin{pmatrix} & & 0 \\ A' & & \vdots \\ & & 0 \\ 0 \cdots 0\, 1 \end{pmatrix},$$

we have an isomorphism $\mathrm{Iso}_{e_n}(\mathrm{O}(n)) \cong \mathrm{O}(n-1)$. Thus we have a homeomorphism $S^{n-1} \cong \mathrm{O}(n)/\mathrm{O}(n-1)$. □

In the above example, we chose a special element $e_n \in S^{n-1}$ to find an explicit description of the isotropy subgroup $\mathrm{Iso}_{e_n}(\mathrm{O}(n))$. This is guaranteed by the following fact.

Exercise 3.6.43. Suppose a topological group G acts on a topological space X transitively. Then for any $x, y \in X$, show that $\mathrm{Iso}_x(G)$ and $\mathrm{Iso}_y(G)$ conjugate to each other in G.

Exercise 3.6.44. Prove an analogue of Example 3.6.42 for the unitary group $U(n) = \{A \in M_n(\mathbb{C}) \mid A^t\overline{A} = E_n\}$ and the $(2n-1)$-dimensional sphere $S^{2n-1} = \{(z_1, \ldots, z_n) \in \mathbb{C}^n \mid |z_1|^2 + \cdots + |z_n|^2 = 1\}$.

In other words, if we let $U(n)$ act on S^{2n-1} by the matrix multiplication, show that

- $U(n)$ is compact,
- the action of $U(n)$ on S^{2n-1} is transitive, and
- we have a homeomorphism $U(n)/U(n-1) \cong S^{2n-1}$.

We have seen a lot of examples of group actions and their quotients. We end this section by stating an important relation between fiber bundles and quotients by group actions.

Before we state a general theorem, let us investigate a special case. Let $G = O(n)$ and $H = O(n-1)$. Notice that the composition $p : O(n) \to O(n)/O(n-1) \cong S^{n-1}$ of the projection and a homeomorphism $O(n)/O(n-1) \cong S^{n-1}$ is given by $p(A) = Ae_n$.

Proposition 3.6.45. *Define a map $p : O(n) \to S^{n-1}$ by $p(A) = Ae_n$ for $A \in O(n)$. Then p is a fiber bundle with fiber $O(n-1)$.*

Proof. We need to find local trivializations. Our idea is the following:

Step 1 Find an open covering $\{U_\alpha\}$ of S^{n-1} and continuous maps $s_\alpha : U_\alpha \to p^{-1}(U_\alpha)$ satisfying $p \circ s_\alpha = 1_{U_\alpha}$.

Step 2 For each α, define a map $\varphi_\alpha : p^{-1}(U_\alpha) \to U_\alpha \times O(n-1)$ by $\varphi_\alpha(A) = (p(A), s_\alpha(p(A))^{-1}A)$.

Step 3 In order to show that φ_α is a homeomorphism, define a map $\psi_\alpha : U_\alpha \times O(n-1) \to p^{-1}(U_\alpha)$ by $\psi_\alpha(x, A) = s_\alpha(x)A$ and show that this is an inverse to φ_α.

We postpone **Step 1** and assume that we have an open covering $\{U_\alpha\}$ and maps $\{s_\alpha\}$.

Step 2: We first need to show that the map φ_α is well-defined and continuous. Since s_α is continuous, φ_α is continuous as a map to $U_\alpha \times O(n)$.

In order to show $s_\alpha(p(A))^{-1}A \in O(n-1)$ or $s_\alpha(p(A))^{-1}A \in \text{Iso}_{e_n}(O(n))$, we compare $s_\alpha(p(A))e_n$ and Ae_n. By the definition of p and $p \circ s_\alpha = 1_{U_\alpha}$, we have $s_\alpha(p(A))e_n = p(s_\alpha(p(A))) = p(A) = Ae_n$, which implies $s_\alpha(p(A))^{-1}Ae_n = e_n$ and we have

$$s_\alpha(p(A))^{-1}A \in O(n-1) = \text{Iso}_{e_n}(O(n)).$$

Step 3: The map ψ_α is obviously continuous. Since

$$p(\psi_\alpha(x, A)) = s_\alpha(x)Ae_n = s_\alpha(x)e_n = p(s_\alpha(x)) = x,$$

we have $\psi_\alpha(x, A) \in p^{-1}(U_\alpha)$.

For $(x, A) \in U_\alpha \times O(n-1)$, we have

$$\varphi_\alpha \circ \psi_\alpha(x, A) = \varphi_\alpha(s_\alpha(x)A)$$
$$= (p(s_\alpha(x)A), s_\alpha(p(s_\alpha(x)A)^{-1}s_\alpha(x)A)).$$

By $A \in O(n-1) = \text{Iso}_{e_n}(O(n))$, we have

$$p(s_\alpha(x)A) = s_\alpha(x)Ae_n$$
$$= s_\alpha(x)e_n$$
$$= p(s_\alpha(x)) = x.$$

Hence

$$\varphi_\alpha \circ \psi_\alpha(x, A) = (x, s_\alpha^{-1}s_\alpha(x)A)$$
$$= (x, A).$$

On the other hand, for $A \in p^{-1}(U_\alpha)$,

$$\psi_\alpha \circ \varphi_\alpha(A) = \psi_\alpha(p(A), s_\alpha(p(A))^{-1}A)$$
$$= s_\alpha(p(A))s_\alpha(p(A))^{-1}A$$
$$= A$$

and thus φ_α is a homeomorphism.

Step 1: Take

$$U_+ = S^{n-1} \setminus \{e_n\}$$
$$U_- = S^{n-1} \setminus \{-e_n\}$$

as an open covering of S^{n-1}. Define a map $s_+ : U_+ \to p^{-1}(U_+)$ by

$$s_+(\boldsymbol{x}) = \begin{pmatrix} 1 - \frac{x_1 x_1}{1-x_n} & -\frac{x_1 x_2}{1-x_n} & \cdots & -\frac{x_1 x_{n-1}}{1-x_n} & x_1 \\ -\frac{x_2 x_1}{1-x_n} & 1 - \frac{x_2 x_2}{1-x_n} & & & x_2 \\ \vdots & & \ddots & & \vdots \\ -\frac{x_{n-1}x_1}{1-x_n} & & & 1 - \frac{x_{n-1}x_{n-1}}{1-x_n} & x_{n-1} \\ x_1 & x_2 & \cdots & x_{n-1} & x_n \end{pmatrix}$$

for $\boldsymbol{x} = (x_1, \ldots, x_n) \in U_+$. It is left to the reader to verify that $s_+(\boldsymbol{x}) \in O(n)$. An easy calculation shows that $p \circ s_+(x) = s_+(x)e_n = x$ and thus the map s_+ satisfies the required condition. We define s_- by using s_+. For $x \in U_-$, define $s_-(x) = C^{-1}s_+(Cx)$, where

$$C = \begin{pmatrix} -1 & & & \\ & 1 & & 0 \\ & & \ddots & \\ & 0 & & 1 \\ & & & & -1 \end{pmatrix}.$$

$\square$

The reader might wonder where do maps $s_\alpha : U_\alpha \to p^{-1}(U_\alpha)$ satisfying $p \circ s_\alpha = 1_{U_\alpha}$ come from. An interpretation of such maps can be given by using the following fact from group theory.

Lemma 3.6.46. *Let G be a group and H a subgroup. Denote the projection by $p : G \to G/H$. Suppose a map $s : G/H \to G$ with $p \circ s = 1_{G/H}$ is given. Define a map $\varphi : G \to G/H \times H$ by $\varphi(g) = (p(g), s(p(g))^{-1}g)$. Then φ is a bijection.*

Furthermore, if G is a topological group and s is continuous, φ is a homeomorphism.

Proof. Follows from the same argument as in the proof of Proposition 3.6.45. $\square$

Definition 3.6.47. For a continuous map $p : E \to B$, a continuous map $s : B \to E$ satisfying $p \circ s = 1_B$ is called a *cross section* of p.

Lemma 3.6.46 says that the projection $p : G \to G/H$ be comes a trivial bundle, if it has a cross section. In Proposition 3.6.45, we constructed "local cross sections" $s_\pm : U_\pm \to p^{-1}(U_\pm)$, since all we needed was a locally trivial bundle.

Definition 3.6.48. Let $p : E \to B$ be a continuous map. We say p has *local cross sections* if, for any $x \in B$, there exists an open neighborhood U_x and a continuous map $s_x : U_x \to p^{-1}(U_x)$ such that $p \circ s_x = 1_{U_x}$. The map s_x is called a *local cross section* at x.

The proof of Proposition 3.6.45 can be extended as follows without a change.

Theorem 3.6.49. *Let G be a topological group and H a subgroup. If the projection $p : G \to G/H$ has local cross sections, it becomes a fiber bundle with fiber H.*

The proof is left to the reader.

It should be noted that, for a large class of topological groups G and its closed subgroup H, the projection $G \to G/H$ is known to have local cross sections. Hence it is a fiber bundle with fiber H. For example, this is the case when G is a Lie group. A proof can be found in a book by Toda and Mimura [Mimura and Toda (1991)].

3.7 Principal Bundles

Let G be a topological group and H a subgroup. Suppose that the projection $p : G \to G/H$ has local cross sections. By Theorem 3.6.49, $p : G \to G/H$ is a fiber bundle with fiber H. Let us find a structure group of this fiber bundle.

By the proof of Proposition 3.6.45 or Theorem 3.6.49, local trivializations of $p : G \to G/H$ is given by local cross sections. Namely, if $s_\alpha : U_\alpha \to p^{-1}(U_\alpha)$ is a local cross section over an open set $U_\alpha \subset G/H$. A local trivialization $\varphi_\alpha : p^{-1}(U_\alpha) \to U_\alpha \times H$ is defined by $\varphi_\alpha(y) = (p(y), s_\alpha(p(y))^{-1}y)$ for $x \in U_\alpha \cap U_\beta$ and $h \in H$, $y \in p^{-1}(U_\alpha \cap U_\beta)$. Furthermore we also know that its inverse is given by $\varphi_\alpha^{-1}(x, h) = s_\alpha(x)h$.

Let U_β be another open set with $U_\alpha \cap U_\beta \neq \emptyset$ having a local cross section s_β and the associated local trivialization φ_β. Then we have

$$\begin{aligned}
\varphi_\beta \circ \varphi_\alpha^{-1}(x, h) &= \varphi_\beta(s_\alpha(x)h) \\
&= (p(s_\alpha(x)h), s_\beta(p(s_\alpha(x)h))^{-1}s_\alpha(x)h) \\
&= (x, s_\beta(x)^{-1}s_\alpha(x)h).
\end{aligned}$$

In other words, the coordinate transformation

$$\Phi^{\alpha\beta} : U_\alpha \cap U_\beta \to \mathrm{Homeo}(H)$$

is given by $\Phi^{\alpha\beta}(x)(h) = s_\beta(x)^{-1}s_\alpha(x)h$.

Let $\mathrm{ad}(\mu_H) : H \to \mathrm{Homeo}(H)$ be the adjoint to the group multiplication μ_H of H. Define a map $\overline{\Phi}^{\alpha\beta} : U_\alpha \cap U_\beta \to H$ by $\overline{\Phi}^{\alpha\beta}(x) = s_\beta(x)^{-1}s_\alpha(x)$. Then the diagram

$$
\begin{array}{ccc}
H & \xrightarrow{\quad \mathrm{ad}(\mu_H) \quad} & \mathrm{Homeo}(H) \\
& \overline{\Phi}^{\alpha\beta} \nwarrow \quad \nearrow \Phi^{\alpha\beta} & \\
& U_\alpha \cap U_\beta &
\end{array}
$$

is commutative. In summary, we obtain the following fact.

Theorem 3.7.1. *Let G be a topological group and H a subgroup. Suppose the projection $p : G \to G/H$ has local cross sections. Then both fiber and structure group of the fiber bundle $G \to G/H$ are H and the action of the structure group is given by the group multiplication $\mu_H : H \times H \to H$ of H.*

Definition 3.7.2. Let G be a topological group. A fiber bundle $p : E \to B$ with fiber G and structure group G is called a *principal G-bundle*, if the action of the structure group on the fiber is given by the group multiplication of G. When G is obvious from the context, it is simply called a *principal bundle*.

With this terminology, Theorem 3.7.1 is restated as follows.

Corollary 3.7.3. *Let G be a topological group and H a subgroup. If the projection $p : G \to G/H$ has local cross sections, it is a principal H-bundle.*

Corollary 3.7.4. *The composition $O(n) \to S^{n-1}$ of the projection $O(n) \to O(n)/O(n-1)$ and the homeomorphism $O(n)/O(n-1) \cong S^{n-1}$ obtained in Example 3.6.42 is a principal $O(n-1)$-bundle.*

Besides this example, principal bundles, in fact, have already appeared in this book. Another important example is the Hopf bundle in Example 3.5.23.

Example 3.7.5. The Hopf bundle $p : S^3 \to S^2$ is a principal S^1-bundle. □

In order to understand principal bundles and general fiber bundles, let us take a look at the Möbius band in detail.

Example 3.7.6. Let $G = S^1 = \{z \in \mathbb{C} \mid |z| = 1\}$ and $H = C_2 = \{1, -1\} \subset G$. By Example 3.6.7, we have a homeomorphism $S^1/C_2 \cong S^1$. Thus the composition $p : S^1 \to S^1/C_2 \cong S^1$ is a principal C_2-bundle. Figure 3.10 describes a geometric meaning of this map p. This is a map given by twisting the circle S^1 and then fold it into a single circle.

Fig. 3.10   twist and fold

Another way to draw this map is Figure 3.11.

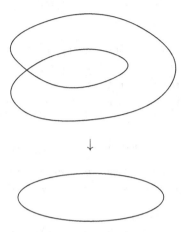

Fig. 3.11   the boundary of the Möbius band

By attaching segments to this figure, we obtain the Möbius band as is shown in Figure 3.12.

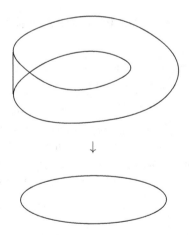

Fig. 3.12   attaching segments to Figure 3.11

Conversely, we obtain the C_2-bundle by removing the interior of the Möbius band. □

It should be note that, in this example, the Möbius band is also a fiber bundle with structure group C_2 and base space S^1. More generally, we can expect an analogous correspondence between principal G-bundles and fiber bundles with structure group G when they have the same base space. In order to study this problem, let us investigate the construction in Theorem 3.5.31, which allows us to reconstruct a fiber bundle $p : E \to B$ with fiber F and structure group G from an open covering $\{U_\alpha\}$ of B, the action of G on F, and coordinate transformations $\Phi^{\alpha\beta} : U_\alpha \cap U_\beta \to G$ by gluing trivial bundles $U_\alpha \times F$.

Thus a fiber bundle $p : E \to B$ with structure group G and fiber F is determined by the following data:

(1) an open covering $\{U_\alpha\}$ of B,
(2) maps $\Phi^{\alpha\beta} : U_\alpha \cap U_\beta \to G$, and
(3) an action of G on F, i.e. $G \to \mathrm{Homeo}(F)$.

On the other hand, in the case of a principal G-bundle, the last data in the above list is given by the group multiplication of G, and is already contained in the group structure of G. Thus a principal G-bundle $p : P \to B$ is determined by the following data:

(1) an open covering $\{U_\alpha\}$ of B, and
(2) maps $\Phi^{\alpha\beta} : U_\alpha \cap U_\beta \to G$.

By removing the third data from a fiber bundle $E \to B$ with structure group G, we obtain a principal G-bundle. Conversely, given a principal G-bundle $P \to B$, by adding the third data, namely an action of G on F, we should be able to construct a fiber bundle with fiber F on the same base space.

In summary, we have the following correspondence.

Fact 3.7.7. *Let G be a topological group and fix an action of G on a space F. Suppose that an open covering $\{U_\alpha\}$ of B and a family of maps $\Phi = \{\Phi^{\alpha\beta} : U_\alpha \cap U_\beta \to G\}$ are given.*

If the action of G on F induces an injective map[3] $G \to \mathrm{Homeo}(F)$. We obtain a one-to-one correspondence

$$
\begin{Bmatrix} \text{principal} \quad G\text{-bundles} \\ \text{over } B \text{ with coordinate} \\ \text{transformations } \Phi \end{Bmatrix} \longleftrightarrow \begin{Bmatrix} \text{fiber bundles over } B \\ \text{with fiber } F, \text{ structure} \\ \text{group } G, \text{ and coordi-} \\ \text{nate transformations } \Phi \end{Bmatrix}. \tag{3.4}
$$

---

[3]Such an action is called *faithful*.

This correspondence is given by coordinate transformations and local trivializations and is quite complicated. Fortunately, there is a much simpler construction of "$\longrightarrow$" in (3.4).

We first need to define an action of G on the total space of a principal G-bundle.

Definition 3.7.8. Let $p : P \to B$ be a principal G-bundle. Define a right action $P \times G \to P$ of G on P as follows: let $\{\varphi_\alpha : p^{-1}(U_\alpha) \to U_\alpha \times G\}_{\alpha \in A}$ be a local trivialization. Take $x \in P$ and $g \in G$. If $x \in p^{-1}(U_\alpha)$ write $\varphi_\alpha(x) = (p(x), \overline{\varphi}_\alpha(x))$ and define an action of g on x by

$$x \cdot g = \varphi_\alpha^{-1}(\varphi_\alpha(x)g) = \varphi_\alpha^{-1}(p(x), \overline{\varphi}_\alpha(x)g).$$

This action $p^{-1}(U_\alpha) \times G \to p^{-1}(U_\alpha)$ is also given by the composition

$$p^{-1}(U_\alpha) \times G \overset{\varphi_\alpha \times 1_G}{\longrightarrow} (U_\alpha \times G) \times G = U_\alpha \times (G \times G) \overset{1_{U_\alpha} \times \mu}{\longrightarrow} U_\alpha \times G \overset{\varphi_\alpha^{-1}}{\longrightarrow} p^{-1}(U_\alpha).$$

Since $P = \bigcup_{\alpha \in A} p^{-1}(U_\alpha)$, we obtain a map $P \times G \to P$.

Lemma 3.7.9. *The action of G on P is well-defined. Namely, when $x \in p^{-1}(U_\alpha) \cap p^{-1}(U_\beta)$, two ways to define $x \cdot g$ by using φ_α and φ_β agree.*

Proof. Suppose $x \in p^{-1}(U_\alpha) \cap p^{-1}(U_\beta), g \in G$. Let φ_α and φ_β be local trivializations over U_α and U_β and write $\varphi_\alpha(x) = (p(x), \overline{\varphi}_\alpha(x))$. By definition, we have $x \cdot g = \varphi_\alpha^{-1}(p(x), \overline{\varphi}_\alpha(x)g)$ over $p^{-1}(U_\alpha)$ and $x \cdot g = \varphi_\beta^{-1}(p(x), \overline{\varphi}_\beta(x)g)$ over $p^{-1}(U_\beta)$. Thus it suffices to show the following

$$\varphi_\beta \varphi_\alpha^{-1}(p(x), \overline{\varphi}_\alpha(x)g) = (p(x), \overline{\varphi}_\beta(x)g).$$

Let $\Phi^{\alpha\beta} : U_\alpha \cap U_\beta \to G$ be the coordinate transformation. Then we have $\varphi_\beta \varphi_\alpha^{-1}(y, h) = (y, \Phi^{\alpha\beta}(y)h)$ for $(y, h) \in (U_\alpha \cap U_\beta) \times G$. Namely we have

$$\varphi_\beta \varphi_\alpha^{-1}(p(x), \overline{\varphi}_\alpha(x)g) = (p(x), \Phi^{\alpha\beta}(p(x))\overline{\varphi}_\alpha(x)g).$$

And we are left to show $\overline{\varphi}_\beta(x) = \Phi^{\alpha\beta}(p(x))\overline{\varphi}_\alpha(x)$, which follows from the following calculation

$$\varphi_\beta \varphi_\alpha^{-1}(p(x), \overline{\varphi}_\alpha(x)) = \varphi_\beta \varphi_\alpha^{-1}\varphi_\alpha(x) = \varphi_\beta(x) = (p(x), \overline{\varphi}_\beta(x))$$
$$\varphi_\beta \varphi_\alpha^{-1}(p(x), \overline{\varphi}_\alpha(x)) = (p(x), \Phi^{\alpha\beta}(p(x))\overline{\varphi}_\alpha(x)).$$

$\square$

It is left to the reader to verify that the above definition is an action of G. The continuity of the action follows from the following well-known fact.

Lemma 3.7.10. *Let $f : X \to Y$ be a map between topological spaces. If there exists an open covering $X = \bigcup_{\alpha \in A} U_\alpha$ of X such that the restriction $f|_{U_\alpha}$ of f to each U_α, f is continuous.*

Remark 3.7.11. The above action of G is defined in such a way to make the local trivialization $\varphi_\alpha : p^{-1}(U_\alpha) \to U_\alpha \times G$ compatible with G actions, where the action of G, where the action of G on $U_\alpha \times G$ is given by $(x, h) \cdot g = (x, hg)$.

Let us make a digression and study the quotient space P/G.

Theorem 3.7.12. *For any principal G-bundle $p : P \to B$, the projection p induces a homeomorphism $\bar{p} : P/G \xrightarrow{\cong} B$ which makes the following diagram commutative.*

$$
\begin{array}{ccc}
P & \xrightarrow{\;=\;} & P \\
\Big\downarrow{\scriptstyle \pi} & & \Big\downarrow{\scriptstyle p} \\
P/G & \xrightarrow{\;\bar{p}\;} & B.
\end{array}
\tag{3.5}
$$

Proof. Define a map $\bar{p} : P/G \to B$ by $\bar{p}([x]) = p(x)$, where $[x]$ is the equivalence class of $x \in P$ in P/G. In order to verify that this map satisfies the requirements, we need to show the following:

(1) $\bar{p}$ is well-defined,
(2) $\bar{p}$ is continuous,
(3) $\bar{p}$ is a bijection,
(4) $\bar{p}$ is an open map, and
(5) $\bar{p}$ makes the diagram (3.5) commutative.

In order to show that $\bar{p}$ is well-defined, suppose $[x] = [y]$. Then there exists $g \in G$ with $x = yg$. Let

$$
\varphi_\alpha = p \times \overline{\varphi}_\alpha : p^{-1}(U_\alpha) \longrightarrow U_\alpha \times G
$$

be a local trivialization on a neighborhood U_α of $p(y)$. By the definition of yg, we have $yg = \varphi_\alpha^{-1}(p(y), \overline{\varphi}_\alpha(y)g)$. By the commutativity of the diagram

$$
\begin{array}{ccc}
p^{-1}(U_\alpha) & \xleftarrow{\;\;\varphi_\alpha^{-1}\;\;} & U_\alpha \times G \\
& {\scriptstyle p} \searrow \quad \swarrow {\scriptstyle \mathrm{pr}_1} & \\
& U_\alpha, &
\end{array}
$$

we have

$$p(x) = p(yg) = p \circ \varphi_\alpha^{-1}(p(y), \overline{\varphi}_\alpha(y)g) = \mathrm{pr}_1(p(y), \overline{\varphi}_\alpha(y)g) = p(y),$$

where pr_1 is the projection onto the first coordinate. Hence $\bar{p}$ is well-defined.

It is immediate to verify that $\bar{p}$ makes the diagram (3.5) commutative. The continuity of $\bar{p}$ follows from the commutativity of (3.5) and Lemma 3.6.14.

We next show that $\bar{p}$ is a bijection. Since p is surjective, so is $\bar{p}$ by the commutativity of (3.5). Suppose $\bar{p}([x]) = \bar{p}([y])$ for $x, y \in P$. Then $p(x) = p(y)$. Denote this common point by $b = p(x) = p(y)$ and let

$$\varphi_\alpha = p \times \overline{\varphi}_\alpha : p^{-1}(U_\alpha) \longrightarrow U_\alpha \times G$$

be a local trivialization around b. Let $g = \overline{\varphi}_\alpha(y)^{-1}\overline{\varphi}_\alpha(x)$. Then we have

$$\varphi_\alpha(yg) = (p(yg), \overline{\varphi}_\alpha(yg)) = (p(y), \overline{\varphi}_\alpha(y)g) = (b, \overline{\varphi}_\alpha(x)) = \varphi_\alpha(x).$$

Since φ_α is a homeomorphism, we have $x = yg$, which implies $[x] = [y]$.

Finally the fact that $\bar{p}$ is an open map follows from the next Lemma. □

Lemma 3.7.13. *For any fiber bundle $p : E \to B$, the projection p is an open map.*

Exercise 3.7.14. Prove this.

Example 3.7.15. Since the Hopf bundle $p : S^3 \to S^2$ is a principal S^1-bundle, there should be an action of S^1 on S^3 whose quotient space is S^2. Let us find this action.

By Example 3.1.11, local trivializations $\varphi_\pm : p^{-1}(U_\pm) \to U_\pm \times S^1$ are given by

$$\varphi_+(z_1, z_2) = \left(2|z_1|^2 - 1, 2z_1\bar{z}_2, \tfrac{z_2}{|z_2|}\right)$$

$$\varphi_-(z_1, z_2) = \left(2|z_1|^2 - 1, 2z_1\bar{z}_2, \tfrac{z_1}{|z_1|}\right),$$

where $U_\pm = S^2 \setminus \{(\pm 1, 0)\}$. Furthermore, according to Definition 3.7.8, the action of the structure group S^1 of the Hopf bundle on the total space S^3 is given by

$$x \cdot w = \varphi_\pm^{-1}(p(x), \overline{\varphi}_\pm(x)w)$$

for $x \in p^{-1}(U_\pm)$ and $w \in S^1$, where $\overline{\varphi}_\pm$ is the map given by $\varphi_\pm(x) = (p(x), \overline{\varphi}_\pm(x))$. Let us find an explicit description of $x \cdot w$ when $x \in p^{-1}(U_+)$. For $x = (z_1, z_2) \in \mathbb{C}^2$, $\overline{\varphi}_+(z_1, z_2) = \tfrac{z_2}{|z_2|}$ implies

$$\varphi_+^{-1}(x, z, w) = \left(\frac{zw}{2\sqrt{\frac{1-x}{2}}}, w\sqrt{\frac{1-x}{2}}\right)$$

and we have

$$(z_1, z_2) \cdot w = \varphi_+^{-1}(p(z_1, z_2), \overline{\varphi}_+(z_1)w)$$

$$= \varphi_+^{-1}\left(2|z_1|^2 - 1, 2z_1\overline{z}_2, \tfrac{z_2}{|z_2|}w\right)$$

$$= \left(\frac{2z_1\overline{z}_2\tfrac{z_2}{|z_2|}w}{2\sqrt{\tfrac{1-(2|z_1|^2-1)}{2}}}, \frac{z_2}{|z_2|}w\sqrt{\frac{1-(2|z_1|^2-1)}{2}}\right)$$

$$= \left(\frac{z_1|z_2|w}{\sqrt{1-|z_1|^2}}, \frac{z_2}{|z_2|}w\sqrt{1-|z_1|^2}\right)$$

$$= (z_1w, z_2w)$$

for $(z_1, z_2) \in p^{-1}(U_+)$ and $w \in S^1$. In other words, the action of S^1 on $p^{-1}(U_+)$ is given by the multiplications of complex numbers in both coordinates. The same calculation shows that the action of S^1 on $p^{-1}(U_-)$ is also given by the multiplications of complex numbers. This is exactly the action in Example 3.6.18. By the definition of $\mathbb{C}P^1$, we have $S^3/S^1 = \mathbb{C}P^1$. On the other hand, we have $S^3/S^1 \cong S^2$ by Theorem 3.7.12. And we have a homeomorphism $\mathbb{C}P^1 \cong S^2$. □

Let us go back to the discussion on Fact 3.7.7, which claimed

(a principal G bundle) + (an action of G on F) =

(a fiber bundle with structure group G and fiber F).

We have been looking for a simple construction of a fiber bundle with structure group G and fiber F from a principal G-bundle. The action defined in Definition 3.7.8 can be used to give such a construction.

Definition 3.7.16. Let G be a topological group. Suppose that G acts on topological spaces P and F continuously from the right and the left, respectively. Define a left action $\mu : G \times (P \times F) \to P \times F$ of G on $P \times F$ by $\mu(g, x, y) = (xg^{-1}, gy)$.

The quotient space by this action is denoted by $P \times_G F$. The element of $P \times_G F$ represented by $(x, y) \in P \times F$ is denoted by $[x, y]$.

Exercise 3.7.17. Verify that this is actually an action of G on $P \times G$.

Theorem 3.7.18. *For a principal G-bundle $p : P \to B$ and space F with an action of G, define a map $\bar{p} : P \times_G F \to B$ by $\bar{p}([x, y]) = p(x)$. This is a fiber bundle with structure group G and fiber F.*

Proof. Let $\{\varphi_\alpha : p^{-1}(U_\alpha) \to U_\alpha \times G\}$ be local trivializations of a principal G-bundle $p : P \to B$. We construct local trivializations $\{\psi_\alpha : \bar{p}^{-1}(U_\alpha) \to U_\alpha \times F\}$ of $\bar{p} : P \times_G F \to B$ by using $\{\varphi_\alpha\}$. We have

$$\bar{p}^{-1}(U_\alpha) = \{[x,y] \in P \times_G F \mid \bar{p}([x,y]) \in U_\alpha\}$$
$$= \{[x,y] \in P \times_G F \mid p(x) \in U_\alpha\}$$
$$= p^{-1}(U_\alpha) \times_G F.$$

Under this identification, the maps φ_α induce maps

$$\bar{p}^{-1}(U_\alpha) = p^{-1}(U_\alpha) \times_G F \overset{\varphi_\alpha \times_G 1_F}{\longrightarrow} (U_\alpha \times G) \times_G F.$$

On the other hand, the correspondence from $[(x,g),y] \in (U_\alpha \times G) \times_G F$ to $(x,gy) \in U_\alpha \times F$, i.e. the map $(U_\alpha \times G) \times_G F \to U_\alpha \times F$ induced by

$$1_{U_\alpha} \times \mu : U_\alpha \times G \times F \to U_\alpha \times F$$

is a homeomorphism, where μ is the action of G on F. In fact, $(x,y) \mapsto [(x,e),y]$ is the inverse.

Now define a map $\psi_\alpha : \bar{p}^{-1}(U_\alpha) \to U_\alpha \times F$ by the following composition

$$\bar{p}^{-1}(U_\alpha) = p^{-1}(U_\alpha) \times_G F \overset{\varphi_\alpha \times_G 1_F}{\longrightarrow} (U_\alpha \times G) \times_G F \overset{1_{U_\alpha} \times \mu}{\longrightarrow} U_\alpha \times F.$$

Since both φ_α and $1_{U_\alpha} \times \mu$ are homeomorphisms, so is ψ_α. It is also easy to verify that ψ_α makes the following diagram commutative

$$\begin{array}{ccc} \bar{p}^{-1}(U_\alpha) & \overset{\psi_\alpha}{\longrightarrow} & U_\alpha \times F \\ & \underset{\bar{p}}{\searrow} \quad \underset{pr_1}{\swarrow} & \\ & U_\alpha. & \end{array}$$

Thus $\bar{p} : P \times_G F \to B$ is a fiber bundle with fiber F.

In order to show that G is a structure group, we need to find coordinate transformations. For $(x,y) \in (U_\alpha \cap U_\beta) \times F$, we have

$$\psi_\beta \circ \psi_\alpha^{-1}(x,y) = \psi_\beta \circ (\varphi_\alpha^{-1} \times_G 1_F) \circ (1_{U_\alpha} \times \mu)^{-1}(x,y)$$
$$= \psi_\beta \circ (\varphi_\alpha^{-1} \times_G 1_F)(x,e,y)$$
$$= \psi_\beta(\varphi_\alpha^{-1}(x,e),y)$$
$$= (1_{U_\beta} \times \mu) \circ (\varphi_\beta \times_G 1_F)(\varphi_\alpha^{-1}(x,e),y)$$
$$= (1_{U_\beta} \times \mu)(\varphi_\beta \circ \varphi_\alpha^{-1}(x,e),y)$$
$$= (1_{U_\beta} \times \mu)(x, \Phi^{\alpha\beta}(x)e,y)$$
$$= (x, \Phi^{\alpha\beta}(x)y),$$

where $\Phi^{\alpha\beta} : U_\alpha \cap U_\beta \to G$ is a coordinate transformation of $p : P \to B$. This calculation shows that coordinate transformations of $\bar{p} : P \times_G F \to B$ are given by the same maps $\Phi^{\alpha\beta}$ and it has G as a structure group. $\qquad\square$

Remark 3.7.19. The above proof show that the correspondence

$$(P \to B) \longmapsto (P \times_G F \to B)$$

is given by attaching F to P by using the coordinate transformations of $p : E \to B$. This is exactly the correspondence in Fact 3.7.7

$$\left\{ \begin{matrix} \text{principal} & G\text{-bundles} \\ \text{over } B \end{matrix} \right\} \longrightarrow \left\{ \begin{matrix} \text{fiber bundles over } B \text{ with} \\ \text{structure group } G \text{ and fiber } F \end{matrix} \right\}.$$

Let us take a look at the correspondence in Theorem 3.7.18 in the case of the Möbius band.

Example 3.7.20. Let $p : S^1 \to S^1/C_2 \cong S^1$ be the principal C_2-bundle in Example 3.7.6. When we regard $S^1 = \{ z \in \mathbb{C} \, | \, |z|^2 = 1 \}$, p is given by $p(z) = z^2$. The action of $C_2 = \{1, -1\}$ on the total space S^1 is given by the multiplication of complex numbers.

On the other hand, the action of C_2 on the interval $[-\frac{1}{2}, \frac{1}{2}]$ is given by the multiplication of real numbers $(\pm 1) \cdot t = \pm t$. According to Theorem 3.7.18, the quotient space $S^1 \times_{C_2} [\frac{1}{2}, \frac{1}{2}]$ should be homeomorphic to the Möbius band M. Let us verify this. More precisely, if we define an action of C_2 on $S^1 \times [-\frac{1}{2}, \frac{1}{2}]$ by $(-1)(z, t) = (-z, -t)$, we want show that the quotient space is homeomorphic to the Möbius band.

Fig. 3.13   splitting of a circle

As is the case of Example 1.1.5, we split the circle as in Figure 3.13. We also split the annulus $S^1 \times [-\frac{1}{2}, \frac{1}{2}]$ according to the splitting of the circle as is shown in Figure 3.14. The action of C_2 identifies the left and right halves. The two fibers on the boundaries are identified with a rotation of 180 degree.

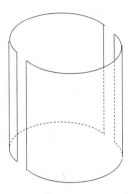

Fig. 3.14　splitting of an annulus

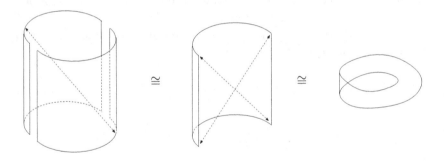

Fig. 3.15　Möbius bundle as an associated bundle

Thus we obtain the space in Figure 3.15, which is the Möbius band.
Furthermore $S^1 \times_{C_2} [-\frac{1}{2}, \frac{1}{2}]$ and the Möbius band are not only homeomorphic but also "isomorphic" as fiber bundles over S^1. In order to give
a precise statement, however, we need to define isomorphisms of fiber bundles, which is the subject of §4.1.　　　　　　　　　　　　　　　□

Exercise 3.7.21. Give an explicit construction of a homeomorphism from
$S^1 \times_{C_2} [-\frac{1}{2}, \frac{1}{2}]$ to the Möbius band following the idea described by Figures
in Example 3.7.20.

We conclude this section with an important property of principal bundles. By Corollary 3.7.3, the projection $G \to G/H$ is a principal H-bundle, if it has local cross sections. If it has a global cross section, it becomes a trivial fiber bundle. This fact can generalized to principal G-bundles.

Theorem 3.7.22. *Let G a topological group and $p : P \to B$ a principal G-bundle. If p has a cross section, then it is a trivial bundle.*

We leave it to the reader to modify the proof of Theorem 3.6.49, hence of Proposition 3.6.41 to prove this theorem.

Remark 3.7.23. In the above theorem, the assumption that p essential. For example, the Möbius band has a cross section, as is shown in Figure 3.16. But it is not trivial. This is because the Möbius band is not principal.

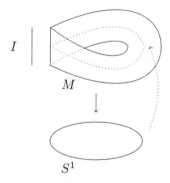

Fig. 3.16   the cross section to the center of the Möbius band

Chapter 4

Classification of Fiber Bundles

In this chapter, we seek for descriptions of fiber bundles over a fixed base space B. We have seen in §3.7 that, when coordinate transformations are fixed, there is a one-to-one correspondence

$$\left\{ \begin{matrix} \text{principal } G\text{-bundles} \\ \text{over } B \end{matrix} \right\} \longleftrightarrow \left\{ \begin{matrix} \text{fiber bundles over } B \text{ with} \\ \text{structure group } G \text{ and fiber } F \end{matrix} \right\}.$$

Thus it suffices to find a description of the set of principal G-bundles over B.

This set looks like a "function of B", since the set is determined when a space B is given. In fact, the aim of this chapter is to describe this set in terms of B. Precisely speaking, what we are going to do is to give a description of the set of "isomorphism classes" of fiber bundles over B. As we identify two homeomorphic spaces, we would like to identify two "isomorphic" fiber bundles. We need a way to compare two fiber bundles. In modern mathematics, we use maps or morphism to compare two objects. Let us begin with the definition of maps between fiber bundles.

4.1 Maps between Fiber Bundles

A fiber bundle is a complicated object consisting of base space, total space, fiber, local trivializations, coordinate transformations, and structure group. A "map" between two fiber bundles should give correspondences between these data. Since it is a "map", it should consists of continuous maps between base spaces, total spaces, and fibers. Let us forget structure groups for a while to concentrate on maps between these three kinds of spaces.

Given two fiber bundles

$$\xi = (F \longrightarrow E \xrightarrow{\ p\ } B)$$

$$\xi' = (F' \longrightarrow E' \xrightarrow{\ p'\ } B')$$

a map from ξ to ξ' should consists of a map $B \to B'$ between base spaces, a map $E \to E'$ between total spaces, and a map $F \to F'$ between fibers. However, the first two suffice.

Definition 4.1.1. Let

$$\xi = (p : E \longrightarrow B)$$
$$\xi' = (p' : E' \longrightarrow B')$$

be fiber bundles. A *fiber-preserving map* from ξ to ξ' is a pair of maps

$$\tilde{f} : E \longrightarrow E'$$
$$f : B \longrightarrow B'$$

which makes the diagram

$$
\begin{array}{ccc}
E & \xrightarrow{\ \tilde{f}\ } & E' \\
{\scriptstyle p}\big\downarrow & & \big\downarrow{\scriptstyle p'} \\
B & \xrightarrow{\ f\ } & B'
\end{array}
\tag{4.1}
$$

commutative.

A fiber-preserving map is denoted by

$$(\tilde{f}, f) : (E, B) \longrightarrow (E', B')$$

or

$$\boldsymbol{f} = (\tilde{f}, f) : \xi \longrightarrow \xi'.$$

It is easy to verify that, for any $x \in B$, $\tilde{f}$ maps the fiber $p^{-1}(x)$ over x to the fiber $p'^{-1}(f(x))$ over $f(x)$. This fact can be expressed by the following diagram

$$
\begin{array}{ccc}
p^{-1}(x) & \xrightarrow{\ \tilde{f}|_{p^{-1}(x)}\ } & p'^{-1}(f(x)) \\
\big\uparrow & & \big\uparrow \\
\big\downarrow & & \big\downarrow \\
E & \xrightarrow{\ \tilde{f}\ } & E' \\
{\scriptstyle p}\big\downarrow & & \big\downarrow{\scriptstyle p'} \\
B & \xrightarrow{\ f\ } & B'.
\end{array}
$$

We need to impose more conditions in order to call $\boldsymbol{f} = (\tilde{f}, f)$ to be a map of fiber bundles. For example, we need conditions on such structure groups and coordinate transformations. We introduced and named fiber-preserving maps since they are fundamental. Another reason is that fiber-preserving maps are used as maps between fibrations (Chapter 5) and quasifibrations (§5.11).

We first consider conditions related to local trivializations.

Definition 4.1.2. Let $p : E \to B$ and $p' : E' \to B'$ be fiber bundles with fiber F and F', respectively. Let

$$\boldsymbol{f} = (\tilde{f}, f) : (E, B) \longrightarrow (E', B')$$

be a fiber-preserving map. For $x \in B$, choose local trivializations $\varphi_\alpha : p^{-1}(U_\alpha) \to U_\alpha \times F$ and $\psi_\beta : p'^{-1}(V_\beta) \to V_\beta \times F'$ around x and $f(x)$, respectively. We denote the adjoint to the composition

$$(U_\alpha \cap f^{-1}(V_\beta)) \times F \xrightarrow{\varphi_\alpha^{-1}} p^{-1}(U_\alpha \cap f^{-1}(V_\beta)) \xrightarrow{\tilde{f}}$$

$$p'^{-1}(f(U_\alpha) \cap V_\beta) \xrightarrow{\psi_\beta} (f(U_\alpha) \cap V_\beta) \times F' \xrightarrow{\mathrm{pr}_2} F'$$

by

$$L^{\boldsymbol{f}}_{\alpha\beta} : U_\alpha \cap f^{-1}(V_\beta) \longrightarrow \mathrm{Map}(F, F').$$

Since $(\tilde{f}, f)$ is a fiber-preserving map, $\tilde{f}$ maps the fiber over x to the fiber over $f(x)$. If we identify $p^{-1}(x)$ with F and $p'^{-1}(f(x))$ with F' through local trivializations, we have $L^{\boldsymbol{f}}_{\alpha\beta}(x) \in \mathrm{Map}(F, F')$ by Lemma 3.4.2. Furthermore $L^{\boldsymbol{f}}_{\alpha\beta}$ is continuous by Lemma 3.4.6.

We use these maps $L^{\boldsymbol{f}}_{\alpha\beta}$ to discuss relations between structure groups and fiber-preserving maps. For simplicity, we only consider maps between fiber bundles that share the same fiber and structure group.

Definition 4.1.3. Let $\xi = (p : E \to B)$ and $\xi' = (p' : E' \to B')$ be fiber bundles with the same fiber F and the same structure group G. We also assume that the actions of G on F are given by the same map $\mu_G : G \times F \to F$.

Under these assumptions, a *bundle map* from ξ to ξ' is a fiber-preserving map from ξ to ξ' such that the maps

$$L^{\boldsymbol{f}}_{\alpha\beta} : U_\alpha \cap f^{-1}(V_\beta) \longrightarrow \mathrm{Map}(F, F)$$

in Definition 4.1.2 take values in G. In other words, there exists a dotted continuous map $\bar{L}^{\boldsymbol{f}}_{\alpha\beta}$ which makes the following diagram commutative

$$
\begin{array}{ccc}
G & \xrightarrow{\;\mathrm{ad}(\mu_G)\;} & \mathrm{Map}(F, F) \\
& \nwarrow \quad \nearrow & \\
\exists \bar{L}^{\boldsymbol{f}}_{\alpha\beta} & \quad L^{\boldsymbol{f}}_{\alpha\beta} & \\
& U_\alpha \cap f^{-1}(V_\beta). &
\end{array}
$$

Remark 4.1.4. It should be noted that, if $\boldsymbol{f} = (\tilde{f}, f)$ is a bundle map from ξ to ξ', then for any point x in the base space of ξ, the restriction $\tilde{f}|_{p^{-1}(x)} : p^{-1}(x) \to p'^{-1}(f(x))$ is a continuous map.

Bundle maps between principal bundles have the following property. The proof is left to the reader.

Proposition 4.1.5. *Let* $\xi = (p : E \to B)$ *and* $\xi' = (p' : E' \to B')$ *be principal G-bundles for G be a topological group. If* $\boldsymbol{f} = (\tilde{f}, f) : (E, B) \to (E', B')$ *is a bundle map, the following diagram is commutative*

$$
\begin{array}{ccc}
E \times G & \xrightarrow{\tilde{f} \times 1_G} & E' \times G \\
\downarrow & & \downarrow \\
E & \xrightarrow{\quad \tilde{f} \quad} & E',
\end{array}
\tag{4.2}
$$

where the vertical maps are the action of G on total spaces given by Definition 3.7.8.

We use the following terminology for such maps.

Definition 4.1.6. Suppose a topological group G acts on topological spaces X and Y from the right. We say a continuous map $f : X \to Y$ is G-*equivariant* if it makes the diagram

$$
\begin{array}{ccc}
X \times G & \xrightarrow{1_G \times f} & Y \times G \\
\downarrow & & \downarrow \\
X & \xrightarrow{\quad f \quad} & Y
\end{array}
$$

commutative, where the vertical maps are the actions of G. The same terminology is used for left actions.

We use the following definition for isomorphisms of fiber bundles.

Definition 4.1.7. Let $\xi = (p : E \to B)$ and $\xi' = (p' : E' \to B)$ be fiber bundles with the same fiber F, the same structure group G, and the same base space B. We say that ξ and ξ' are *isomorphic* or *equivalent* and denote by $\xi \cong \xi'$ if there exists a bundle map $\boldsymbol{f} = (\tilde{f}, f) : (E, B) \to (E', B)$ with $f = 1_B$. Such a bundle map $\boldsymbol{f}$ is called a *bundle isomorphism*.

In the rest of this chapter, we always assume that a structure group is specified in any fiber bundle. There is no loss of generality, since we may take $G = \mathrm{Homeo}(F)$.

We first need to verify the following.

Proposition 4.1.8. *The relation $\cong$ among fiber bundles is an equivalence relation.*

Proof. We need to verify the following:

(reflexivity) For any ξ, $\xi \cong \xi$.
(symmetry) For any ξ and η, $\xi \cong \eta \Rightarrow \eta \cong \xi$.
(transitivity) For any ξ, ζ, and η, $\xi \cong \zeta$, $\zeta \cong \eta \Rightarrow \xi \cong \eta$.

It is easy to prove the reflexivity and transitivity. In fact, when $\xi = (p : E \to B)$, the identity maps

$$
\begin{array}{ccc}
E & \xrightarrow{\ =\ } & E \\
\downarrow{\scriptstyle p} & & \downarrow{\scriptstyle p} \\
B & \xrightarrow{\ =\ } & B
\end{array}
$$

define an isomorphism from ξ to itself. For three fiber bundles $\xi = (p : E \to B)$, $\zeta = (p' : E' \to B)$, and $\eta = (p'' : E'' \to B)$, suppose we have bundle isomorphisms

$$
\begin{array}{ccc}
E & \xrightarrow{\ \tilde{f}\ } & E' \\
\downarrow{\scriptstyle p} & & \downarrow{\scriptstyle p'} \\
B & \xrightarrow{\ =\ } & B
\end{array}
\qquad
\begin{array}{ccc}
E' & \xrightarrow{\ \tilde{g}\ } & E'' \\
\downarrow{\scriptstyle p'} & & \downarrow{\scriptstyle p''} \\
B & \xrightarrow{\ =\ } & B.
\end{array}
$$

Then the composition

$$
\begin{array}{ccc}
E & \xrightarrow{\ \tilde{g}\circ\tilde{f}\ } & E'' \\
\downarrow{\scriptstyle p} & & \downarrow{\scriptstyle p''} \\
B & \xrightarrow{\ =\ } & B
\end{array}
$$

is a bundle isomorphism from ξ to η.

In order to show the symmetry, suppose

$$
\begin{array}{ccc}
E & \xrightarrow{\ \tilde{f}\ } & E' \\
\downarrow{\scriptstyle p} & & \downarrow{\scriptstyle p'} \\
B & \xrightarrow{\ =\ } & B
\end{array}
$$

is a bundle isomorphism from $\xi = (p : E \to B)$ to $\eta = (p' : E' \to B)$. We need to construct a bundle isomorphism

$$
\begin{array}{ccc}
E' & \xrightarrow{\;\tilde{g}\;} & E \\
\downarrow{\scriptstyle p'} & & \downarrow{\scriptstyle p} \\
B & \xrightarrow{\;=\;} & B.
\end{array}
$$

An obvious idea is to use $\tilde{g} = \tilde{f}^{-1}$. Since $\tilde{f}$ is a homeomorphism when restricted to each fiber, it is a bijection on the whole total space. Thus the inverse $\tilde{f}^{-1}$ exists and makes the above diagram commutative. Let us show that $\tilde{f}^{-1}$ is continuous.

Choose an open covering $B = \bigcup_{\alpha \in A} U_\alpha$ of B such that both ξ and η have local trivializations

$$
\varphi_\alpha : p^{-1}(U_\alpha) \longrightarrow U_\alpha \times F
$$
$$
\psi_\alpha : p'^{-1}(U_\alpha) \longrightarrow U_\alpha \times F
$$

over U_α. Since $E' = \bigcup_{\alpha \in A} p'^{-1}(U_\alpha)$ is also an open covering, it suffices to show that $\tilde{f}^{-1}$ is continuous on each $p'^{-1}(U_\alpha)$. By assumption, the composition

$$
U_\alpha \times F \xrightarrow{\varphi_\alpha^{-1}} p^{-1}(U_\alpha) \xrightarrow{\tilde{f}} p'^{-1}(U_\alpha) \xrightarrow{\psi_\alpha} U_\alpha \times F
$$

defines a continuous map $\Phi : U_\alpha \to \mathrm{Homeo}(F)$ by $\psi_\alpha \circ f \circ \varphi_\alpha^{-1}(x, y) = (x, \Phi(x)(y))$. For each point $x \in U_\alpha$, define $\Psi(x) = \Phi(x)^{-1}$. Then Ψ is continuous, since it is given by the composition

$$
\Psi : U_\alpha \xrightarrow{\Phi} G \xrightarrow{\nu} G \subset \mathrm{Homeo}(F),
$$

where ν the inverse in G. Define $g_\alpha : U_\alpha \times F \to U_\alpha \times F$ by $g_\alpha(x, y) = (x, \Psi(x)(y))$. Then this is continuous and is inverse to the map given by $\psi_\alpha \circ f \circ \varphi_\alpha^{-1}(x, y) = (x, \Phi(x)(y))$. Thus the diagram

$$
\begin{array}{ccc}
p'^{-1}(U_\alpha) & \xrightarrow{\;\psi_\alpha\;} & U_\alpha \times F \\
\downarrow{\scriptstyle \tilde{f}^{-1}} & & \downarrow{\scriptstyle g_\alpha} \\
p^{-1}(U_\alpha) & \xrightarrow{\;\varphi_\alpha\;} & U_\alpha \times F
\end{array}
$$

is commutative and $\tilde{f}^{-1}$ is continuous. It is left to the reader to verify that $\tilde{f}^{-1}$ defines a bundle map. $\qquad\square$

Corollary 4.1.9. *If* $(\tilde{f}, 1_B) : (E, B) \to (E', B)$ *is a bundle isomorphism,* $\tilde{f} : E \to E'$ *is a homeomorphism.*

Example 4.1.10. Let M_0 be an annulus and M_2 the space obtained from a rectangle by gluing the side edges after twisting by 2π, as is shown in Figure 4.1.

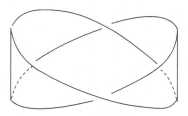

Fig. 4.1 M_2

These spaces can be explicitly defined by

$$M_0 = \left\{ (2\cos\varphi, 2\sin\varphi, t) \,\middle|\, -\tfrac{1}{2} \le t \le \tfrac{1}{2}, 0 \le \varphi \le 2\pi \right\}$$

$$M_2 = \left\{ ((2+t\cos\varphi)\cos\varphi, (2+t\cos\varphi)\sin\varphi, t\sin\varphi) \,\middle|\, -\tfrac{1}{2} \le t \le \tfrac{1}{2}, 0 \le \varphi \le 2\pi \right\}.$$

The projections onto the central circles are denoted by $p_0 : M_0 \to S^1$ and $p_2 : M_2 \to S^1$, respectively. Then we obtain fiber bundles with fiber $[-\tfrac{1}{2}, \tfrac{1}{2}]$. The spaces M_0 and M_2 look different but they are isomorphic as fiber bundles. In particular, M_0 and M_2 are homeomorphic. A homeomorphism can be defined as follows.

Cut M_2 along the vertical lines AB and CD as is shown in Figure 4.2. Denote the front side by E_1 and the back side by E_2. Cut M_0 similarly along the lines $A'B'$ and $C'D'$ and denote the front side by E_1' and the back side by E_2'.

There are homeomorphisms $f_1 : E_1 \to E_1'$ and $f_2 : E_2 \to E_2'$ which map vertices A, B, D, C to A', B', C', D', respectively. These two maps can be glued together along AB and CD to give a continuous map, which is a bundle isomorphism.

$$
\begin{array}{ccc}
M_2 & \xrightarrow{\ f_1 \cup f_2\ } & M_0 \\
\downarrow{\scriptstyle p_2} & & \downarrow{\scriptstyle p_0} \\
S^1 & \xrightarrow{\ =\ } & S^1.
\end{array}
$$

$\square$

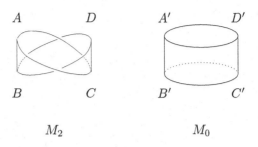

$$M_2 \qquad\qquad M_0$$

Fig. 4.2 cutting M_2 and M_0

Exercise 4.1.11. Find explicit descriptions of maps used in Example 4.1.10 and show that they are continuous.

Exercise 4.1.12. Let M_n be a space obtained from a rectangle by gluing the side edges after the twist of $n\pi$. The projection onto the central circle $p_n : M_n \to S^1$ is a fiber bundle. Show that we have bundle isomorphisms

$$M_{2n} \cong M_0$$
$$M_{2n+1} \cong M_1.$$

Remark 4.1.13. In Example 4.1.10, we have shown that M_2 and M_0 are homeomorphic. A homeomorphism is given by $f_1 \cup f_2$. Since we identify homeomorphic spaces in topology, we should regard M_2 and M_0 to be the "same". On the other hand, most people who do not know topology would think that they are different.

This is because people can only recognize spaces embedded in the 3-dimensional Euclidean space $\mathbb{R}^3$. The notion of homeomorphism has nothing to do with embedding.

In this sense, the notion of homeomorphism is quite different from the intuition of "same spaces".

Let us write the conditions for two bundles to be isomorphic by using local trivializations.

Proposition 4.1.14. *Let* $E \xrightarrow{p} B$ *and* $E' \xrightarrow{p'} B$ *be fiber bundles with fiber* F *and structure group* G. *We also assume that the actions of* G *on* F *are the same. Let* $\{U_\alpha\}_{\alpha \in A}$ *be an open covering of* B *on which both* E *and* E' *have local trivializations. Denote the coordinate transformations of* E *and* E' *by* $\{\Phi^{\alpha\beta} : U_\alpha \cap U_\beta \to G\}$ *and* $\{\Psi^{\alpha\beta} : U_\alpha \cap U_\beta \to G\}$, *respectively.*

Then E and E' are isomorphic if and only if there exists a map λ_α : $U_\alpha \to G$ for each $\alpha \in A$ such that, for $\alpha, \beta \in A$ and $x \in U_\alpha \cap U_\beta$,

$$\lambda_\beta(x)^{-1}\Psi^{\alpha\beta}(x)\lambda_\alpha(x) = \Phi^{\alpha\beta}(x).$$

Proof. Suppose there exists a bundle isomorphism $\tilde{f} : E \to E'$. Let

$$\varphi_\alpha : p^{-1}(U_\alpha) \longrightarrow U_\alpha \times F$$
$$\psi_\alpha : p'^{-1}(U_\alpha) \longrightarrow U_\alpha \times F$$

be local trivialization of E and E', respectively. For each $x \in U_\alpha \cap U_\beta$ and $y \in F$, we may write

$$\psi_\beta \tilde{f} \varphi_\alpha^{-1}(x, y) = (x, \lambda_{\alpha\beta}(x)(y)).$$

By the definition of bundle map, there exists a continuous map

$$\bar{\lambda}_{\alpha\beta} : U_\alpha \cap U_\beta \to G$$

such that $\lambda_{\alpha\beta} = \mathrm{ad}(\beta) \circ \bar{\lambda}_{\alpha\beta}$. Define $\lambda_\alpha = \bar{\lambda}_{\alpha\alpha}$. Consider the composition $\psi_\beta \circ \tilde{f} \circ \varphi_\alpha^{-1}$ in the diagram

$(U_\alpha \cap U_\beta) \times F$ $\qquad\qquad\qquad\qquad\qquad$ $(U_\alpha \cap U_\beta) \times F$

φ_α $\qquad\qquad\qquad\qquad\qquad\qquad\qquad$ ψ_β

$p^{-1}(U_\alpha \cap U_\beta) \xrightarrow{\ \tilde{f}\ } p'^{-1}(U_\alpha \cap U_\beta)$

φ_β $\qquad\qquad\qquad\qquad\qquad\qquad\qquad$ ψ_α

$(U_\alpha \cap U_\beta) \times F$ $\qquad\qquad\qquad\qquad\qquad$ $(U_\alpha \cap U_\beta) \times F.$

We have

$$\psi_\beta \circ \tilde{f} \circ \varphi_\alpha^{-1}(x, y) = \psi_\beta \circ \psi_\alpha^{-1} \circ \psi_\alpha \circ \tilde{f} \circ \varphi_\alpha^{-1}(x, y)$$
$$= \psi_\beta \circ \psi_\alpha^{-1}(x, \lambda_\alpha(x)(y))$$
$$= (x, \Psi^{\alpha\beta}(x)\lambda_\alpha(x)(y))$$

by using the right bottom. On the other hand, by using the left bottom, we have

$$\psi_\beta \circ \tilde{f} \circ \varphi_\alpha^{-1}(x, y) = \psi_\beta \circ \tilde{f} \circ \varphi_\beta^{-1} \circ \varphi_\beta \circ \varphi_\alpha^{-1}(x, y)$$
$$= \psi_\beta \circ \tilde{f} \circ \varphi_\beta^{-1}(x, \Phi^{\alpha\beta}(x)(y))$$
$$= (x, \lambda_\beta(x)\Phi^{\alpha\beta}(x)(y))$$

and we have $\lambda_\beta(x)\Phi^{\alpha\beta}(x) = \Psi^{\alpha\beta}(x)\lambda_\alpha(x)$ and thus

$$\Phi^{\alpha\beta}(x) = \lambda_\beta(x)^{-1}\Psi^{\alpha\beta}(x)\lambda_\alpha(x).$$

Conversely, suppose such maps $\lambda_\alpha : U_\alpha \to G$ are given. For each α, define $\tilde{f}_\alpha : p^{-1}(U_\alpha) \to p'^{-1}(U_\alpha)$ by the composition

$$p^{-1}(U_\alpha) \xrightarrow{\varphi_\alpha} U_\alpha \times F \xrightarrow{\Lambda_\alpha} U_\alpha \times F \xrightarrow{\psi_\alpha^{-1}} p'^{-1}(U_\alpha),$$

where $\Lambda_\alpha(x, y) = (x, \lambda_\alpha(x)(y))$.

When $p^{-1}(U_\alpha \cap U_\beta) \neq \emptyset$, both $\tilde{f}_\alpha(x)$ and $\tilde{f}_\beta(x)$ are defined for $x \in p^{-1}(U_\alpha \cap U_\beta) \neq \emptyset$. By the condition

$$\lambda_\beta(x)\Phi^{\alpha\beta}(x) = \Psi^{\alpha\beta}(x)\lambda_\alpha(x),$$

it is easy to check that the maps coincide on the intersection. Thus by Lemma 3.7.10, we obtain bundle map $\tilde{f} : E \to E'$, which is a bundle isomorphism. $\qquad\square$

We have only studied maps between fiber bundles when they have the same structure group and the fiber. It is possible to define maps between fiber bundles when the structure groups or the fibers are different. In fact, we need such maps when we study vector bundles, which is the subject of §A.2.

4.2 Pullbacks

Pullback is an important concept which is closely related to bundle maps. This is an operation which produces a fiber bundle by pulling back an existing fiber bundle via a continuous map between base spaces.

Definition 4.2.1. Let $\xi = (p : E \to B)$ be a fiber bundle. For a continuous map $f : X \to B$, define

$$f^*(E) = \{(x, e) \in X \times E \mid f(x) = p(e)\}.$$

Define also maps

$$f^*(p) : f^*(E) \longrightarrow X$$
$$p^*(f) : f^*(E) \longrightarrow E$$

by $f^*(p)(x, e) = x$ and $p^*(f)(x, e) = e$, respectively. This operation of making $f^*(E)$ from E is called *pullback*. By Theorem 4.2.3, $f^*(p) : f^*(E) \to X$ is a fiber bundle. This bundle is denoted by $f^*(\xi)$ and is called the *pullback* of ξ along f.

Remark 4.2.2. Note that we have a commutative diagram

$$
\begin{array}{ccc}
f^*(E) & \xrightarrow{\ p^*(f)\ } & E \\
\downarrow{\scriptstyle f^*(p)} & & \downarrow{\scriptstyle p} \\
X & \xrightarrow{\ \ f\ \ } & B.
\end{array}
\tag{4.3}
$$

Theorem 4.2.3. *Under the assumption of Definition 4.2.1, the map $f^*(p)$: $f^*(E) \to X$ is a fiber bundle with the same fiber F and the same structure group G as $p : E \to B$.*

Proof. Let $\{\varphi_\alpha : p^{-1}(U_\alpha) \to U_\alpha \times F\}$ be local trivializations of $p : E \to B$. Write $\varphi_\alpha(e) = (p(e), \bar\varphi_\alpha(e))$. In order to find local trivializations of $f^*(E)$, we first need to find an appropriate open covering of X. Denote $V_\alpha = f^{-1}(U_\alpha)$. We use the open covering $\{V_\alpha\}_{\alpha \in A}$ of X.

Define maps $\psi_\alpha : f^*(p)^{-1}(V_\alpha) \to V_\alpha \times F$ by $\psi_\alpha(x, e) = (x, \bar\varphi_\alpha(e))$. Since $f^*(p)^{-1}(V_\alpha) \subset V_\alpha \times p^{-1}(U_\alpha)$, ψ_α is given by the composition

$$
f^*(p)^{-1}(V_\alpha) \hookrightarrow V_\alpha \times p^{-1}(U_\alpha) \xrightarrow{1_{V_\alpha} \times \bar\varphi_\alpha} V_\alpha \times F
$$

and is thus continuous. Let us verify that these maps are local trivializations of $f^*(E)$.

Define maps $\gamma_\alpha : V_\alpha \times F \to f^*(p)^{-1}(V_\alpha)$ by $\gamma_\alpha(x, y) = (x, \varphi_\alpha^{-1}(f(x), y))$, i.e. by the composition

$$
\gamma_\alpha : V_\alpha \times F \xrightarrow{\Delta \times 1_F} V_\alpha \times V_\alpha \times F \xrightarrow{1_{V_\alpha} \times f \times 1_F} V_\alpha \times U_\alpha \times F \xrightarrow{1_{V_\alpha} \times \varphi_\alpha^{-1}} V_\alpha \times p^{-1}(U_\alpha).
$$

Since $p(\varphi_\alpha^{-1}(f(x), y)) = f(x)$, we have $\gamma_\alpha(x, y) \in f^*(E)$ and we obtain a map $\gamma_\alpha : V_\alpha \times F \to f^*(p)^{-1}(V_\alpha)$.

Furthermore, since

$$
\begin{aligned}
\gamma_\alpha \circ \psi_\alpha(x, e) &= \gamma_\alpha(x, \bar\varphi_\alpha(e)) \\
&= (x, \varphi_\alpha^{-1}(f(x), \bar\varphi_\alpha(e))) \\
&= (x, \varphi_\alpha^{-1}(p(e), \bar\varphi_\alpha(e))) \\
&= (x, \varphi_\alpha^{-1} \circ \varphi_\alpha(e)) \\
&= (x, e) \\
\psi_\alpha \circ \gamma_\alpha(x, y) &= \psi_\alpha(x, \varphi_\alpha^{-1}(f(x), y)) \\
&= (x, \bar\varphi_\alpha \circ \varphi_\alpha^{-1}(f(x), y)) \\
&= (x, y),
\end{aligned}
$$

the map γ_α is an inverse to ψ_α. Thus ψ_α is a local trivialization.

We next show that $f^*(E)$ has G as a structure group. For $(x, y) \in$
$(V_\alpha \cap V_\beta) \times F$, we have

$$
\begin{aligned}
\psi_\beta \circ \psi_\alpha^{-1}(x, y) &= \psi_\beta \circ \gamma_\alpha(x, y) \\
&= \psi_\beta(x, \varphi_\alpha^{-1}(f(x), y)) \\
&= (x, \bar{\varphi}_\beta(\varphi_\alpha^{-1}(f(x), y))).
\end{aligned}
$$

Let $\Phi^{\alpha\beta} : U_\alpha \cap U_\beta \to \mathrm{Homeo}(F)$ be a coordinate transformation of $p : E \to$ B. By definition,

$$
\bar{\varphi}_\beta(\varphi_\alpha^{-1}(f(x), y)) = \Phi^{\alpha\beta}(f(x))(y).
$$

Thus

$$
\psi_\beta \circ \psi_\alpha^{-1}(x, y) = (x, \Phi^{\alpha\beta}(f(x))(y)),
$$

which implies that G is a structure group of $f^*(E)$. $\square$

Remark 4.2.4. The above proof shows that the coordinate transformations of $f^*(E) \to X$ are given by $\{\Phi^{\alpha\beta} \circ f\}$.

The reader is recommended to work out the following exercise in order to get intuitions on pullbacks.

Exercise 4.2.5. Consider the example in Exercise 4.1.12. Recall that M_0 is an annulus, M_1 is a Möbius band, and, more generally, M_n is obtained from a rectangle by gluing the side edges with the twist of $n\pi$. Let us denote the fiber bundle $M_n \to S^1$ by ξ_n.

Define a map $\varphi_n : S^1 \to S^1$ by $\varphi_n(z) = z^n$ for $z \in S^1 = \{z \in \mathbb{C} \,|\, |z| = 1\}$. Show that the pullback of ξ_1 along φ_n is isomorphic to ξ_n.

The following two properties of pullbacks are fundamental.

Proposition 4.2.6. *For a fiber bundle $p : E \to B$ and a continuous map $f : X \to B$, the map $p^*(f) : f^*(E) \to E$ in Definition 4.2.1 is a bundle map.*

Proof. It is immediate to verify that the diagram (4.3) is commutative. Let us compute the fiber over $x \in X$. We have

the fiber over $x = f^*(p)^{-1}(x)$
$$= \{(x,e) \in f^*(E) \mid f^*(p)(x,e) = x\}$$
$$= \{(x,e) \in X \times E \mid f(x) = p(e)\}$$
$$\stackrel{\cong}{\longrightarrow} \{e \in E \mid f(x) = p(e)\} \qquad (4.4)$$
$$= p^{-1}(f(x))$$
$$= \text{the fiber over } f(x).$$

Here the correspondence (4.4) is given by $(x,e) \mapsto e$, i.e. $p^*(f)$. Thus the map $p^*(f)$ defines a homeomorphism between the fiber over x and the fiber over $f(x)$.

If the bundle $p : E \to B$ has G as a structure group, the coordinate transformation of $f^*(p)$ is given by the composition

$$V_\alpha \cap V_\beta = f^{-1}(U_\alpha) \cap f^{-1}(U_\beta) \stackrel{f}{\longrightarrow} U_\alpha \cap U_\beta \stackrel{\Phi^{\alpha\beta}}{\longrightarrow} \text{Homeo}(F)$$

by Remark 4.2.4. Thus if G is a structure group of p and if the coordinate transformations factor as

$$U_\alpha \cap U_\beta \stackrel{\bar{\Phi}^{\alpha\beta}}{\longrightarrow} G \stackrel{\text{ad}(\mu)}{\longrightarrow} \text{Homeo}(F),$$

the coordinate transformations of $f^*(p)$ is given by

$$V_\alpha \cap V_\beta \stackrel{f}{\longrightarrow} U_\alpha \cap U_\beta \stackrel{\bar{\Phi}^{\alpha\beta}}{\longrightarrow} G \stackrel{\text{ad}(\mu)}{\longrightarrow} \text{Homeo}(F).$$

Thus $p^*(f)$ satisfies the condition for structure group of bundle map. $\square$

Proposition 4.2.7. *The converse to the above proposition holds. Namely, if*

$$p : E \longrightarrow B$$
$$p' : E' \longrightarrow X$$

are fiber bundles having the same fiber and the structure group and if

$$\begin{array}{ccc} E' & \stackrel{\tilde{f}}{\longrightarrow} & E \\ {\scriptstyle p'}\downarrow & & \downarrow{\scriptstyle p} \\ X & \stackrel{f}{\longrightarrow} & B \end{array}$$

is a bundle map, the bundle $E' \stackrel{p'}{\longrightarrow} X$ is isomorphic to the pullback $f^*(E) \stackrel{f^*(p)}{\longrightarrow} X$.

Proof. Since $(\tilde{f}, f) : (E', X) \to (E, B)$ is a bundle map, we have $p(\tilde{f}(e)) = f(p'(e))$ for any $e \in E'$. By the definition of $f^*(E)$, we have $(p'(e), \tilde{f}(e)) \in f^*(E)$. Define $\bar{f} : E' \to f^*(E)$ by $\bar{f}(e) = (p'(e), \tilde{f}(e))$. This is continuous since each factor is continuous. We are going to show that this is a bundle isomorphism. It is easy to show that the diagram

$$
\begin{array}{ccc}
E' & \xrightarrow{\bar{f}} & f^*(E) \\
{\scriptstyle p'}\downarrow & & \downarrow{\scriptstyle f^*(p)} \\
X & \xrightarrow{=} & X
\end{array}
$$

is commutative.

In order to show that $\bar{f}$ induces a homeomorphism on each fiber, consider the composition

$$
E' \xrightarrow{\bar{f}} f^*(E) \xrightarrow{p^*(f)} E,
$$

which is can be easily seen to coincide with $\tilde{f}$. Since both $p^*(f)$ and $\tilde{f}$ are homeomorphisms on each fiber, so is $\bar{f}$.

Let us consider the structure group. Let $\{\varphi_\lambda : p^{-1}(U_\lambda) \to U_\lambda \times F\}_{\lambda \in \Lambda}$ and $\{\varphi'_\gamma : p'^{-1}(V_\gamma) \to V_\gamma \times F\}_{\gamma \in \Gamma}$ be local trivializations of $p : E \to B$ and $p' : E' \to X$, respectively. We may assume that $f(V_\gamma)$ is contained in one of U_λ by subdividing $\{V_\gamma\}_{\gamma \in \Gamma}$ finely. By the construction of the local trivializations of $f^*(E) \to X$, it has local trivializations over the open covering $\{f^{-1}(U_\lambda)\}_{\lambda \in \Lambda}$. Let us denote the local trivializations by $\{\psi_\lambda : p^*(f)^{-1}(U_\lambda) \to f^{-1}(U_\lambda) \times F\}_{\lambda \in \Lambda}$.

Let G the structure group of these bundles. We need to show that, when $f(V_\gamma) \subset U_\lambda$, we have a factorization

$$
\begin{array}{ccc}
U_\lambda & \xrightarrow{\text{ad}(\text{pr}_2 \circ \psi_\lambda \circ \bar{f} \circ \varphi'^{-1}_\gamma)} & \text{Homeo}(F) \\
 & \searrow \qquad \nearrow{\scriptstyle \text{ad}(\mu)} & \\
 & G. &
\end{array}
$$

Write

$$
\varphi_\lambda(e) = (p(e), (\text{pr}_2 \circ \varphi_\lambda)(e)).
$$

By the proof of Theorem 4.2.3, local trivializations $\psi_\lambda : f^*(p)^{-1}(U_\lambda) \to U_\lambda \times F$ of $f^*(E)$ are given by $\psi_\lambda(x, e) = (x, (\text{pr}_2 \circ \varphi_\lambda)(e))$. Then the composition

$$
V_\gamma \times F \xrightarrow{\varphi'^{-1}_\lambda} p'^{-1}(U_\lambda) \xrightarrow{\bar{f}} f^*(p)^{-1}(U_\lambda) \xrightarrow{\psi_\lambda} U_\lambda \times F
$$

is given by

$$(x, y) \longmapsto (x, \mathrm{pr}_2(\varphi_\lambda(\tilde{f}(\varphi_\gamma'^{-1}(x, y))))) = (x, \mathrm{ad}(\mathrm{pr}_2 \circ \varphi_\lambda \circ \tilde{f} \circ \varphi_\gamma'^{-1})(x)(y)).$$

Thus we have

$$\mathrm{ad}(\mathrm{pr}_2 \circ \psi_\lambda \circ \tilde{f} \circ \varphi_\gamma'^{-1}) = \mathrm{ad}(\mathrm{pr}_2 \circ \varphi_\lambda \circ \tilde{f} \circ \varphi_\gamma'^{-1}).$$

On the other hand, since $\tilde{f}$ is a bundle map, the map $\mathrm{ad}(\mathrm{pr}_2 \circ \varphi_\lambda \circ \tilde{f} \circ \varphi_\gamma'^{-1})$ factors as

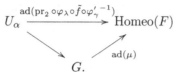

And thus $\bar{f}$ is a bundle map. Since it is the identity on X, it is an isomorphism of fiber bundles. $\qquad\square$

We conclude this section by a couple of more useful properties of pullbacks.

Lemma 4.2.8. *Let $\xi = (p : E \to B)$ be a fiber bundle. For continuous maps $f : X \to B$ and $g : Y \to X$, we have a bundle isomorphism $(f \circ g)^*(\xi) \cong g^*(f^*(\xi))$.*

Lemma 4.2.9. *For a fiber bundle $\xi = (p : E \to B)$, we have a bundle isomorphism $1_B^*(\xi) \cong \xi$.*

Proofs of these two facts easily follow from the definition of pullbacks and are left to the reader.

Definition 4.2.10. Let $p : E \to B$ be a fiber bundle and $A \subset B$ a subspace. The inclusion map is denoted by $i : A \hookrightarrow B$. The bundle $i^*(E)$ is called the *restriction* of E to A and is denoted by $E|_A$.

Note that the map $p : E \to B$ does not have to be a fiber bundle in order to define the space $f^*(E)$. In general, given a pair of maps

$$\begin{array}{ccc} & & Y \\ & & \downarrow g \\ X & \xrightarrow{f} & Z, \end{array}$$

we denote

$$X \times_Z Y = \{(x, y) \in X \times Y \mid f(x) = g(y)\}$$

and call it the pullback or the *fiber product* of f and g. This construction is characterized by the following universal property.

Proposition 4.2.11. *Given a commutative diagram*

$$
\begin{array}{ccc}
W & \xrightarrow{\ \tilde{f}\ } & Y \\
{\scriptstyle \tilde{g}}\downarrow & & \downarrow{\scriptstyle g} \\
X & \xrightarrow{\ f\ } & Z
\end{array}
$$

There exists a unique map $\tilde{f} \times_Z \tilde{g} : W \to X \times_Z Y$ which makes the diagram commutative.

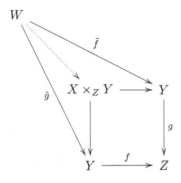

The proof of this fact is also left to the reader.

4.3   Fiber Bundles and Homotopy

The pullback operation allow us to construct new fiber bundles from fiber bundles by using maps between base spaces. For example, we have seen in Exercise 4.2.5 that we obtain infinitely many fiber bundles $M_n \to S^1$ from a fiber bundle $M_1 \to S^1$. It should be also noted that we do not need to deal with local trivializations when we take pullbacks.

An obvious question is the following.

Problem 4.3.1. Let ξ be a fiber bundle over B. Given continuous maps $f, g : X \to B$, when are pullbacks $f^*(\xi)$ and $g^*(\xi)$ isomorphic?

It turns out that, as we will see later (Theorem 4.3.25), if f can be deformed into g continuously, then the pullbacks are isomorphic $f^*(\xi) \cong g^*(\xi)$. Recall that, in Definition 2.2.13, we introduced the notion of homotopy as a continuous deformation of paths. Homotopy is, in fact, the notion for continuous deformations of continuous maps.

Definition 4.3.2. Let $f_0, f_1 : X \to Y$ be continuous maps. If there exists a continuous map $H : X \times [0, 1] \to Y$ such that

(1) $H(x,0) = f_0(x)$

(2) $H(x,1) = f_1(x)$

for each $x \in X$, we say f_0 and f_1 are *homotopic* and denote $f_0 \simeq f_1$. The map H is called a *homotopy* from f_0 to f_1.

Exercise 4.3.3. Verify that the relation $\simeq$ is an equivalent relation on $\text{Map}(X, Y)$.

Remark 4.3.4. When H is a homotopy from f_0 to f_1, define a map $f_t : X \to Y$ for $t \in [0,1]$ by $f_t(x) = H(x,t)$. In other words, $f_t = \text{ad}(H)(t)$. By Lemma 3.4.2, $f_t : X \to Y$ is a continuous map for each $t \in [0,1]$. When t changes from 0 to 1, the map f_t is deformed from f_0 to f_1 continuously. This is a meaning of homotopy.

Remark 4.3.5. The homotopy of paths (Definition 2.2.13) used in Chapter 2 is also a kind of homotopy but it is required to fix the end points paths. Definition 4.3.2 the most basic definition of homotopy and, by adding various conditions, we obtain variants such as homotopy of paths.

Remark 4.3.6. As we will see in this book, the notion of homotopy plays a central role in the study of fiber bundles, and in topology. The reader might have had an impression that continuous deformations of maps is a natural idea. Probably this is because the notion of map is well known as a foundation of mathematics. It turns out that homotopies appeared in the literature rather recently. According to Dieudonné's book [Dieudonné (1989)], The term "homotopy" first appeared in the paper [Dehn and Heegaard (1907)] by Dehn and Heegaard published in 1907. And the current definition first appeared in Brouwer's paper [Brouwer (1976)] in 1911.

Since then, the notion of homotopy gained more and more popularity, as people discovered that various geometric phenomena and concepts can be described by using homotopy. Now we have a research field called homotopy theory, in which generalizations of fiber bundles, called fibrations[1] and the dual notion, called cofibrations[2] are used to develop abstract and unique techniques. Some aspects of homotopy theory are described in Chapter 6 from by personal point of view.

Let's get back to the discussion on homotopy. The following is one of the most fundamental relations between homotopy and fiber bundles.

---

[1] §5.2.

[2] §5.9.

Theorem 4.3.7 (Lift of Homotopy). *Suppose we have*

(1) fiber bundles $p : E \to Y$ and $p' : E' \to X$,
(2) a fiber preserving map $(\tilde{f}_0, f_0) : (E', X) \to (E, Y)$, and
(3) a homotopy $H : X \times [0,1] \to Y$ satisfying $H(x,0) = f_0(x)$ for all $x \in X$.

If X is compact Hausdorff, there exists a lift of H to a bundle map

$$
\begin{array}{ccc}
E' \times [0,1] & \xrightarrow{\ \tilde{H}\ } & E \\
\Big\downarrow{\scriptstyle p' \times 1_{[0,1]}} & & \Big\downarrow{\scriptstyle p} \\
X \times [0,1] & \xrightarrow{\ H\ } & Y
\end{array}
$$

such that $\tilde{H}(e,0) = \tilde{f}_0(e)$ for each $e \in E'$.

The assumptions and the conclusion of the theorem can be described as a diagram:

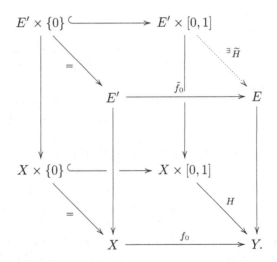

Theorem 4.3.7 says that, if the vertical maps are projections of fiber bundles and the horizontal solid arrows are given in such a way that they are bundle maps and make the squares commutative, then there exists a dotted bundle map $\tilde{H}$ making the diagram commutative.

In Theorem 4.3.7, we have assumed that X is compact Hausdorff for simplicity. As we will see later, this condition can be relaxed. The case of compact Hausdorff spaces is general enough to understand the idea of the proof of Theorem 4.3.7.

Unfortunately, even in the case of compact Hausdorff spaces, the proof is quite complicated. Let us take a look at a rough idea of the proof before we go into the details of the proof of Theorem 4.3.7.

Suppose $p : E \to Y$ is a trivial bundle, it is easy to construct $\widetilde{H}$. In fact, since $E = Y \times F$, we can write $\tilde{f}_0(e) = (\tilde{f}_0'(e), \tilde{f}_0''(e))$. Then define

$$\widetilde{H}(e,t) = (H(p'(e), t), \tilde{f}_0''(e)).$$

An obvious idea is to cut $E \to Y$ into small trivial bundles. And then decompose. Then we can construct $\widetilde{H}$ on each small fiber bundle. If we can glue these maps together, we are done. Since X is compact, we may take the decomposition as a decomposition into a finite number of small fiber bundles. Then we can glue these small fiber bundles inductively.

In order to make this idea work, we need to make use of various properties of compact Hausdorff spaces. Let us recall them before the proof of Theorem 4.3.7. Proofs can be found in textbooks on pointset topology such as [Kelley (1975)].

Theorem 4.3.8. *Any compact Hausdorff space is normal.*

Normal spaces have the following nice property.

Theorem 4.3.9 (Urysohn's Lemma). *When X is normal and $W \subset W' \subset X$ are open subsets of X satisfying $\overline{W} \subset W'$, then there exists a continuous map $u : X \to [0,1]$ satisfying the following conditions*

(1) $u(\overline{W}) = 1$
(2) $u(X \setminus W') = 0$.

In order to prove Theorem 4.3.7, we need to use local trivialization of the fiber bundle $E' \times [0,1]$ over $X \times [0,1]$. It would be nice if we take an open covering of $X \times [0,1]$ whose members are products of open sets. In fact, it is possible. It can be done by using the Lebesgue number of compact metric spaces used in §2.2. The following is an extension of Corollary 2.2.9.

Lemma 4.3.10. *Suppose X is compact. For any open covering $\{V_\alpha\}_{\alpha \in A}$ of $X \times [0,1]$, there exist a partition of $[0,1]$ $0 < \frac{1}{\ell} < \frac{2}{\ell} < \cdots < \frac{\ell-1}{\ell} < 1$ open sets $U_{1,1}, \ldots, U_{n_1,1}, \ldots, U_{1,\ell}, \ldots, U_{n_\ell,\ell} \subset X$ of X satisfying the following conditions:*

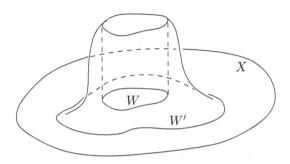

Fig. 4.3   the graph of the map $u : X \to [0,1]$

(1) $\displaystyle\bigcup_{k=1}^{\ell}\bigcup_{i=1}^{n_k} U_{i,k} \times W_k = X \times [0,1]$, *and*

(2) for each (i,k), there exists $\alpha \in A$ such that $U_{i,k} \times W_k \subset V_\alpha$,

where $W_1 = [0, \frac{4}{3\ell})$, $W_2 = (\frac{2}{3\ell}, \frac{7}{3\ell})$, ..., $W_\ell = (\frac{3\ell-1}{3\ell}, 1]$.

Proof. The idea of the proof is basically the same as Corollary 2.2.9.

For each $(x,t) \in X \times [0,1]$, there exists an α and open subsets $U_{x,t} \subset X$ and $V_{x,t} \subset [0,1]$ satisfying

$$(x,t) \in U_{x,t} \times V_{x,t} \subset V_\alpha.$$

For each $t \in [0,1]$, $\{U_{x,t}\}_{x\in X}$ is an open covering X. Since X is compact, we may choose a finite number of open sets $U_{x_1,t}, \ldots, U_{x_{n_t},t}$ which cover X. Define $V_t = \displaystyle\bigcap_{i=1}^{n_t} V_{x_i,t}$. Then

$$\bigcup_{i=1}^{n_t} U_{x_i,t} \times V_t = X \times V_t.$$

Note that $\{V_t\}_{t\in[0,1]}$ is an open covering of $[0,1]$. Since $[0,1]$ is a compact metric space, Lemma 2.2.8 gives us a number σ. Choose $\ell \in \mathbb{N}$ with $\frac{5}{3\ell} < \sigma$ and define $W_1 = [0, \frac{4}{3\ell})$, $W_2 = (\frac{2}{3\ell}, \frac{7}{3\ell})$, ..., $W_\ell = (\frac{3\ell-4}{3\ell}, 1]$. Then we have $d(W_k) \leq \frac{5}{3\ell}$ and thus, for each k, there exists t_k such that $W_k \subset V_{t_k}$. For each k, we have

$$\bigcup_{i=1}^{n_{t_k}} U_{x_i,t_k} \times V_{t_k} = X \times V_{t_k}$$

and, for $1 \le i \le n_{t_k}$, we have

$$U_{x_i,t_k} \times W_k \subset U_{x_i,t_k} \times V_{t_k} \subset V_\alpha.$$

These are partitions of $[0,1]$ $0 < \frac{1}{\ell} < \frac{2}{\ell} < \cdots < \frac{\ell-1}{\ell} < 1$ and an open covering of $X\{U_{x_i,t_k}\}_{1 \le i \le n_t, 1 \le t \le \ell}$ that we wanted. $\square$

Now we are ready to prove Theorem 4.3.7.

Proof of Theorem 4.3.7. Let $\{V_\alpha\}_{\alpha \in A}$ be an open covering of Y such that the bundle $E \to Y$ has a local trivialization $\varphi_\alpha : p^{-1}(V_\alpha) \to V_\alpha \times F$ on each V_α. Then $\{H^{-1}(V_\alpha)\}_{\alpha \in A}$ is an open covering of $X \times [0,1]$.

By Lemma 4.3.10, there exists a partition of $[0,1]$ $0 < \frac{1}{\ell} < \frac{2}{\ell} < \cdots < \frac{\ell-1}{\ell} < 1$ and open sets $U_{1,1}, \ldots, U_{n_1,1}, \ldots, U_{1,\ell}, \ldots, U_{n_\ell,\ell} \subset X$ such that

$$\bigcup_{k=1}^{\ell} \bigcup_{i=1}^{n_k} U_{i,k} \times W_k = X \times [0,1]$$

and, for each (i,k), there exists $\alpha \in A$ such that $U_{i,k} \times W_k \subset H^{-1}(V_\alpha)$. Recall that $W_1 = [0, \frac{4}{3\ell})$, $W_2 = (\frac{2}{3\ell}, \frac{7}{3\ell})$, ..., $W_\ell = (\frac{3\ell-4}{3\ell}, 1]$. Denote $t_k = \frac{k}{\ell}$. Then $t_k \in W_k \cap W_{k+1}$ for each k.

We are going to construct $\tilde{H}$ on $E' \times [0, t_k]$ by induction on k. When $k = 0$, we may use $\tilde{f}_0$ on $E' \times \{0\}$. Suppose we have extended this map to $E' \times [0, t_k]$. In order to extend it further to $E' \times [0, t_{k+1}]$, we decompose $X \times [t_k, t_{k+1}]$ into small pieces and then define $\tilde{H}$ on $E' \times [t_k, t_{k+1}]$.

By Theorem 4.3.8, X is a normal space. When $x \in U_{i,k+1}$, there exists a pair of open sets (O', O) of X satisfying

$$x \in O, \ \overline{O} \subset O', \ \overline{O}' \subset U_{i,k+1}.$$

Since X is compact, we may choose a finite number of pairs $\{(O'_j, O_j)\}_{j=1,\ldots,s}$ so that $X = \bigcup_{j=1}^{s} O_j$.

Apply Urysohn's Lemma (Theorem 4.3.9) to each (O'_j, O_j) with $[0,1]$ replaced by $[t_k, t_{k+1}]$ to obtain continuous maps $u_j : X \to [t_k, t_{k+1}]$ satisfying

$$u_j(\overline{O_j}) = t_{k+1}$$
$$u_j(X \setminus O'_j) = t_k.$$

With these functions, define

$$\tau_0(x) = t_k$$
$$\tau_j(x) = \max\{u_1(x), \ldots, u_j(x)\}$$

for each $x \in X$. Then we have

$$t_k = \tau_0(x) \le \tau_1(x) \le \ldots \le \tau_s(x) = t_{k+1}.$$

Define

$$X_j = \{(x,t) \in X \times [t_k, t_{k+1}] \,|\, t \le \tau_j(x)\}$$

so that we obtain an increasing sequence of closed subsets

$$X_0 = X \times \{t_k\} \subset X_1 \subset X_2 \subset \cdots \subset X_s = X \times [t_k, t_{k+1}]$$

in $X \times [t_k, t_{k+1}]$.

Let E'_j be the restriction[3] of the bundle $E' \times [0,1]$ to X_j. Then we obtain an increasing sequence of fiber bundles

$$E'_0 = E' \times \{t_k\} \subset E'_1 \subset \cdots \subset E'_s = E' \times [t_k, t_{k+1}].$$

We construct homotopies on E'_j by induction on j. Suppose we have extended the homotopy to E'_{j-1}. Denote it by $\widetilde{H}_{j-1}$. In order to extend it to $\widetilde{H}_j : E'_j \times [0,1] \to E$, we decompose X_j into X_{j-1} and its complement. For each $1 \le j \le s$, define

$$X_{j-1,j} = \left\{(x,t) \in \overline{O'_j} \times [t_k, t_{k+1}] \,\Big|\, \tau_{j-1}(x) \le x \le \tau_j(x)\right\}.$$

See Figure 4.4. Then $X_j = X_{j-1} \cup X_{j-1,j}$. By construction, there exist $U_{i,k+1}$ and V_α such that

$$X_{j-1,j} \subset \overline{O'_j} \times [t_k, t_{k+1}]$$

$$\overline{O'_j} \times [t_k, t_{k+1}] \subset U_{i,k+1} \times [t_k, t_{k+1}]$$

$$H(U_{i,k+1} \times [t_k, t_{k+1}]) \subset V_\alpha.$$

Define $E'_{j-1,j} = E'|_{X_{j-1,j}}$, then $E'_j = E'_{j-1} \cup E'_{j-1,j}$. Write the local trivialization of $E \to Y$ on V_α as $\varphi_\alpha(e) = (p(e), \bar{\varphi}_\alpha(e))$ and define

$$\widetilde{H}_j(e,t) = \varphi_\alpha^{-1}(H(p'(e),t), \bar{\varphi}_\alpha(\widetilde{H}_{j-1}(e, \tau_{j-1}(p'(e)))))$$

for $(e,t) \in E'_{j-1,j} \times [0,1]$. The map $\widetilde{H}_j$ is continuous on $E'_{j-1,j}$. We also define $\widetilde{H}_j|_{E'_{j-1} \times [0,1]} = \widetilde{H}_{j-1}$. On the common boundary of $E'_{j-1,j}$ and E'_{j-1}, i.e. when $t = \tau_{j-1}(p'(e))$, we have

$$\begin{aligned}
\text{Right Hand Side} &= \varphi_\alpha^{-1}(H(p'(e),t), \bar{\varphi}_\alpha(\widetilde{H}_{j-1}(e,t))) \\
&= \varphi_\alpha^{-1}(p \circ \widetilde{H}_{j-1}(e,t), \bar{\varphi}_\alpha(\widetilde{H}_{j-1}(e,t))) \\
&= \varphi_\alpha^{-1} \circ \varphi_\alpha(\widetilde{H}_{j-1}(e,t)) \\
&= \widetilde{H}_{j-1}(e,t)
\end{aligned}$$

and it agrees with $\widetilde{H}_{j-1}$. Thus we have succeeded in extending $\widetilde{H}_{j-1}$ to E'_j. This completes the construction of $\widetilde{H}$ on $E' \times [0,1]$. $\square$

[3]Definition 4.2.10.

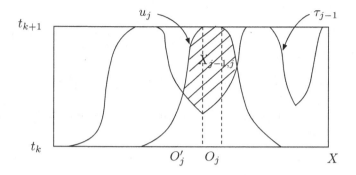

Fig. 4.4   a partition of $X \times [0, 1]$

Exercise 4.3.11. Verify that the homotopy $\widetilde{H}$ satisfies the conditions in Theorem 4.3.7.

An important idea in the above proof is to decompose $X \times [0, 1]$ into a union of closed sets whose boundaries are given by graphs of continuous functions. When continuous maps are defined on such closed sets and they agree on the boundaries, we may glue these maps together to obtain a continuous map. More precisely, we used the following fact, whose proof is left to the reader.

Lemma 4.3.12. *Let A and B be closed subsets of a topological space X. Given continuous maps $f : A \to Y$ and $g : B \to Y$ satisfying $f|_{A \cap B} = g|_{A \cap B}$, there exists a unique continuous map $h : A \cup B \to Y$ such that $h|_A = f$ and $h|_B = g$.*

Definition 4.3.13. We denote the map h in Lemma 4.3.12 by $f \cup g$.

The assumption in Theorem 4.3.7 that X is compact Hausdorff is used to obtain functions τ_j by applying Urysohn's Lemma. In other words, we may replace the compact-Hausdorffness assumption by the existence of these functions to obtain the same conclusion.

Definition 4.3.14. Let $\{U_\alpha\}_{\alpha \in A}$ be a covering of a topological space X. A *partition of unity subordinate to* this covering is a family of maps $\{f_\gamma : X \to [0, 1]\}_{\gamma \in \Gamma}$ satisfying the following conditions:

(1) Define $\mathrm{Supp}(f_\gamma) = \overline{\{x \in X \mid f_\gamma(x) \neq 0\}}$ for each γ. Then $\{\mathrm{Supp}(f_\gamma)\}_{\gamma \in \Gamma}$ is a locally finite covering of X, which means that

$X = \bigcup_{\gamma \in \Gamma} \mathrm{Supp}(f_\gamma)$ and, for any $x \in X$, there exists an open neighborhood U of x which intersect only a finite number of $\mathrm{Supp}(f_\gamma)$.

(2) For each $x \in X$, $\displaystyle\sum_{\gamma \in \Gamma} f_\gamma(x) = 1$.

(3) For each $\gamma \in \Gamma$, there exists $\alpha \in A$ such that $\mathrm{Supp}(f_\gamma) \subset U_\alpha$.

When a covering has a partition of unity subordinate to it, it is called a *numerable covering*.

Note that numerable coverings do not need to be open coverings. Dold used this concept of numerable coverings to generalize Theorem 4.3.7.

Theorem 4.3.15. *Let $\{U_\alpha\}_{\alpha \in A}$ be a numerable covering of a topological space B, and $p : E \to B$ a fiber bundle. If p is trivial on each U_α, then Theorem 4.3.7 holds for any topological space X.*

This can be proved by following the argument of the proof of Theorem 4.3.7 and is omitted. The interested reader is recommend to take a look at the original paper [Dold (1963)].

Precisely speaking, what Dold proved is not Theorem 4.3.15, but the following equivalent form.

Corollary 4.3.16 (Covering Homotopy Theorem). *Let $p : E \to B$ be a fiber bundle. Suppose continuous maps $f : X \to B$ and $\tilde{f} : X \to E$ make the diagram*

commutative. Let $H : X \times [0,1] \to B$ be a homotopy with $H(x,0) = f(x)$.

If B has a numerable covering, then there exists a homotopy $\tilde{H} : X \times [0,1] \to E$ such that $\tilde{H}(x,0) = \tilde{f}(x)$ and makes the diagram commutative.

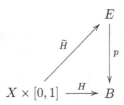

Remark 4.3.17. We may state this theorem by the following diagram

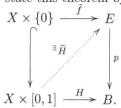

The Covering Homotopy Theorem says that when the square is commutative, there exists a dotted arrow $\widetilde{H}$ making the two triangles commutative.

As we mentioned in Remark 2.2.4, this is a generalization (without uniqueness) of the existences of lifts of paths and homotopies in the case of covering spaces.

Proof of 4.3.16. Take $E' \to X$ to be the trivial bundle $X \xrightarrow{=} X$ in Theorem 4.3.15. $\qquad\square$

It turns out that Theorem 4.3.15 is equivalent to Corollary 4.3.16, whose proof is left to the reader as an exercise.

An important class of spaces to which Theorem 4.3.15 can be applied is the following.

Definition 4.3.18. A topological space X is called *paracompact*, if any open covering $\{U_\alpha\}_{\alpha \in A}$ of X has a locally finite subdivision. In other words, there exists an open covering $\{V_\beta\}_{\beta \in B}$ of X satisfying the conditions that

(1) for any $\beta \in B$, there exists $\alpha \in A$ with $V_\beta \subset U_\alpha$, and
(2) for any $x \in X$, there exists an open neighborhood U of x such that only a finite number of V_β's intersect with U nontrivially.

The following fact can be found in Chapter VIII §4 in Dugundi's book [Dugundji (1978)].

Theorem 4.3.19. *Any paracompact Hausdorff space has a partition of unity subordinate to any open covering.*

Corollary 4.3.20. *If $p : E \to B$ is a fiber bundle over a paracompact Hausdorff space B. Then for any topological space X, Corollary 4.3.16, hence Theorem 4.3.7 holds.*

We end this section by some facts that follow from Theorem 4.3.7 and Corollary 4.3.16, which will be used in the proof of Theorem 4.5.21).

We first need the notion of deformation retraction.

Definition 4.3.21. Let A be a subspace of a topological space X. Denote the inclusion map by $i : A \hookrightarrow X$. If there exists a continuous map $r : X \to A$ with

(1) $r \circ i = 1_A$
(2) $i \circ r \simeq 1_X$,

then A is called a *deformation retract* of X. A homotopy for $i \circ r \simeq 1_X$ is called a *deformation retraction*.

Remark 4.3.22. When only the first condition $r \circ i = 1_A$ holds, A is called a *retract* of X and r is called a *retraction*.

Corollary 4.3.23. *Let $p : E \to X$ be a fiber bundle over a paracompact Hausdorff space X. Suppose $A \subset X$ is a deformation retract of X. Then there exists a bundle map*

$$
\begin{array}{ccc}
E & \xrightarrow{\tilde{r}} & E|_A \\
\downarrow{\scriptstyle p} & & \downarrow{\scriptstyle p} \\
X & \xrightarrow{r} & A.
\end{array}
$$

Proof. Let $h : X \times [0,1] \to X$ be a homotopy from 1_X to $i \circ r$. Then, by Theorem 4.3.7, h has a lift

$$
\begin{array}{ccc}
E \times [0,1] & \xrightarrow{\tilde{h}} & E \\
\downarrow{\scriptstyle p \times 1_{[0,1]}} & & \downarrow{\scriptstyle p} \\
X \times [0,1] & \xrightarrow{h} & X.
\end{array}
$$

Define $\tilde{r} = \tilde{h}|_{X \times \{1\}}$. Then we obtain a bundle map

$$
\begin{array}{ccc}
E & \xrightarrow{\tilde{r}} & E \\
\downarrow{\scriptstyle p} & & \downarrow{\scriptstyle p} \\
X & \xrightarrow{r} & X.
\end{array}
$$

Since the image of $\tilde{r}$ is $E|_A = p^{-1}(A)$, the proof is completed. $\quad\square$

We also need the following fact.

Corollary 4.3.24. *Let X be a paracompact Hausdorff space. If $p : E \to X \times [0,1]$ is a fiber bundle, there exists a fiber bundle $p' : E' \to X$ over X such that $p : E \to X \times [0,1]$ is isomorphic to $p' \times 1_{[0,1]} : E' \times [0,1] \to X \times [0,1]$.*

Proof. Let $i_0 : X = X \times \{0\} \hookrightarrow X \times [0,1]$ be the inclusion map and define $E' = i_0^*(E)$. By definition we have a bundle map

$$
\begin{array}{ccc}
E' & \longrightarrow & E \\
{\scriptstyle p'}\downarrow & & \downarrow{\scriptstyle p} \\
X & \xrightarrow{\ i_0\ } & X \times [0,1].
\end{array}
$$

Here we regard E' as a subspace of E and p' is a restriction of p. In order to apply Theorem 4.3.7, we need a homotopy between base spaces. We use the identity map

$$ H = 1_{X \times [0,1]} : X \times [0,1] \xrightarrow{=} X \times [0,1] $$

as a homotopy. Since $H(x,0) = (x,0) = i_0(x)$, it satisfies the assumption of Theorem 4.3.7 and we obtain a bundle map

$$
\begin{array}{ccc}
E' \times [0,1] & \xrightarrow{\ \widetilde{H}\ } & E \\
{\scriptstyle p' \times 1_{[0,1]}}\downarrow & & \downarrow{\scriptstyle p} \\
X \times [0,1] & \xrightarrow{\ H\ } & X \times [0,1]
\end{array}
$$

Since H is the identity map, this is an isomorphism of fiber bundles by definition. $\qquad\square$

An important consequence of this corollary is the following theorem, which is one of the most significant relations between fiber bundles and homotopy and plays an important role in the classification of fiber bundles.

Theorem 4.3.25. *Let $p : E \to Y$ be a fiber bundle. Let $f_0, f_1 : X \to Y$ be continuous maps with $f_0 \simeq f_1$. If X is paracompact Hausdorff, then the two bundles $f_0^*(E) \to X$ and $f_1^*(E) \to X$ obtained by pulling back p along f_0 and f_1 are isomorphic.*

Proof. Let $H : X \times [0,1] \to Y$ be a homotopy from f_0 to f_1. The pullback $H^*(E)$ of E along H is a fiber bundle over $X \times [0,1]$. Thus by Corollary 4.3.24, there exists a fiber bundle $E' \to X$ over X and an isomorphism of fiber bundles $H^*(E) \cong E' \times [0,1]$.

On the other hand, for each $t \in [0,1]$, let $i_t : X = X \times \{t\} \hookrightarrow X \times [0,1]$ be the inclusion map. Since $f_0 = H \circ i_0$ and $f_1 = H \circ i_1$, we have

$$f_0^*(E) = (H \circ i_0)^*(E) = i_0^*(H^*(E)) \cong i_0^*(E' \times [0,1])$$
$$f_1^*(E) = (H \circ i_1)^*(E) = i_1^*(H^*(E)) \cong i_1^*(E' \times [0,1])$$

by Proposition 4.2.8.

Note that, for each $t \in [0,1]$,

$$
\begin{array}{ccccc}
E' & =\!=\!= & E' \times \{t\} & \hookrightarrow & E' \times [0,1] \\
\downarrow & & & & \downarrow \\
X & =\!=\!= & X \times \{t\} & \xrightarrow[i_t]{} & X \times [0,1]
\end{array}
$$

is a bundle map. By Proposition 4.2.7, we obtain an isomorphism $E' \cong i_t^*(E' \times [0,1])$. Thus we have a sequence of isomorphism

$$f_0^*(E) \cong i_0^*(E' \times [0,1]) \cong E' \cong i_1^*(E' \times [0,1]) \cong f_1^*(E)$$

and the proof is completed. $\qquad\square$

4.4 Classification of Fiber Bundles: Simple Cases

Before we dive into the classification problem of general fiber bundles, let us consider simple cases. Simple in the sense of homotopy. The simplest spaces in the sense of homotopy are called contractible spaces.

Definition 4.4.1. A topological space X is called *contractible* if there exist a point $x_0 \in X$ and a homotopy $H : X \times [0,1] \to X$ such that, for any $x \in X$, $H(x,0) = x$ and $H(x,1) = x_0$. In other words, the identity map 1_X of X and the constant map c_{x_0} to x_0 are homotopic. This homotopy H is called a *contraction* of X to x_0.

Example 4.4.2. The n-dimensional unit disk

$$D^n = \left\{ (x_1,\ldots,x_n) \in \mathbb{R}^n \,\big|\, x_1^2 + \cdots + x_n^2 \leq 1 \right\}$$

is a typical example of a contractible space. More generally, any convex set in $\mathbb{R}^n$ is contractible. $\qquad\square$

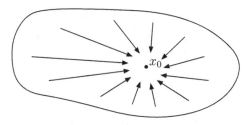

Fig. 4.5 contractible space

It is reasonable to expect from the relation between homotopy and fiber bundles studied in §4.3 that fiber bundles over a contractible space are not so complicated.

Theorem 4.4.3. *Let* $p : E \to X$ *be a fiber bundle with* F *over a paracompact Hausdorff space* X. *If* X *is contractible,* p *is isomorphic to a trivial bundle. In other words, we have a bundle map*

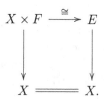

Proof. Since X is contractible, there exist a point $x_0 \in X$ and a homotopy $H : X \times [0,1] \to X$ between 1_X and c_{x_0}. By Theorem 4.3.25, we obtain an isomorphism $1_X^*(E) \cong c_{x_0}^*(E)$. By the definition of pullback,

$$
\begin{aligned}
1_X^*(E) &= \{(x,e) \in X \times E \mid 1_X(x) = p(e)\} \\
&= \{(x,e) \in X \times E \mid x = p(e)\} \\
&= \{(p(e),e) \in X \times E \mid e \in E\} \\
&\cong E
\end{aligned}
$$

and

$$
\begin{aligned}
c_{x_0}^*(E) &= \{(x,e) \in X \times E \mid c_{x_0}(x) = p(e)\} \\
&= \{(x,e) \in X \times E \mid x_0 = p(e)\} \\
&= X \times \{e \in E \mid x_0 = p(e)\} \\
&= X \times p^{-1}(x_0) \\
&= X \times F.
\end{aligned}
$$

And we have an isomorphism of fiber bundles $E \cong X \times F$.   □

This theorem classifies fiber bundles over a contractible space completely. In particular, any fiber bundle over the n-dimensional disk D^n is always trivial.

The next simplest case would be spaces that are obtained by gluing two contractible spaces such as a sphere. Denote the upper and the lower hemispheres of S^n by

$$
\begin{aligned}
S_+^n &= \{(x_0,\ldots,x_n) \in S^n \mid x_n \geq 0\} \\
S_-^n &= \{(x_0,\ldots,x_n) \in S^n \mid x_n \leq 0\},
\end{aligned}
$$

respectively. Then we have

$$
S^n = S_+^n \cup S_-^n. \tag{4.5}
$$

Since both S_+^n and S_-^n are homeomorphic to D^n and hence are contractible.

Let $p : E \to S^n$ a fiber bundle with fiber F. Denote the inclusion map by $i_\pm : S_\pm^n \hookrightarrow S^n$. Since S_+^n and S_-^n are contractible, the restrictions $i_+^*(E)$ and $i_-^*(E)$ of E to S_+^n and S_-^n are trivial. Let

$$
\varphi_+ : i_+^*(E) \xrightarrow{\ \cong\ } S_+^n \times F
$$

$$
\varphi_- : i_-^*(E) \xrightarrow{\ \cong\ } S_-^n \times F
$$

be trivializations.

Definition 4.4.4. Under the above situation, denote the adjoint of the composition

$$
(S_+^n \cap S_-^n) \times F \xrightarrow{\varphi_+^{-1}} p^{-1}(S_+^n \cap S_-^n) \xrightarrow{\varphi_-} (S_+^n \cap S_-^n) \times F \xrightarrow{\mathrm{pr}_2} F \tag{4.6}
$$

by $B(p) : S^{n-1} \cong S_+^n \cap S_-^n \to \mathrm{Homeo}(F)$. This is called the *classifying map* of p.

Suppose p has a structure group G. The classifying map $B(p)$ is close to a coordinate transformation but is not quite, since S_+^n and S_-^n are not open. We need to enlarge them slightly. Take $0 < \varepsilon < 1$ and define

$$U_{+,\varepsilon} = \{(x_0, \ldots, x_n) \in S^n \mid x_n > -\varepsilon\}$$
$$U_{-,\varepsilon} = \{(x_0, \ldots, x_n) \in S^n \mid x_n < \varepsilon\}.$$

Let $i_{\pm,\varepsilon} : U_{\pm,\varepsilon} \hookrightarrow S^n$ be inclusion maps. Since $U_{\pm,\varepsilon}$ are contractible, we have isomorphisms of fiber bundles $\varphi_{\pm,\varepsilon} : i_{\pm,\varepsilon}^*(E) \xrightarrow{\cong} U_{\pm,\varepsilon} \times F$. As is the case of (4.6), we obtain a map $\Phi_{+-} : U_{+,\varepsilon} \cap U_{-,\varepsilon} \to \mathrm{Homeo}(F)$. The classifying map $B(p)$ agrees with the composition

$$S^{n-1} = S_+^n \cap S_-^n \hookrightarrow U_{+,\varepsilon} \cap U_{-,\varepsilon} \xrightarrow{\overline{\Phi}_{+-}} \mathrm{Homeo}(F).$$

When $\varphi_{\pm,\varepsilon}$ are local trivializations of p, there exists a map $\overline{\Phi}_{+-}$ making the diagram

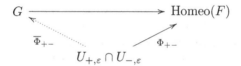

commutative. Let us call the composition

$$S^{n-1} = S_+^n \cap S_-^n \hookrightarrow U_{+,\varepsilon} \cap U_{-,\varepsilon} \xrightarrow{\overline{\Phi}_{+-}} G$$

the classifying map of p and denote it by the same symbol $B(p)$.

The bundle E is obtained by gluing trivial bundles $S_+^n \times G$ and $S_-^n \times G$ under the decomposition (4.5). Since G is the structure group, the coordinate transformation $\overline{\Phi}_{+-}$ is determined by $B(p)$. In fact, this map classifies bundles over S^n.

Theorem 4.4.5. *Let G be a arcwise connected topological group. Let $p : E \to S^n$ and $p' : E' \to S^n$ be fiber bundles with fiber F and structure group G. We also assume that the actions of G on F are the same.*

Suppose that there exists $0 < \varepsilon < 1$ such that both E and E' have local trivializations

$$\varphi_{\pm,\varepsilon} : p^{-1}(U_{\pm,\varepsilon}) \longrightarrow U_{\pm,\varepsilon} \times F$$
$$\psi_{\pm,\varepsilon} : (p')^{-1}(U_{\pm,\varepsilon}) \longrightarrow U_{\pm,\varepsilon} \times F.$$

Then E and E' are isomorphic if and only if the classifying maps $B(p)$ and $B(p')$ are homotopic $B(p) \simeq B(p')$.

We need the following refinement of Proposition 4.1.14.

Proposition 4.4.6. *Let* $E \xrightarrow{p} B$ *and* $E' \xrightarrow{p'} B$ *be fiber bundles with fiber* F *and structure group* G. *Suppose that the actions of* G *and* F *are the same. Let* $\{A_i\}_{i=1,\ldots,n}$ $\{U_i\}_{i=1,\ldots,n}$ *be a finite closed covering and a finite of open covering of* B *such that* $A_i \subset U_i$ *and both* E *and* E' *have local trivializations over* U_i *for each* i. *For* $1 \leq i,j \leq n$, *let* $\{\Phi^{ij} : A_i \cap A_j \to G\}$ *and* $\{\Psi^{ij} : A_i \cap A_j \to G\}$ *be restrictions of coordinate transformations of* E *and* E', *respectively.*

Then E *and* E' *are isomorphic if and only if there exists a continuous map* $\lambda_i : A_i \to G$ *for each* $1 \leq i \leq n$ *such that, for any* $1 \leq i,j \leq n$ *and* $x \in A_i \cap A_j$,

$$\lambda_j(x)^{-1}\Psi^{ij}(x)\lambda_i(x) = \Phi^{ij}(x). \tag{4.7}$$

Proof. If E and E' are isomorphic, Proposition 4.1.14 gives us maps $\lambda_i : U_i \to G$. The restrictions of these maps to A_i's satisfy (4.7).

Suppose we have maps $\lambda_i : A_i \to G$ satisfying (4.7). In the proof of Proposition 4.1.14, we constructed maps $\tilde{f}_\alpha : p^{-1}(U_\alpha) \to p'^{-1}(U_\alpha)$ for each α and the glue them together to obtain a continuous map $\tilde{f} : E \to E'$ by using Lemma 3.7.10.

By using the same argument, we obtain maps $\tilde{f}_i : p^{-1}(A_i) \to p'^{-1}(A_i)$. Since $\{A_i\}$ is a closed covering, we make use of Lemma 4.3.12 to glue these maps inductively to obtain a bundle map $\tilde{f} : E \to E'$. $\qquad\square$

Proof of Theorem 4.4.5. Suppose E and E' are isomorphic. Let $\tilde{f} : E \to E'$ be a bundle isomorphism. By Proposition 4.1.14, $\tilde{f}$ is determined by two maps $\lambda_+ : U_{+,\varepsilon} \to G$ and $\lambda_- : U_{-,\varepsilon} \to G$. Since $B(p)$ and $B(p')$ are restrictions of coordinate transformations of E and E', respectively, we have $B(p)(x) = \lambda_-(x)^{-1}B(p')(x)\lambda_+(x)$ for each $x \in S^{n-1} = S^n_+ \cap S^n_-$. We define a homotopy from $B(p)$ to $B(p')$ by using these maps $\lambda_\pm$.

Since S^n_+ is contractible, there exists a homotopy $H_+ : S^n_+ \times [0,1] \to S^n_+$ which shrinks S^n_+ to $x_0 = (1,0,\ldots,0) \in S^{n-1}$. Note that $H_+(x,0) = x$ and $H_+(x,1) = x_0$ for $x \in S^n_+$. Define $h_+(x,t) = \lambda_+(H_+(x,t))$ for $(x,t) \in S^n_+ \times [0,1]$. Similarly we define a map $h_- : S^n_- \times [0,1] \to G$ by $h_-(x,t) = \lambda_-(H_-(x,t))$, where $H_- : S^n_- \times [0,1] \to S^n_-$ is a homotopy which shrinks S^n_- to x_0. With these maps, define $F(x,t) = h_-(x,t)^{-1}B(p')(x)h_+(x,t)$ for

$(x, t) \in S^{n-1} \times [0, 1]$. Then

$$F(x, 0) = h_-(x, 0)^{-1} B(p')(x) h_+(x, 0)$$
$$= \lambda_-(x)^{-1} B(p')(x) \lambda_+(x)$$
$$= B(p)(x)$$
$$F(x, 1) = h_-(x, 1)^{-1} B(p')(x) h_+(x, 1)$$
$$= \lambda_-(x_0)^{-1} B(p')(x) \lambda_+(x_0).$$

Thus $B(p)$ is homotopic to the map given by $x \mapsto \lambda_-(x_0)^{-1} B(p')(x) \lambda_+(x_0)$. The next step is to construct a homotopy from this map to $B(p')$.

Since G is arcwise connected, there exist paths $\left\{ \begin{matrix} w_- : [0, 1] \to G \\ w_+ : [0, 1] \to G \end{matrix} \right\}$ which connect $\left\{ \begin{matrix} \lambda_-(x_0) \text{ and } e \\ \lambda_+(x_0) \text{ and } e \end{matrix} \right\}$, respectively, where e is the unit of G. Define $G(x, t) = w_-(t)^{-1} B(p')(x) w_+(t)$ for $(x, t) \in S^{n-1} \times [0, 1]$. Then we have $G(x, 0) = F(x, 1)$ and $G(x, 1) = B(p')(x)$ and hence $B(p) \simeq B(p')$.

Conversely, suppose $\bar{H}$ is a homotopy from $B(p)$ to $B(p')$. Define $L(x) = B(p')(x)^{-1} B(p)(x)$ for $x \in S^{n-1}$. Then $H(x, t) = \bar{H}(x, t)^{-1} B(p)(x)$ is a homotopy $H : S^{n-1} \times [0, 1] \to G$ from L to c_e, where c_e is the constant map to the unit e. Since

$$H(S^{n-1} \times \{1\}) = c_e(S^{n-1}) = \{e\},$$

H induces a map $\widetilde{H} : S^{n-1} \times [0, 1]/S^{n-1} \times \{1\} \to G$. As is shown in Figure 4.6, $S^{n-1} \times [0, 1]/S^{n-1} \times \{1\}$ is homeomorphic to S_+^n. (See Exercise 4.4.10.)

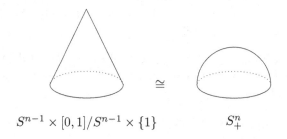

$$S^{n-1} \times [0, 1]/S^{n-1} \times \{1\} \qquad\qquad S_+^n$$

Fig. 4.6   cone and hemisphere

Thus the map $\widetilde{H}$ can be regarded as a map $\lambda_+ : S_+^n \to G$. Define $\lambda_- = c_e : U_{-,\varepsilon} \to G$. If $\overline{\Phi}^{+-}$ and $\overline{\Phi'}^{+-}$ are coordinate transformations of

E and E', we have

$$\begin{aligned}
\lambda_-(x)^{-1}\overline{\Phi'}^{+-}(x)\lambda_+(x) &= e \cdot \overline{\Phi'}^{+-}(x)\tilde{H}(x,1)\\
&= \overline{\Phi'}^{+-}(x)L(x)\\
&= B(p')(x)B(p')(x)^{-1}B(p)(x)\\
&= B(p)(x)
\end{aligned}$$

for $x \in S_+^n \cap S_-^n$. By Proposition 4.4.6, E and E' are isomorphic.   □

An essential argument used in the above proof can be generalized as follows.

Lemma 4.4.7. *Let X and Y be topological spaces. Choose $y_0 \in Y$. A continuous map $f : X \to Y$ is homotopic to a constant map c_{y_0} if and only if there exists an extension $\tilde{f} : X \times [0,1]/X \times \{1\} \to Y$ of f such that $\tilde{f}([x,1]) = y_0$. Here we regard $X = X \times \{0\} \subset X \times [0,1]/X \times \{1\}$.*

We study related topics in §4.6.

Definition 4.4.8. For a topological space X, the quotient space $X \times [0,1]/X \times \{1\}$ is called *cone* of X and is denoted by CX.

Exercise 4.4.9. Show that the cone CX is contractible for any X.

Exercise 4.4.10. Prove that CS^{n-1} is homeomorphic to D^n, hence to S_+^n.

4.5   Classifying Fiber Bundles over CW Complexes

We have studied fiber bundles over a disk D^n and a sphere S^n which is obtained by gluing two disks. In this section, we study fiber bundles over spaces obtained by gluing disks.

Definition 4.5.1. Let X be a topological space. A *cell decomposition* of X consists a family of continuous maps $\{\varphi_\lambda : D^{n_\lambda} \to X\}_{\lambda \in \Lambda}$ satisfying the following conditions:

   (1) For each λ, the restriction $\varphi_\lambda : \mathrm{Int}D^{n_\lambda} \to X$ is a homeomorphism onto its image.

   (2) Denote $e_\lambda = \varphi_\lambda(\mathrm{Int}D^{n_\lambda})$. Then

$$X = \bigcup_{\lambda \in \Lambda} e_\lambda$$

and $e_\lambda \cap e_\mu = \emptyset$ if $\lambda \neq \mu$.

(3) Define $\dim e_\lambda = n$ if $e_\lambda = \varphi_\lambda(\mathrm{Int}\, D^n)$. For a nonnegative integer r, define

$$X^{(r)} = \bigcup_{\substack{\mu \in \Lambda \\ \dim e_\mu \leq r}} e_\mu.$$

Then if $\dim e_\lambda = n$, we have

$$\varphi_\lambda(\partial D^n) = \varphi_\lambda(S^{n-1}) \subset X^{(n-1)}.$$

A Hausdorff space equipped with a cell decomposition is called a *cell complex*. We also use the collection $\{e_\lambda\}_{\lambda \in \Lambda}$ or the expression

$$X = \bigcup_{\lambda \in \Lambda} e_\lambda$$

to denote the cell decomposition. The subspace $X^{(r)}$ is called the *r-skeleton* of X. We also denote $X^{(-1)} = \emptyset$. Each e_λ is called a *cell* and the map φ_λ is called the *characteristic map* of e_λ. When $\dim e_\lambda = n$, it is called an *n-cell*.

Remark 4.5.2. Each characteristic map $\varphi_\lambda : D^{n_\lambda} \to X$ is a map onto $\overline{e_\lambda}$. Since D^{n_λ} is compact and we assume that X is Hausdorff, this map is a closed map by Lemma 3.6.8 hence is closed. This fact will be used in the proof the proof of Proposition 4.5.14.

Denote $\partial e_\lambda = \overline{e_\lambda} \setminus e_\lambda$ and call it the *boundary* of e_λ. Then the third condition for cell decomposition is equivalent to

$$\partial e_\lambda \subset \bigcup_{\substack{\mu \in \Lambda \\ \dim e_\mu < \dim e_\lambda}} e_\mu,$$

since $\partial e_\lambda = \varphi_\lambda(\partial D^{n_\lambda})$.

Remark 4.5.3. In the definition of cell complex, we assumed that the domain of each characteristic map is a closed disk, although each cell is homeomorphic to an open disks. The use of closed disks is sometimes to strong. For example, hyperplanes $H_1, \ldots, H_k$ in $\mathbb{R}^n$ decomposes $\mathbb{R}^n$ into a union of small pieces, each of which is homeomorphic to an open disk. This decomposition is very close to a cell decomposition. However, unbounded cells do not admit characteristic maps from closed disks.

A generalization of cell complexes which includes such a decomposition is recently introduced and studied by the author in [Tamaki (2018)].

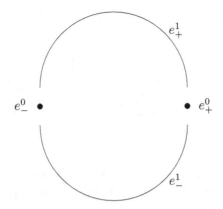

Fig. 4.7   a cell decomposition of S^1

Example 4.5.4. Decompose S^1 as is shown in Figure 4.7. This decomposition $S^1 = e_+^0 \cup e_-^0 \cup e_+^1 \cup e_-^1$ is a cell decomposition of S^1.

This cell decomposition can be extended to S^2

$$S^2 = e_+^0 \cup e_-^0 \cup e_+^1 \cup e_-^1 \cup e_+^2 \cup e_-^2$$

as is shown in Figure 4.8.

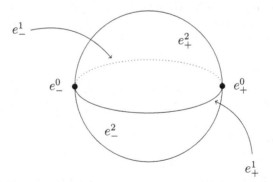

Fig. 4.8   a cell decomposition of S^2

More generally the n-dimensional sphere S^n has a cell decomposition

$$S^n = e_+^0 \cup e_-^0 \cup e_+^1 \cup e_-^1 \cup \cdots \cup e_+^n \cup e_-^n$$

having two cells in each dimension. $\square$

Exercise 4.5.5. Give explicit characteristic maps of S^n and prove that it is a cell decomposition.

Example 4.5.6. Add a vertical segment e^1 to the cell decomposition of S^1 in Example 4.5.4 as in Figure 4.9. This decomposition

$$e^0_- \cup e^0_+ \cup e^1_- \cup e^1_+ \cup e_1$$

is not a cell decomposition, since the boundary of e^1 is not included in the 0-skeleton $e^0_- \cup e^0_+$.

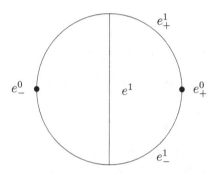

Fig. 4.9 S^1 with a vertical line

□

Notice that the classification of fiber bundles over S^n made an essential use of the cell decomposition of S^n in Example 4.5.4. It should also be noted that there exists a simpler cell decomposition of S^n.

Example 4.5.7. Choose a point e^0 in S^n and define $e^n = S^n \setminus \{e^0\}$. See Figure 4.10.

Since e^n is homeomorphic to an n-dimensional open disk, the decomposition $S^n = e^0 \cup e^n$ is a cell decomposition of S^n.

□

Exercise 4.5.8. Prove that, for any point x in S^n, the complement $S^n \setminus \{x\}$ is homeomorphic to $\mathbb{R}^n$, and hence to $\mathrm{Int}(D^n)$.

Let us discuss the geometric meaning of cell decompositions. The first two conditions mean that X can be decomposed into a union of small pieces that are homeomorphic to open disks. What is the meaning of the third condition? Suppose that X has a cell complex having only one n-cell e^n

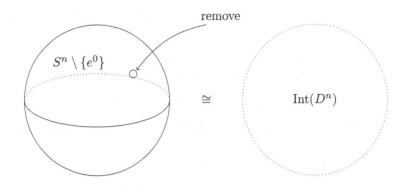

Fig. 4.10   the minimal cell decomposition of S^n

and all other cells are of dimension less than or equal to $n-1$. Namely we may write

$$X = X^{(n-1)} \cup e^n = X^{(n-1)} \cup \overline{e^n}.$$

We interpret X to be a space obtained from the $(n-1)$-skeleton by attaching e^n, $\overline{e}^n$, or D^n. Let us take a close look at the way e^n is attached. By the third condition of cell decomposition, the characteristic map φ of e^n gives rise to a map

$$\varphi|_{S^{n-1}} : \partial D^n = S^{n-1} \longrightarrow X^{(n-1)}.$$

(See Figure 4.11.) Since $\varphi(\text{Int}(D^n)) = e^n$ does not intersect with $X^{(n-1)}$, X is obtained by attaching the boundary of D^n to $X^{(n-1)}$ by the map $\varphi|_{S^{n-1}}$.

If X has more n-cells, other n-cells are also attached in a similar way. Thus the third condition in cell decomposition allows us to reconstruct a cell complex starting from 0-cells by attaching n-cells to the $(n-1)$-skeleton inductively. If this were to be true, we would be able to study properties of cell complexes by induction on dimensions. However, we have to be a little bit careful. The above argument only gives us a bijection as sets. In order to compare topologies, we need a precise definition of spaces obtained by attaching other spaces by continuous maps.

Definition 4.5.9. Let X and Y be topological spaces, A a subspace of Y, and $f : A \to X$ a continuous map. Define a space $X \cup_f Y$ as a quotient space by

$$X \cup_f Y = (X \amalg Y)/_{\sim_f},$$

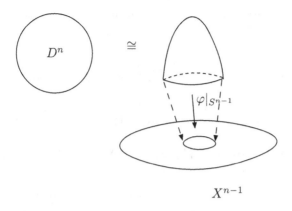

X^{n-1}

Fig. 4.11   attaching a cell

where the relation $\sim_f$ is the equivalence relation generated by the relation $a \sim_f f(a)$ for $a \in A$.

This space $X \cup_f Y$ is called the *space obtained by attaching Y to X by f*.

Example 4.5.10. Let X be a single point set $*$, $Y = D^n$, and $A = \partial D^n = S^{n-1}$. The unique map from S^{n-1} to $*$ is denoted by $f : S^{n-1} \to *$. Then we have

$$X \cup_f Y = * \cup_f D^n = D^n / \partial D^n.$$

The fact that this space is homeomorphic to S^n can be shown as follows. Let $\varphi_n : D^n \to S^n$ be the characteristic map of the n-cell in Example 4.5.7. Since $\varphi_n(\partial D^n) = e^0$, we obtain a continuous map $\tilde{\varphi}_n : D^n / \partial D^n \to S^n$. By the condition on characteristic map, this is a bijection. Since $D^n / \partial D^n$ is compact and S^n is Hausdorff, this is a homeomorphism. $\square$

Let X be a cell complex with cell decomposition $X = \bigcup_{\lambda \in \Lambda} e_\lambda$ and characteristic maps of n-cells e_λ given by $\varphi_\lambda : D^n \to X$. The question is if we obtain a homeomorphism

$$X^{(n)} \cong X^{(n-1)} \cup_{\amalg \varphi_\lambda |_{S_\lambda^{n-1}}} \left(\coprod_{\dim e_\lambda = n} D_\lambda^n \right)$$

for each n. Here D_λ^n is a copy of D^n. The inclusion map of $X^{(n-1)}$ into

$X^{(n)}$ and characteristic maps φ_λ of n-cells induce a continuous bijection

$$X^{(n-1)} \cup_{\amalg \varphi_\lambda|_{S^{n-1}}} \left(\coprod_{\dim e_\lambda = n} D_\lambda^n \right) \longrightarrow X^{(n)}.$$

Unfortunately, this map is not a homeomorphism in general.

Example 4.5.11. Any topological space X has a decomposition $X = \bigcup_{x \in X} \{x\}$, which can be regarded as a cell decomposition in to 0-cells. The space obtained by attaching $\coprod_{x \in X} D_x^0$ to $X^{(-1)} = \emptyset$ is X made discrete and is not homeomorphic to X in general. $\square$

We need to impose conditions on topology of cell complexes.

Definition 4.5.12. A cell complex $X = \bigcup_{\lambda \in \Lambda} e_\lambda$ is called a *CW complex* if the following two conditions are satisfied:

(C) For any cell e_λ the closure $\bar{e}_\lambda$ intersects with only a finite number of cells.

(W) A subset $A \subset X$ is closed in X if and only if $A \cap \bar{e}_\lambda$ is closed in $\bar{e}_\lambda$ for all $\lambda \in \Lambda$.

Here the conditions (C) and (W) stand for *closure finiteness* and *weak topology*, respectively.

Remark 4.5.13. It is known that many spaces naturally appearing in various kinds of geometry have structures of CW complexes. For example, smooth manifolds in differential geometry, algebraic varieties over $\mathbb{R}$ or $\mathbb{C}$ in algebraic geometry, and analytic spaces can be regarded as CW complexes. See [Lefschetz and Whitehead (1933); Cairns (1934); Hironaka (1975); Milnor (1963)], for example.

As we expect, we have the following description.

Proposition 4.5.14. *Let X be a CW complex with cell decomposition $X = \bigcup_{\lambda \in \Lambda} e_\lambda$. For each cell e_λ, the characteristic map is denoted by $\varphi_\lambda : D^n \to X$. For each $n \geq 0$, let*

$$\bar{p}_n : X^{(n-1)} \cup_{\amalg \varphi_\lambda|_{S^{n-1}}} \left(\coprod_{\dim e_\lambda = n} D_\lambda^n \right) \longrightarrow X^{(n)}$$

be the map induced by the inclusion of the $(n-1)$-skeleton $X^{(n-1)}$ and the characteristic maps of n-cells. Then this is a homeomorphism.

Proof. Since $\bar{p}_n$ is a surjective continuous map, it remains to show that it is a closed map. Let

$$p : X^{(n-1)} \amalg \left(\coprod_{\dim e_\lambda = n} D_\lambda^n \right) \longrightarrow X^{(n-1)} \cup_{\amalg \varphi_\lambda|_{S^{n-1}}} \left(\coprod_{\dim e_\lambda = n} D_\lambda^n \right)$$

be the projection. By the definition of quotient topology, we have

$$A \subset X^{(n-1)} \cup_{\amalg \varphi_\lambda|_{S^{n-1}}} \left(\coprod_{\dim e_\lambda = n} D_\lambda^n \right) : \text{closed} \Longleftrightarrow p^{-1}(A) : \text{closed}.$$

On the other hand,

$$p^{-1}(A) = (X^{(n-1)} \cap A) \amalg \left(\coprod_{\dim e_\lambda = n} D_\lambda^n \cap p^{-1}(A) \right).$$

If A is closed, both $X^{(n-1)} \cap A$ and $D_\lambda^n \cap p^{-1}(A)$ are closed for each λ. We want to show that $\bar{p}_n(A)$ is closed in $X^{(n)}$ by using these conditions. By the condition (W), it suffices to show that $\bar{p}_n(A) \cap \bar{e}_\mu$ is closed in $\bar{e}_\mu$ for each cell e_μ in $X^{(n)}$. Let $p_n = \bar{p}_n \circ p$, then we have

$$\bar{p}_n(A) \cap \bar{e}_\mu = \left(p_n(X^{(n-1)} \cap A) \cup \bigcup_\lambda p_n(D_\lambda \cap p^{-1}(A)) \right) \cap \bar{e}_\mu.$$

By the condition (C), $\bar{e}_\mu$ intersects only a finite number of n-cells $e_{\lambda_1}, \ldots, e_{\lambda_k}$ nontrivially. Then we have

$$\bar{p}_n(A) \cap \bar{e}_\mu$$
$$= \left(p_n(X^{(n-1)} \cap A) \cup p_n(D_{\lambda_1}^n \cap p^{-1}(A)) \cup \cdots \cup p_n(D_{\lambda_k}^n \cap p^{-1}(A)) \right) \cap \bar{e}_\mu$$
$$= \left(p_n(X^{(n-1)} \cap A) \cup \varphi_{\lambda_1}(D_{\lambda_1}^n \cap p^{-1}(A)) \cup \cdots \cup \varphi_{\lambda_k}(D_{\lambda_k}^n \cap p^{-1}(A)) \right) \cap \bar{e}_\mu.$$

Since p_n is a homeomorphism onto $X^{(n-1)}$ when restricted to $X^{(n-1)}$, the image $p_n(X^{(n-1)} \cap A)$ is a closed set. The fact that each $\varphi_{\lambda_i}(D_{\lambda_i}^n \cap p^{-1}(A))$ is closed is already discussed in Remark 4.5.2. Hence $\bar{p}_n(A) \cap \bar{e}_\mu$ is a closed set. □

Exercise 4.5.15. Show that, for a CW complex X, its n-skeleton $X^{(n)}$ is also a CW complex and is a closed subset of X for all $n \geq 0$.

Example 4.5.16. The n-dimensional sphere S^n becomes a CW complex by the cell decomposition in Examples 4.5.4. It is also a CW complex by the cell decomposition in Example 4.5.7. □

More generally, we have the following useful fact.

Proposition 4.5.17. *If the number of cells in a cell complex X is finite, then it is a CW complex.*

Proof. See [Lundell and Weingram (1969)], for example. □

We also use the following terminologies.

Definition 4.5.18. For a cell complex X, define

$$\dim X = \begin{cases} \max\left\{\dim e_\lambda \mid e_\lambda \subset X : \text{cell}\right\}, & \text{if max exists} \\ \infty, & \text{otherwise} \end{cases}.$$

This is called the *dimension* of X.

When the number of cells in X is finite, it is called a *finite complex*.

A closed subset A of X is called a *subcomplex* if it can be expressed as a union of cells in X.

In the rest of this chapter, we study the classification of fiber bundles over CW complexes. Thanks to Theorem 3.7.7, it suffices to consider principal bundles.

Definition 4.5.19. For a topological group G, the set of isomorphism classes of principal G-bundles over X is denoted by $P_G(X)$.

With this notation, what we are going to do is to find a description of this set $P_G(X)$ in terms of G and X. In order to state the aim of this chapter, we need the following notation.

Definition 4.5.20. The quotient of the set $\mathrm{Map}(X, Y)$ of continuous maps from X to Y by the relation $\simeq$ (Definition 4.3.2) is denoted by $[X, Y]$. This is called the *homotopy set* from X to Y.

Here is our classification theorem.

Theorem 4.5.21 (Classification Theorem). *For topological group G (satisfying certain conditions), there exists a topological space BG and a principal G-bundle $p : EG \to BG$ such that, for any CW complex X, the map $B : [X, BG] \to P_G(X)$ given by $[f] \mapsto f^*(EG)$ is a bijection.*

In other words, for any principal G-bundle $E \to X$ over X, there exists a continuous map $f : X \to BG$ such that $f^*(EG) \cong E$.

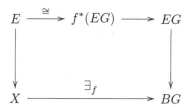

Furthermore if f and g are such maps, they are homotopic $f \simeq g$.

Remark 4.5.22. The conditions on topological groups will be made clear during the proof, which occupies the rest of this chapter besides §4.11. We are going to see two ways to construct the principal G-bundle $p : EG \to BG$ and the conditions on topological groups depend on the construction. This is why we cannot state the conditions explicitly now.

Definition 4.5.23. The bundle $EG \to BG$ in Theorem 4.5.21 is called a *universal G-bundle* and the base space BG is called the *classifying space* of G.

4.6   CW Complexes and Homotopy

In order to prove Theorem 4.5.21, we need to understand homotopy theoretic properties of CW complexes, which is the subject of this section.

As we have seen in Proposition 4.5.14, a CW complex is made of disks of various dimensions by attaching them to lower dimensional disks. Note that disks are contractible, which means that they are trivial up to homotopy. When we attach disks, however, we use restrictions of characteristic maps to the boundaries. Thus we need to understand continuous maps defined on the boundaries of disks, namely spheres.

Definition 4.6.1. A topological space X is said to be *n-connected* if, for each $k \leq n$, the homotopy set from the k-dimensional sphere to X consists of a single element $[S^k, X] = \{*\}$. In other words, if $k \leq n$, they any two maps $f, g : S^k \to X$ are homotopic $f \simeq g$.

Lemma 4.6.2. *A topological space X is 0-connected if and only if it is arcwise connected.*

Proof. Suppose X is 0-connected. For any points $x_0, x_1 \in X$, define maps $f, g : S^0 \to X$ by $f(1) = g(1) = x_0$ and $f(-1) = x_0$, $g(-1) = x_1$. Here we regard

$$S^0 = \left\{ x \in \mathbb{R} \mid x^2 = 1 \right\} = \{1, -1\}.$$

Since f and g are continuous and X is 0-connected, f and g are homotopic. Let $H : S^0 \times [0,1] \to X$ be a homotopy from f to g. Define a map $w : [0,1] \to X$ by $w(t) = H(-1, t)$. Then

$$w(0) = H(-1, 0) = f(-1) = x_0$$
$$w(1) = H(-1, 1) = g(-1) = x_1.$$

And w is a path from x_0 to x_1.

Conversely, suppose X is arcwise connected. Given a map $f : S^0 \to X$, a path from $f(1)$ to $f(-1)$ can be used to define a homotopy from $f : S^0 \to X$ to the constant map to $f(1)$. Given another map $g : S^0 \to X$, it is also homotopic to the constant map to $g(1)$. A path from $f(1)$ to $g(1)$ defines a homotopy between these two constant maps. Thus f and g are homotopic.   □

Recall that, when we introduced the fundamental group in §2.3, the notion of based spaces was also introduced (Definition 2.3.3). We should require maps between based spaces to preserve base points.

Definition 4.6.3. For based spaces (X, x_0) and (Y, y_0), a *base point preserving map* or a *based map* from (X, x_0) to (Y, y_0) is a continuous map $f : X \to Y$ with $f(x_0) = y_0$.

For two based maps

$$f : (X, x_0) \longrightarrow (Y, y_0)$$
$$g : (X, x_0) \longrightarrow (Y, y_0),$$

a *based point preserving homotopy* or a *based homotopy* from f to g is a homotopy $H : X \times [0,1] \to Y$ such that, for any $t \in [0,1]$, $H(x_0, t) = y_0$. If such a homotopy H exists, we say f and g are *homotopic* and denote $f \simeq g$.

The set of based maps from X to Y is denoted by $\mathrm{Map}((X, x_0), (Y, y_0))$. When the base points are obvious from the context, it is denoted by $\mathrm{Map}_*(X, Y)$. The set of homotopy classes of based maps from (X, x_0) to (Y, y_0) is denoted by

$$[(X, x_0), (Y, y_0)] = [X, Y]_* = \mathrm{Map}_*(X, Y)/ \simeq .$$

This is called the *based homotopy set* from X to Y. It contains a special element, the homotopy class of the constant map to the base point, which is denoted by $*$.

Example 4.6.4. The homotopy set $[(S^1, e_0), (X, x_0)]$ agrees with the fundamental group $\pi_1(X, x_0)$ introduced in Definition 2.3.19. We will see later that they are isomorphic as groups, but let us verify that there is a bijection between them. Define a map

$$\pi^* : \mathrm{Map}((S^1, e_0), (X, x_0)) \longrightarrow \Omega(X, x_0)$$

by $\pi^*(f) = f \circ \pi$, where $\pi : [0, 1] \to S^1$ is the restriction of the map exp in Exercise 2.1.6 and is a map which glues the end points of $[0, 1]$ together to make it into a circle. It is easy to verify that based homotopies induce homotopies of loops and we obtain a map

$$\pi^* : [(S^1, e_0), (X, x_0)] \longrightarrow \pi_1(X, x_0).$$

A loop $\ell : [0, 1] \to X$ in X based on x_0 induces a map $\bar{\ell} : [0, 1]/\{0, 1\} \to X$ since $\ell(0) = \ell(1)$. By identifying $[0, 1]/\{0, 1\}$ with S^1, we may regard it as a based map $\bar{\ell} : S^1 \to X$. This construction defines an inverse to π^* and thus π^* is a bijection. □

This correspondence suggests the following generalization of the fundamental group.

Definition 4.6.5. Regard S^n as a based space with $e_0 = (1, 0, \ldots, 0) \in S^n$ the base point. For a based space (X, x_0), define $\pi_n(X, x_0) = [(S^n, e_0), (X, x_0)]$. As we will see later, it has a structure of group when $n \geq 1$ and it is called the n-th *homotopy group* of X. When the base point is obvious from the context, it is denoted by $\pi_n(X)$.

The n-connectivity can be stated by using homotopy groups as follows.

Proposition 4.6.6. *Let (X, x_0) be an arcwise connected based space. Then the following three conditions are equivalent:*

(1) X is n-connected, i.e. for any $k \leq n$, $[S^k, X] = \{*\}$.
(2) For any $k \leq n$, $\pi_k(X, x_0) = \{*\}$.
(3) If $k \leq n$, for any continuous map $f : S^k \to X$, there exists a continuous map $F : D^{k+1} \to X$ such that $F|_{S^k} = f$. In other words, any map $f : S^k \to X$ can be extended to a map on D^{k+1}.

The idea of the proof is essentially the same as Lemma 4.4.7 but we need to take care of base points.

Proof. We prove that (1) and (3) are equivalent. The case of (2) and (3) is left to the reader.

(1) $\implies$ (3): Let x_0 be the base point of X. For $k \le n$, let $c_{x_0} : S^k \to X$ be the constant map to x_0. Since X is n-connected, any map $f : S^k \to X$ is homotopic to c_{x_0}. Let $H : S^k \times [0,1] \to X$ be a homotopy from c_{x_0} to f. Since $H(S^k \times \{0\}) = c_{x_0}(S^k) = \{x_0\}$, H induces a map $\tilde{H} : S^k \times [0,1]/S^k \times \{0\} \to X$. As we have seen in Exercise 4.4.10, we may identify

$$S^k \times [0,1]/S^k \times \{0\} \cong D^{k+1}$$

and the map $\tilde{H}$ can be regarded as a map $F : D^{k+1} \to X$ with $F|_{S^k} = f$.

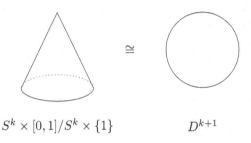

$$S^k \times [0,1]/S^k \times \{1\} \qquad\qquad D^{k+1}$$

Fig. 4.12   cone and disk are homeomorphic

(3) $\implies$ (1): Given continuous maps $f, g : S^k \to X$, we construct a homotopy from f to g.

By assumption, f can be extended to a map $F : D^{k+1} \to X$. Let $p : S^k \times [0,1] \to S^k \times [0,1]/S^k \times \{0\}$ be the projection and define $H : S^k \times [0,1] \to X$ by the following composition

$$S^k \times [0,1] \xrightarrow{p} S^k \times [0,1]/S^k \times \{0\} \cong D^{k+1} \xrightarrow{F} X.$$

Then we have

$$H(x,0) = F|_{S^k}(x) = f(x)$$
$$H(x,1) = F(x)$$

for any $x \in S^k$. In particular, $H(x,0)$ is a constant map. Let $x_0 = F(0)$. Then H is a homotopy from f to the constant map c_{x_0}. Similarly there exists a homotopy G from g to a constant map c_{x_1}. If we can construct a homotopy K from c_{x_0} to c_{x_1}, then web obtain a homotopy from f to g by concatenating homotopies H, G, and K.

In order to construct a homotopy from c_{x_0} to c_{x_1}, we define a map $h : S^0 \to X$ by $h(1) = x_0$ and $h(-1) = x_1$. Its extension $\tilde{h} : D^1 = [0,1] \to X$ can be used to define $K : S^k \times [0,1] \to X$ by $K(x,t) = \tilde{h}(t)$. This is a homotopy from c_{x_0} to c_{x_1}. $\qquad\square$

Exercise 4.6.7. Found a homeomorphism

$$S^n \times [0,1]/(S^n \times \{0\} \cup \{e_0\} \times [0,1]) \cong D^{n+1}$$

where e_0 is the base point of S^n and show that (2) and (3) are equivalent.

Definition 4.6.8. For a based space (X, x_0), the space $X \times [0,1]/(X \times \{1\} \cup \{x_0\} \times [0,1])$ is called the *reduced cone* of X and is denoted by CX.

Exercise 4.6.9. Show that for any based space X, its reduced cone is contractible via based homotopy.

Lemma 2.3.22 can be proved by using the identification in Proposition 4.6.6.

Proof of Lemma 2.3.22. Suppose the fundamental group of X is trivial. In order to show that X is simply-connected, let $\ell, \ell' : [0,1] \to X$ which share the end points. Denote $\ell(0) = \ell'(0) = x_0$ and $\ell(1) = \ell'(1) = x_1$. Then we have $\ell * \nu(\ell') \in \Omega(X, x_0)$. By the correspondence in Example 4.6.4, we regard it as an element of $\mathrm{Map}_*(S^1, X)$. Since the fundamental group is trivial, by Proposition 4.6.6, we have a map $L : D^2 \to X$ such that $L|_{S^1} = \ell * \nu(\ell')$.

Let $\psi : [0,1] \times [0,1] \to D^2$ be the quotient map which switches two factors and then collapses $\{0\} \times [0,1]$ and $\{1\} \times \{0,1\}$ to $(-1,0)$ and $(1,0)$, respectively. Then $L \circ \psi$ is a homotopy from ℓ to ℓ' which preserves two end points. Thus X is simply connected.

The converse is left to the reader. □

Exercise 4.6.10. Show that, if an arcwise connected space X is simply connected, then $\pi_1(X, x_0)$ consists of a single point for each $x_0 \in X$.

Recall that the fundamental group is made into a group via the loop product. This group structure can be extended to $\pi_n(X, x_0)$ for $n \geq 2$. In order to define the group structure, we first need the following construction.

Definition 4.6.11. Let X and Y be based spaces. The *wedge sum*) of X and Y is defined by

$$X \vee Y = X \times \{y_0\} \cup \{x_0\} \times Y \subset X \times Y,$$

where x_0 and y_0 are base points of X and Y, respectively.

Although the wedge sum $X \vee Y$ is defined as a subspace of the product $X \times Y$, it can be regarded as a space obtained by gluing X and Y by identifying x_0 and y_0 as is shown in Figure 4.13.

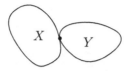

Fig. 4.13   wedge sum

Lemma 4.6.12. *For an integer $n \geq 1$, let E be the equator of S^n, i.e.*

$$E = \{(x_0, \ldots, x_n) \in S^n \mid x_0 = 0\}.$$

Then we have a homeomorphism $S^n/E \cong S^n \vee S^n$.

The following homeomorphism is used in the proof.

Exercise 4.6.13. Denote the one-point compactification of $\mathbb{R}^n$ by $\mathbb{R}^n \cup \{\infty\}$. Define a map $\tau : [0,1]^n/\partial[0,1]^n \to \mathbb{R}^n \cup \{\infty\}$ by

$$\tau([t_1, \ldots, t_n]) = \begin{cases} \left(\tan \frac{2\pi t_1 - \pi}{2}, \ldots, \tan \frac{2\pi t_n - \pi}{2}\right), & (t_1, \ldots, t_n) \notin \partial[0,1]^n \\ \infty, (t_1, \ldots, t_n) \in \partial[0,1]^n. \end{cases}$$

Show that this is a homeomorphism.

Proof of Lemma 4.6.12. The homeomorphism $\sigma : \mathbb{R}^n \xrightarrow{\cong} S^n \setminus \{x\}$ obtained in Exercise 4.5.8 can be extended to a homeomorphism $\tilde{\sigma} : \mathbb{R}^n \cup \{\infty\} \to S^n$, where $\mathbb{R}^n \cup \{\infty\}$ is the one-point compactification of $\mathbb{R}^n$. Let $x = e_n = (0 \ldots, 0, 1)$. Define a homeomorphism $\varphi : [0,1]^n/\partial[0,1]^n \xrightarrow{\cong} S^n$ as the composition of this map and the homeomorphism τ in Exercise 4.6.13. Then we have

$$\varphi^{-1}(E) = \left\{[(t_1, \ldots, t_n)] \in [0,1]^n/\partial[0,1]^n \mid t_1 = \tfrac{1}{2}\right\}.$$

Consider the commutative diagram

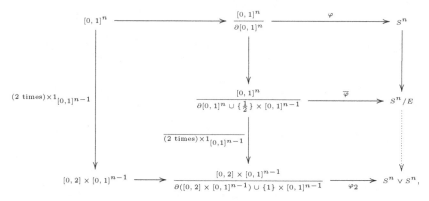

where $\overline{\varphi}$ is the map induced from φ, φ_2 is the map obtained by concatenating two copies of φ by using the decomposition $[0,2] \cong [0,1] \cup [1,2]$ and a homeomorphism $[1,2] \cong [0,1]$, and $\overline{(2 \text{ times})} \times 1_{[0,1]^{n-1}}$ is the homeomorphism induced by the map given by multiplying 2 to the first coordinate.

Then we obtain the dotted homeomorphism $S^n/E \to S^n \vee S^n$, since both $\overline{\varphi}$ are φ_2 homeomorphisms. $\square$

Definition 4.6.14. For an integer $n \geq 1$, define pinch : $S^n \to S^n \vee S^n$ to be the composition

$$S^n \longrightarrow S^n/E \xrightarrow{\cong} S^n \vee S^n$$

of the projection and the homeomorphism in Lemma 4.6.12. This is called the *pinching map*

By using the homeomorphism $\varphi : [0,1]^n/\partial[0,1]^n \xrightarrow{\cong} S^n$ used in the proof of Lemma 4.6.12, it can be described as

$$\text{pinch}(\varphi([t_1, t_2, \ldots, t_n])) = \begin{cases} \varphi([2t_1, t_2, \ldots, t_n], e_0), & 0 \leq t_1 \leq \frac{1}{2} \\ (e_0, \varphi([2t_1 - 1, t_2, \ldots, t_n])), & \frac{1}{2} \leq t_1 \leq 1. \end{cases}$$
$$(4.8)$$

We also need a folding map.

Definition 4.6.15. For a based space X, define a map fold : $X \vee X \to X$ by the identity map on each X, namely

$$\text{fold}(x, x') = \begin{cases} x, & x' = x_0 \\ x', & x = x_0. \end{cases}$$

This is called the *folding map*.

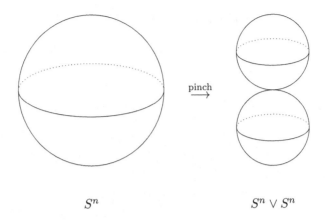

$$S^n \qquad\qquad S^n \vee S^n$$

Fig. 4.14   collapsing the equator

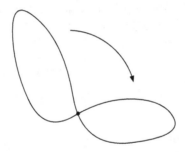

Fig. 4.15   folding map

Now we are ready to define a group operation on homotopy groups.

Definition 4.6.16. For based maps $f, g : S^n \to X$, define a map $f + g : S^n \to X$ by the composition

$$S^n \xrightarrow{\text{pinch}} S^n \vee S^n \xrightarrow{f \vee g} X \vee X \xrightarrow{\text{fold}} X,$$

where $f \vee g$ is the map defined by

$$(f \vee g)(x, y) = \begin{cases} (f(x), x_0), & y = e_0 \\ (x_0, g(y)), & x = e_0. \end{cases}$$

Here x_0 and e_0 are the base points of X and S^n, respectively.

Lemma 4.6.17. *If $n \geq 1$, the operation $+$ defines a map*

$$+ : \pi_n(X, x_0) \times \pi_n(X, x_0) \longrightarrow \pi_n(X, x_0).$$

Proof. We need to show that, for based maps $f_0, f_1, g_0, g_1 : S^k \to X$, if $f_0 \simeq f_1$ and $g_0 \simeq g_1$, then $f_0 + g_0 \simeq f_1 + g_1$.

Let F and G be based homotopies of $f_0 \simeq f_1$ and $g_0 \simeq g_1$, respectively. They induce a based homotopy $H : (S^n \vee S^n) \times [0, 1] \to X$ for $f_0 \vee g_0 \simeq f_1 \vee g_1$ by

$$H((x, y), t) = \begin{cases} F(x, t), & y = e_0 \\ G(y, t), & x = e_0. \end{cases}$$

And we obtain $f_0 + g_0 \simeq f_1 + g_1$. □

Recall that, in the case of fundamental groups, the loop product (Definition 2.3.10) was not associative. It does not satisfy the unit condition, either. We had to use homotopy to show that it induces a group operation on $\pi_1(X, x_0)$. We have the same situation for homotopy groups. In order to construct homotopies, it is better to use the unit cube $[0, 1]^n$ instead of S^n as the domains of elements of $\pi_n(X, x_0)$. In fact, in most textbooks, the group operation of homotopy groups is defined by using the unit cubes. In order to rewrite the definition of homotopy groups by using cubes, we need the notion of pairs of spaces.

Definition 4.6.18. A pair (X, A) of a topological space X and its subspace A is called a *pair of topological spaces*. Let (Y, B) be another pair. A continuous map $f : X \to Y$ is said to be a *map of pairs* if $f(A) \subset B$, in which case we denote $f : (X, A) \to (Y, B)$. Given another map of pairs $g : (X, A) \to (Y, B)$, a *homotopy of pairs* from f to g is a homotopy $H : X \times [0, 1] \to Y$ from f to g which satisfies $H(a, t) \in B$ for any $a \in A$ and $t \in [0, 1]$.

The set of maps of pairs from (X, A) to (Y, B) is denoted by $\mathrm{Map}((X, A), (Y, B))$. Its set of homotopy classes is denoted by $[(X, A), (Y, B)]$ is called the *homotopy set of pairs*.

Remark 4.6.19. When both A and B are single points, $[(X, A), (Y, B)]$ is the based homotopy set.

The correspondence in Example 4.6.4 can be generalized as follows.

Lemma 4.6.20. *For any based space (X, x_0), we have a natural bijection*

$$[([0, 1]^n, \partial[0, 1]^n), (X, x_0)] \cong \pi_n(X, x_0).$$

Proof. We first define a map

$$\pi : \mathrm{Map}\,(([0,1]^n, \partial[0,1]^n), (X, x_0)) \longrightarrow \mathrm{Map}_*(S^n, X).$$

Any map $f : ([0,1]^n, \partial[0,1]^n) \to (X, x_0)$ induces a map

$$\bar{f} : [0,1]^n / \partial[0,1]^n \longrightarrow X,$$

since $f(\partial[0,1]^n) = \{x_0\}$. This is continuous by the definition of quotient topology. Let $\pi(f) : S^n \to X$ be the composition of this map and the inverse of the map φ used in the proof of Lemma 4.6.12. Then this is a based map. This construction can be extended to an analogous correspondence from homotopies of pairs to homotopies of based maps. Thus we obtain a map

$$\pi : [([0,1]^n, \partial[0,1]^n), (X, x_0)] \longrightarrow \pi_n(X, x_0).$$

This map π is surjective, since for any based map $f : S^n \to X$, the composition $[0,1]^n \to [0,1]^n / \partial[0,1]^n \xrightarrow{\varphi} S^n \xrightarrow{f} X$ defines a map of pairs $([0,1]^n, \partial[0,1]^n) \to (X, x_0)$.

In order to show that this map is injective, take $[f], [g] \in [([0,1]^n, \partial[0,1]^n), (X, x_0)]$ and suppose $\pi([f]) = \pi([g])$. Then both $\bar{f}$ and $\bar{g}$ are homotopic as based maps. Let H be the based homotopy. Then the composition

$$[0,1]^n \longrightarrow [0,1]^n / \partial[0,1]^n \xrightarrow{H} X$$

is a homotopy of pairs from f to g, and we have $[f] = [g]$. $\qquad\square$

Remark 4.6.21. The term "natural" used in Lemma 4.6.20 is a mathematical terminology. Here it means that for any based map $h : (X, x_0) \to (Y, y_0)$ the following diagram is commutative:

$$
\begin{array}{ccc}
[([0,1]^n, \partial[0,1]^n), (X, x_0)] & \xrightarrow{\ \pi\ } & \pi_n(X, x_0) \\
{\scriptstyle h_*}\Big\downarrow & & \Big\downarrow{\scriptstyle h_*} \\
[([0,1]^n, \partial[0,1]^n), (Y, y_0)] & \xrightarrow[\ \pi\ ]{} & \pi_n(Y, y_0).
\end{array}
$$

Here the h_*'s are the maps "induced" by h, whose precise definition will be given in Definition 4.8.1.

Lemma 4.6.22. *Under the identification of the bijection in Lemma 4.6.20, the group operation $+$ in $\pi_n(X, x_0)$ can be described as*

$$(f + g)(t_1, \ldots, t_n) = \begin{cases} f(2t_1, \ldots, t_{n-1}, t_n), & 0 \le t_1 \le \tfrac{1}{2} \\ g(2t_1 - 1, \ldots, t_{n-1}, t_n), & \tfrac{1}{2} \le t_1 \le 1 \end{cases}$$

for $f, g : ([0,1]^n, \partial[0,1]^n) \to (X, x_0)$.

Proof. This is an immediate consequence of the formula (4.8) and the definition of the bijection in Lemma 4.6.20. □

Lemma 4.6.23. *When $n \geq 1$, the operation $+$ makes $\pi_n(X, x_0)$ into a group. Its unit is the homotopy class of the constant map $* : S^n \to X$ to the base point.*

The proof is essentially the same as the case of the fundamental group. Details are left to the reader.

Exercise 4.6.24. Prove this Lemma by extending the homotopies used in the proof of the fact that the fundamental group is a group.

Lemma 4.6.25. *When $n \geq 2$, $\pi_n(X, x_0)$ is an Abelian group.*

Proof. Suppose $\alpha, \beta \in \pi_n(X, x_0)$ are represented by maps

$$f, g : ([0,1]^n, \partial[0,1]^n) \to (X, x_0),$$

respectively. We want to construct a homotopy between $f + g$ and $g + f$. We explain the case $n = 2$ by using figures. The case of $n > 2$ can be proved by applying the homotopy of the case $n = 2$ to the first two coordinates.

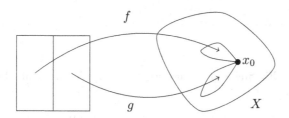

Fig. 4.16　$f + g$

The map $f + g$ is given by Figure 4.16. On the other hand, the map $g + f$ is given by Figure 4.17.

A homotopy connecting these two maps can be constructed, for example, by the following steps as is shown in Figure 4.18:

(1) Divide the domain into rectangles vertically.
(2) Shrink the left rectangle to its lower half and the right rectangle to the upper half so that the domains of f and g are the left bottom and the right top little squares, respectively.
(3) Shift the domain of f to the right and the domain of g to the left.

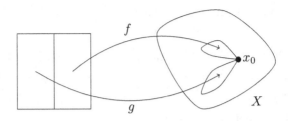

Fig. 4.17 $g + f$

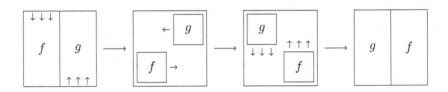

Fig. 4.18   a homotopy between $f + g$ and $g + f$

(4) Enlarge the domain of g and f so that they fill in the left and the right rectangles, respectively.

□

Remark 4.6.26. Since the homotopy group $\pi_n(X, x_0)$ is an Abelian group for $n \geq 2$, the unit is denoted by 0.

One of the most useful relationship between CW complexes and homotopy is the following theorem.

Theorem 4.6.27 (Cellular Approximation Theorem). *Let X and Y be CW complexes. Then for any continuous map $f : X \to Y$, there exists a continuous map $g : X \to Y$ satisfying the following conditions:*

(1) g maps n-cells in X into the n-skeleton of Y. Thus $g\left(X^{(n)}\right) \subset Y^{(n)}$.

(2) g and f are homotopic.

Proof. A proof can be found in most textbooks on algebraic topology. For example, a relatively short proof with geometric intuitions can be found in the book [Fomenko *et al.* (1986)] by Fomenko, Fuks, and Gutenmacher. □

Definition 4.6.28. A continuous map $f : X \to Y$ between CW complexes is called a *cellular map* if, for any $n \geq 0$, $f\left(X^{(n)}\right) \subset Y^{(n)}$.

Remark 4.6.29. With this term, the cellular approximation theorem says that any continuous map between CW complexes can be continuously deformed into a cellular map. However, the author do not like the term "cellular map", since it might give a false impression that it maps cells to cells, which is not the case in general.

Theorem 4.6.27 can be used to determine lower degree part of the homotopy groups of spheres.

Corollary 4.6.30. *The n-dimensional sphere S^n is $(n-1)$-connected, namely, for $k \leq n-1$, $\pi_k(S^n) = 0$.*

Proof. Let $k \leq n - 1$ and $f : S^k \to S^n$ be a continuous map. By Example 4.5.7, both S^k and S^n are CW complexes with cell decompositions

$$S^k = e^0 \cup (S^k \setminus e^0) = e^0 \cup e^k$$
$$S^n = e^0 \cup (S^n \setminus e^0) = e^0 \cup e^n.$$

By the cellular approximation theorem, there exists a cellular map $g : S^k \to S^n$ which is homotopic to f. On the other hand, since $k \leq n - 1$,

$$g\left(S^k\right) = g\left((S^k)^{(k)}\right) \subset (S^n)^{(k)} = e^0$$

and thus g is a constant map. $\qquad\square$

For CW complexes, the covering homotopy theorem (Theorem 4.3.16) can be extended as follows.

Theorem 4.6.31 (Covering Homotopy Extension Theorem). *Let X be a CW complex and A a subcomplex. Suppose $p : E \to Y$ is a fiber bundle and $f : X \to Y$ and $\tilde{f} : X \to E$ are continuous maps making the diagram*

commutative. Furthermore a homotopy $H : X \times [0,1] \to Y$ *with* $H(x,0) =$ $f(x)$ *and a homotopy* $G : A \times [0,1] \to E$ *which makes the diagram*

$$
\begin{array}{ccc}
A \times [0,1] & \xrightarrow{\ G\ } & E \\
\Big\uparrow & & \Big\downarrow{\scriptstyle p} \\
X \times [0,1] & \xrightarrow{\ H\ } & Y
\end{array}
$$

commutative and $G|_{A \times \{0\}} = \tilde{f}|_A$ *are given.*

Then there exists a homotopy $\tilde{H} : X \times [0,1] \to E$ *which makes the triangles in the following diagram commutative*

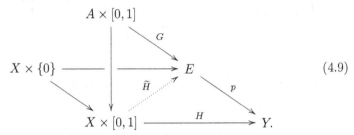

(4.9)

In other words, when we use the covering homotopy theorem, if a lift of homotopy is given on a subcomplex, the we may extended it to a lift $\tilde{H} : X \times [0,1] \to E$ *on* X.

Remark 4.6.32. The above diagram (4.9) can be also expressed as

$$
\begin{array}{ccc}
X \times \{0\} \cup A \times [0,1] & \xrightarrow{\ \tilde{f} \cup G\ } & E \\
\Big\uparrow & \nearrow{\scriptstyle \tilde{H}} & \Big\downarrow{\scriptstyle p} \\
X \times [0,1] & \xrightarrow{\ H\ } & Y,
\end{array}
$$

which looks closer to the covering homotopy theorem.

Proof. This can be proved by essentially the same idea as the proof of Theorem 4.3.7. Starting from the given homotopy $G : A \times [0,1] \to E$, we extend it to $A \cup X^{(n)}$ by induction on n by using the fact that $X^{(n)}$ is obtained from $X^{(n-1)}$ by attaching n-cells. In the proof of Theorem 4.3.7, we cut X into small closed sets by using the fact that X is paracompact Hausdorff. In this case we use cells instead. Since each cell is contractible, the restriction of E to each cell (more precisely, the pull back of E along the characteristic map of each cell) is trivial. The details are omitted. $\square$

By letting $A = \{*\}$, we obtain the following special case of Theorem 4.6.31, which is also useful.

Corollary 4.6.33 (Based Covering Homotopy Theorem). *Let X be a based CW complex, Y a based space, $p : E \to Y$ a fiber bundle, and $f : X \to Y$ and $\tilde{f} : X \to E$ are based maps making the diagram*

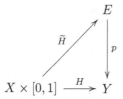

commutative. For a based homotopy $H : X \times [0,1] \to Y$ with $H(x,0) = f(x)$, there exists a based homotopy $\tilde{H} : X \times [0,1] \to E$ such that $\tilde{H}(x,0) = \tilde{f}(x)$ and the diagram

$$
\begin{array}{ccc}
 & & E \\
 & \nearrow^{\tilde{H}} & \downarrow p \\
X \times [0,1] & \xrightarrow{H} & Y
\end{array}
$$

is commutative.

4.7 The First Half of the Proof of the Classification Theorem

Now we are ready to start proving the classification theorem of principal bundles (Theorem 4.5.21). The proof splits in to two parts:

- a construction of a principal G-bundle $EG \to BG$, and
- a proof of the bijectivity of the correspondence $B : [X, BG] \to P_G(X)$.

We introduce two constructions of principal G-bundles $EG \to BG$ in §4.9 and §4.10. In this section, we study the conditions under which the map $[X, B] \to P_G(X)$ obtained by taking the pullback is a surjection and then injection, for a general principal G-bundle $p : E \to B$.

Let us begin with the surjectivity.

Proposition 4.7.1. *Let G be a topological group and $E \to B$ a principal G-bundle. If E is $(n-1)$-connected and X is a CW complex with $\dim X < n$, then the map $[X, B] \to P_G(X)$ given by $f \mapsto f^*(E)$ is surjective.*

Proof. Let $E' \to X$ be a principal G-bundle over X. We need to find a continuous map $f : X \to B$ with $f^*(E) \cong E'$. We construct f on skeletons of X inductively. We are going to construct continuous maps $f_r : X^{(r)} \to B$ and isomorphisms of fiber bundles

$$\tilde{f}_r : i_r^*(E') \xrightarrow{\cong} f_r^*(E) \tag{4.10}$$

by induction on r, where $i_r : X^{(r)} \hookrightarrow X$ is the inclusion map. In order to apply the inductive hypothesis, we also require the maps f_r make the diagram

$$
\begin{array}{ccc}
X^{(r-1)} & \xrightarrow{\ f_{r-1}\ } & B \\
\Big\downarrow{\scriptstyle i_r^{r-1}} & & \Big\| \\
X^{(r)} & \xrightarrow{\ f_r\ } & B
\end{array}
$$

commutative, where $i_r^{r-1} : X^{(r-1)} \hookrightarrow X^{(r)}$ is also the inclusion map.

By Proposition 4.2.7, it suffices to construct a bundle map

$$(f_r, \tilde{f}_r) : (X^{(r)}, i_r^*(E')) \longrightarrow (B, E)$$

to obtain an isomorphism (4.10).

When $r = 0$, the 0-skeleton $X^{(0)}$ is a collection of 0-cells and has the discrete topology. (Exercise 4.7.2.) Since any fiber bundle over a 0-cell is trivial, any map f_0 satisfies the condition

$$f_0^*(E) \cong X^{(0)} \times G \cong i_0^*(E').$$

Suppose we have constructed $\tilde{f}_{r-1}$. Since

$$X^{(r)} = X^{(r-1)} \cup (r\text{-dimensional cells})$$

by Proposition 4.5.14, we obtain $\tilde{f}_r$ if we construct appropriate maps on r-cells and glue them with $\tilde{f}_{r-1}$. Let e^r be an r-cell of X and $\varphi : D^r \to X$ the characteristic map of e^r. By Theorem 4.4.3, $\varphi^*(E')$ is trivial $\varphi^*(E') \cong D^r \times G$, since D^r is trivial. Let $j : S^{r-1} \hookrightarrow D^r$ be the inclusion map. Then we have

$$j^*\varphi^*(E') = (\varphi \circ j)^*(E') \cong S^{r-1} \times G.$$

In the diagram

$$
\begin{array}{ccccccccc}
S^{r-1} \times G & \xrightarrow{\cong} & j^*\varphi^*(E') & \longrightarrow & i_{r-1}^*(E') & \xrightarrow{\cong} & f_{r-1}^*(E) & \longrightarrow & E \\
\downarrow & & \downarrow & & \downarrow & & \downarrow & & \downarrow \\
S^{r-1} & =\!=\!= & S^{r-1} & \xrightarrow{\varphi \circ j} & X^{(r-1)} & =\!=\!= & X^{(r-1)} & \xrightarrow{f_{r-1}} & B
\end{array}
$$

all vertical maps are projections of fiber bundles and horizontal maps are bundle maps. Let $\tilde{f}_{r-1} : S^{r-1} \times G \to E$ be the composition of the maps in the top row. Then this is also a bundle map.

Define a map $h : S^{r-1} \to E$ by $h(x) = \tilde{f}_{r-1}(x, e)$, where $e \in G$ is the unit. By assumption E is $(n-1)$-connected. Since $r < n$, by Proposition 4.6.6, the map h can be extended to $\tilde{h} : D^r \to E$. Define a map $\tilde{F} : D^r \times G \to E$ by $\tilde{F}(x, g) = \mu(\tilde{h}(x), g)$, where $\mu : E \times G \to E$ is the action of G on E in Definition 3.7.8. Define a map $F : D^r \to B$ by $F(x) = p \circ \tilde{F}(x, e)$, then the diagram

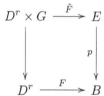

is commutative. In other words, $(F, \tilde{F})$ is a bundle map. Now define a map $f_r : X^{(r)} \to B$ as follows. For each $x \in \overline{e^r}$ in an r-cell, choose $y \in D^r$ with $x = \varphi(y)$ and define $f_r(x) = F(y)$. On $X^{(r-1)}$, it is defined by f_{r-1}. Since $\tilde{h}$ is an extension of h to D^r, F agrees with f_{r-1} on S^{r-1}, which guarantees that it is independent of the choice of y. Thus we obtain a continuous map $f_r : X^{(r)} \to B$. We may also construct a bundle map $i_r^*(E') \to E$ from $\tilde{F}$ which covers f_r by using Proposition 4.1.5 and we obtain an isomorphism $i_r^*(E') \cong f_r^*(E)$.

Now define $f = f_{n-1}$ and we have

$$f^*(E) = f_{n-1}^*(E) \cong i_{n-1}^*(E') = E'.$$

This completes the proof of the surjectivity of the map $[X, B] \to P_G(X)$ given by $f \mapsto f^*(E)$. □

Exercise 4.7.2. Prove that the 0-skeleton $X^{(0)}$ always has the discrete topology for any CW complex X.

Let us consider the surjectivity of this correspondence $[X, B] \to P_G(X)$. We would like to find a condition under which $f^*(E) \cong g^*(E)$ implies $f \simeq g$. We will be able to construct a homotopy $H : X \times [0, 1] \to B$ by using the procedure in the proof of Proposition 4.7.1 if $X \times [0, 1]$ is a CW complex. Note that the decomposition $[0, 1] = \{0\} \cup (0, 1) \cup \{1\}$ defines a structure of 1-dimensional CW complex on $[0, 1]$. We are thus lead to the problem of constructing a structure of CW complex on $X \times Y$ when both X and Y are CW complexes.

Proposition 4.7.3. *For any cell complexes X and Y, the product $X \times Y$ has a structure of cell complex. Furthermore if both X and Y are closure finite, so is $X \times Y$.*

Proof. The decompositions

$$X = \bigcup_{\lambda \in \Lambda} e_\lambda$$

$$Y = \bigcup_{\mu \in M} e'_\mu$$

gives us a decomposition

$$X \times Y = \bigcup_{\lambda \in \Lambda, \mu \in M} e_\lambda \times e'_\mu.$$

For an r-cell e_λ in X and an r'-cell e'_μ in Y, we show that $e_\lambda \times e'_\mu$ is an $(r + r')$-cell in $X \times Y$. Let $\varphi_\lambda : D^r \to X$ and $\varphi'_\mu : D^{r'} \to Y$ be characteristic maps of e_λ and e'_μ, respectively. Choose a homeomorphism $\psi_n : [0,1]^n \to D^n$ for each n. Then the composition

$$\varphi_{\lambda,\mu} : D^{r+r'} \xrightarrow{\psi_{r+r'}^{-1}} [0,1]^{r+r'} = [0,1]^r \times [0,1]^{r'} \xrightarrow{\psi_r \times \psi_{r'}} D^r \times D^{r'} \xrightarrow{\varphi \times \varphi'} X \times Y$$

is continuous with image $\overline{e} \times \overline{e'} = \overline{e \times e'}$. Since $\varphi_{\lambda,\mu}$ maps $\mathrm{Int}(D^{r+r})$ homeomorphically onto $e_\lambda \times e'_\mu$, it is a characteristic map of $e_\lambda \times e'_\mu$.

If both X and Y are closure finite, those cells that intersect with $\overline{e \times e'} = \overline{e} \times \overline{e'}$ are products of cells that intersect with $\overline{e}$ and with $\overline{e'}$. Thus the number is finite. $\square$

Unfortunately, the product $X \times Y$ may not have the weak topology. We need certain assumptions. One of choices to require a finiteness.

Definition 4.7.4. A cell complex X is called *locally finite* if, for each $x \in X$, there exists a finite subcomplex A with $x \in \mathrm{Int}\,A$. It is called *locally countable* if A is taken to be a countable subcomplex.

The following fact can be found as Corollary 5.5 in Chapter II of [Lundell and Weingram (1969)].

Theorem 4.7.5. *Let X and Y be CW complexes. If either X or Y is locally finite[4] or if both X and Y are countable, then $X \times Y$ is a CW complex.*

---

[4]In the Lundell-Weingram book, the locally-finiteness condition is replaced by the locally-compactness condition. But these two conditions are known to be equivalent. See Proposition 3.6 of Chapter II in the same book.

Remark 4.7.6. Liu [Liu (1978)] showed that, under the continuum hypothesis, $X \times Y$ is a CW complex if and only if either one of the conditions in Theorem 4.7.5, i.e.

- either X or Y are locally finite, or
- both X and Y are locally countable

holds.

Another way of dealing with products of CW complexes is to replace the topology by the compactly generated topology. See [Steenrod (1967)], for example.

Now we are ready to discuss the injectivity of the map $[X, B] \to P_G(X)$.

Proposition 4.7.7. *Let G be a topological group and $p : E \to B$ a principal G-bundle with E n-connected. If X is a CW complex with $\dim X < n$, then the map $[X, B] \to P_G(X)$ given by $f \mapsto f^*(E)$ is injective.*

Proof. This is a corollary to Proposition 4.7.1 if we ignore the boundary conditions $H|_{X \times \{0\}} = f$ and $H|_{X \times \{1\}}$, which can be taken care of as follows.

Suppose $f^*(E) \cong g^*(E)$. We want to construct a homotopy $H : X \times [0,1] \to B$ between f and g. Define $E' = f^*(E)$. Then

$$f^*(p) \times 1_{[0,1]} : E' \times [0,1] \longrightarrow X \times [0,1]$$

is a principal G-bundle over $X \times [0,1]$ whose restrictions to $X \times \{0\}$ and $X \times \{1\}$ are isomorphic to $f^*(E)$ and $g^*(E)$, respectively. By the assumption $\dim X < n$, $X \times [0,1]$ is a CW complex with $\dim(X \times [0,1]) < n + 1$. Since E is n-connected, we may construct a homotopy $H : X \times [0,1] \to B$ by using the construction in the proof of Proposition 4.7.1. In this case, however, we start from $f \amalg g : X \times \{0\} \amalg X \times \{1\} \to B$ and extend it on cells. $\square$

Definition 4.7.8. Let G be a topological group. A principal G-bundle $E \to B$ is said to be n-*universal* if the total space E is $(n - 1)$-connected.

Remark 4.7.9. By Proposition 4.7.1 and Proposition 4.7.7, an ∞-universal principal G-bundle, i.e. a principal G-bundle which is n-universal for all n, is a universal G-bundle in the sense of Definition 4.5.23.

By Proposition 4.7.1 and Proposition 4.7.7, it remains to construct a universal G-bundle for the proof of the classification theorem.

4.8   Fiber Bundles and Homotopy Groups

If we want to show a principal G-bundle $p : E \to B$ is universal, we need to show that the homotopy groups of E vanish. We need tools to compute homotopy groups in order to complete the classification theorem.

In this section, we study relations among homotopy groups of the total space E, the base space B, and the fiber F of a given fiber bundle $p :$ $E \to B$. As is stated at the beginning of this chapter, we use maps to compare two objects in modern mathematics. In this case we should use the projection $p : E \to B$ to compare the homotopy groups of E and B. In general we have the following construction.

Definition 4.8.1. Let $f : X \to Y$ be a based map. Define a map $f_* : \pi_n(X) \to \pi_n(Y)$ by $f_*([g]) = [f \circ g]$ for $[g] \in \pi_n(X)$, where $[g]$ is the homotopy class of $g : S^n \to X$. This map f_* is called the *induced homomorphism*, since it is a homomorphism of groups by Exercise 4.8.2.

Exercise 4.8.2. Show that if $n \geq 1$ then f_* is a homomorphism of groups. Also show that, for based maps $f : X \to Y$ and $g : Y \to Z$, we have $(g \circ f)_* = g_* \circ f_*$ and that $(1_X)_* = 1_{\pi_n(X)}$ for the identity map $1_X : X \to X$.

Thus we obtain a homomorphism $p_* : \pi_n(E) \to \pi_n(B)$ with which we can compare $\pi_n(E)$ and $\pi_n(B)$. But what do we mean by comparing two groups? Let us recall linear algebra to recall how to compare two vector spaces.

Proposition 4.8.3. *Let V and W be finite dimensional vector spaces and $f : V \to W$ a linear map. Define*

$$\operatorname{Ker} f = f^{-1}(0)$$
$$\operatorname{Coker} f = W/\operatorname{Im} f.$$

Then the following hold.

(1) f is injective if and only if $\operatorname{Ker} f = 0$.
(2) f is surjective if and only if $\operatorname{Coker} f = 0$.

In particular, f is an isomorphism if and only if $\operatorname{Ker} f = 0$ and $\operatorname{Coker} f = 0$.

In the case of a group homomorphism $f : G \to H$, the differences of G and H that we can measure by f are

$$\operatorname{Ker} f = \{g \in G \mid f(g) = e\}$$
$$\operatorname{Coker} f = H/\operatorname{Im} f.$$

However, in the case of groups, Coker f may not be a group, since Im f is not necessarily a normal subgroup of H. It is better to use Im f instead of Coker f. If we want to describe relations among more that two groups, the notion of exact sequences is convenient.

Definition 4.8.4. A sequence of groups and homomorphisms

$$G_1 \xrightarrow{f_1} G_2 \xrightarrow{f_2} G_3 \tag{4.11}$$

is called *exact* at G_2 if Im f_1 = Ker f_2. In this case, the sequence (4.11) is called an exact sequence.

Furthermore, a sequence of groups and homomorphisms

$$\cdots \longrightarrow G_{k-1} \xrightarrow{f_{k-1}} G_k \xrightarrow{f_k} G_{k+1} \longrightarrow \cdots \tag{4.12}$$

is called an *exact sequence* if

$$G_{k-1} \xrightarrow{f_{k-1}} G_k \xrightarrow{f_k} G_{k+1}$$

is exact at G_k for all k. The sequence (4.12) is also called a *long exact sequence*.

Remark 4.8.5. The sequence (4.11) is exact if and only if

(1) $f_2 \circ f_1 = 0$ and
(2) $f_2(x) = 0 \Rightarrow \exists y \in G_1$ s.t. $x = f_1(y)$.

Now we are ready to compare homotopy groups of the fiber, the total space, and the base space of a fiber bundle.

Proposition 4.8.6. *Let $p : E \to B$ be a fiber bundle with fiber F. Suppose that the base space B is a based CW complex with base point $*$. Then the sequence*

$$\pi_n(F) \xrightarrow{i_*} \pi_n(E) \xrightarrow{p_*} \pi_n(B)$$

is exact for $n \geq 1$, where $i : F \cong p^{-1}(*) \hookrightarrow E$ is the inclusion of the fiber over the base point.

Proof. By Remark 4.8.5, it suffices to show that

(1) $p_* \circ i_* = 0$, and
(2) $p_*([f]) = 0 \Rightarrow \exists [g] \in \pi_n(F)$ s.t. $i_*([g]) = [f]$.

By definition, $F = p^{-1}(*)$, which implies $p \circ i(F) = \{*\}$ and thus $p \circ i = *$, where $*$ is the constant map to the base point. By Exercise 4.8.2, we have $p_* \circ i_* = 0$.

Suppose $p_*([f]) = 0$ for $[f] \in \pi_n(E)$. Then $p \circ f \simeq *$. Let $H : S^n \times [0,1] \to B$ be a based homotopy for this relation. Then $H|_{S^n \times \{0\}} = p \circ f$ and the outside of the diagram

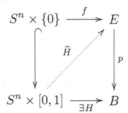

is commutative. Thus the based covering homotopy theorem (Theorem 4.6.33) applies. By using the dotted homotopy $\widetilde{H} : S^n \times [0,1] \to E$ in the diagram, define $g = \widetilde{H}|_{S^n \times \{1\}}$. Then we have

$$p \circ g = p \circ \widetilde{H}|_{S^n \times \{1\}} = H|_{S^n \times \{1\}} = *.$$

Since $p(g(S^n)) = \{*\}$, we have $g(S^n) \subset p^{-1}(\{*\}) = F$. And we obtain an element $[g] \in \pi_n(F)$. On the other hand, $\widetilde{H}$ is a based homotopy for $f \simeq i \circ g$ and we have $[f] = i_*([g])$. $\square$

Note that $p : E \to B$ is surjective and $i : F \to E$ is injective. However, in the exact sequence

$$\pi_n(F) \xrightarrow{i_*} \pi_n(E) \xrightarrow{p_*} \pi_n(B)$$

p_* may not be surjective and i_* is not injective, either.

Example 4.8.7. Recall from Example 2.4.10 that $\pi_1(S^1) \cong \mathbb{Z}$. Since D^2 is contractible and the homotopy groups of a contractible space are trivial, we have $\pi_1(D^2) = 0$.

However, the induced map by the inclusion $i : S^1 \hookrightarrow D^2$ is

$$i_* : \pi_1(S^1) = \mathbb{Z} \longrightarrow 0 = \pi_1(D^2)$$

and thus it is not injective. $\square$

Given a fiber bundle $F \xrightarrow{i} E \xrightarrow{p} B$, let us consider how far the induced map $i_* : \pi_n(F) \to \pi_n(E)$ is from being injective. It is measured by the size of $\operatorname{Ker} i_*$.

Suppose $i_*([f]) = 0$ for $[f] \in \pi_n(F)$. The condition $i_*([f]) = 0$ is equivalent to $f \simeq *$ in E. Following the argument in the proof of

Proposition 4.6.6, it is easily seen to be equivalent to the existence of a based map $\tilde{f} : D^{n+1} \to E$ satisfying $\tilde{f}|_{\partial D^{n+1}} = f$. Since f maps S^n to F, we have $(p \circ \tilde{f})(\partial S^n) = \{*\}$. Thus $p \circ \tilde{f}$ induces a based map $\bar{f} : D^{n+1}/\partial D^{n+1} = S^{n+1} \to B$.

This correspondence seems to define a map $\operatorname{Ker} i_* \to \pi_{n+1}(B)$ but, unfortunately, this is not well-defined. What is well-defined is the other direction.

Definition 4.8.8. Let $p : E \to B$ be a fiber bundle with fiber F. When B is a based CW complex, define a map $\partial : \pi_n(B) \to \pi_{n-1}(F)$ for $n \geq 1$ as follows.

Take $[f] \in \pi_n(B)$. As we have done in Lemma 4.6.20, we may regard $f : D^n \to B$ satisfying $f(\partial D^n) = \{*\}$ under the identification $S^n = D^n/\partial D^n$. We also use the identification

$$D^n = S^{n-1} \times [0,1]/S^{n-1} \times \{0\} \cup \{*\} \times [0,1].$$

Let

$$\pi : S^{n-1} \times [0,1] \longrightarrow S^{n-1} \times [0,1]/S^{n-1} \times \{0\} \cup \{*\} \times [0,1] = D^n$$

be a projection. Define

$$H : S^{n-1} \times [0,1] \xrightarrow{\pi} D^n \xrightarrow{f} B$$

by the composition. For $x \in S^{n-1}$, $H(x,0) = *$ and for $t \in [0,1]$, $H(*,t) = *$. Thus H is a based homotopy. By applying the based covering homotopy theorem, we obtain a homotopy $\tilde{H} : S^{n-1} \times [0,1] \to E$ making the diagram

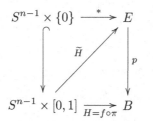

commutative. Now define $g = \tilde{H}|_{S^{n-1} \times \{1\}}$ and regard it as $g : S^{n-1} \to E$. Then we have

$$p \circ g = p \circ \tilde{H}|_{S^{n-1} \times \{1\}} = H|_{S^{n-1} \times \{1\}} = f|_{\partial D^n} = *,$$

which implies $g(S^{n-1}) \subset p^{-1}(*) = F$ and thus defines an element $[g] \in \pi_{n-1}(F)$. The map obtained by this construction is denoted by $\partial : \pi_n(B) \to \pi_{n-1}(F)$ and is called the *connecting homomorphism*.

Exercise 4.8.9. Show that this map ∂ is independent of the choice of $\widetilde{H}$. Also show that ∂ is a group homomorphism if $n \geq 2$.

We have the following relation between ∂ and $\operatorname{Ker} i_*$.

Proposition 4.8.10. *Under the assumption of Proposition 4.8.6, the sequence*

$$\pi_n(B) \xrightarrow{\ \partial\ } \pi_{n-1}(F) \xrightarrow{\ i_*\ } \pi_{n-1}(E)$$

is exact when $n \geq 2$.

Proof. The homotopy $\widetilde{H} : S^{n-1} \times [0,1] \to E$ used in the definition of ∂ is a homotopy in E from the constant map $*$ to the base point and g. And thus

$$i_* \circ \partial([f]) = i_*([g]) = 0$$

or $i_* \circ \partial = 0$.

Conversely, suppose $[g] \in \pi_{n-1}(F)$ is an element with $i_*([g]) = 0$. Let $\widetilde{G} : S^{n-1} \times [0,1] \to E$ be the based homotopy from g to the constant map $*$ in E and define $G = p \circ \widetilde{G} : S^{n-1} \times [0,1] \to B$. Since $\widetilde{H}|_{S^{n-1} \times \{0\}} = *$, $H|_{S^{n-1} \times \{0\}} = *$. On the other hand, H is a based homotopy since $\widetilde{H}$ is. Thus H defines a map

$$f : D^n = S^{n-1} \times [0,1]/S^{n-1} \times \{0\} \cup \{*\} \times [0,1] \longrightarrow B.$$

Since

$$f|_{S^{n-1}} = H|_{S^{n-1} \times \{1\}} = p \circ g = *,$$

f defines an element of $\pi_n(B)$. By the construction of the map ∂, we see $[f] = \partial[g]$. Thus the sequence is exact. $\qquad\square$

We have another exact sequence.

Proposition 4.8.11. *Under the assumption of Proposition 4.8.6,*

$$\pi_n(E) \xrightarrow{\ p_*\ } \pi_n(B) \xrightarrow{\ \partial\ } \pi_{n-1}(F)$$

is exact, if $n \geq 2$.

Proof. Let us first show that $\partial \circ p_* = 0$. Take $[f] \in \pi_n(E)$. We may assume that it is represented by a map $f : D^n \to E$ with $f(\partial D^n) = \{*\}$. Note that the composition

$$S^{n-1} \times [0,1] \xrightarrow{\ \pi\ } D^n \xrightarrow{\ f\ } E$$

can be used as $\widetilde{H}$ in the definition of $\partial([p \circ f])$ and thus

$$\partial([p \circ f]) = [\widetilde{H}|_{S^{n-1} \times \{1\}}]$$
$$= [f|_{S^{n-1}}]$$
$$= [*]$$
$$= 0$$

and we have $\partial \circ p_* = 0$.

Conversely suppose $[f] \in \pi_n(B)$ is an element with $\partial([f]) = 0$. Then $\widetilde{H}|_{S^{n-1} \times \{1\}} \simeq *$ in F for a lift $\widetilde{H} : S^{n-1} \times [0,1] \to E$ of $S^{n-1} \times [0,1] \xrightarrow{\pi} D^n \xrightarrow{f} B$. Let $G : S^{n-1} \times \{1\} \times [0,1] \to F \subset E$ be the homotopy and $G' : (S^{n-1} \times \{0\} \cup \{*\} \times [0,1]) \times [0,1] \to E$ be the constant map to the base point. Then we have maps

$$\widetilde{H} : S^{n-1} \times [0,1] \times \{0\} \longrightarrow E$$
$$G \cup G' : (S^{n-1} \times \{0,1\} \cup \{*\} \times [0,1]) \times [0,1] \longrightarrow E.$$

Denote

$$X = S^{n-1} \times [0,1]$$
$$A = S^{n-1} \times \{0,1\} \cup \{*\} \times [0,1]$$
$$K = G \cup G'$$

for simplicity. Then $\widetilde{H}|_{A \times \{0\}} = K|_{A \times \{0\}}$. Define a map $H : X \times [0,1] \to B$ by $H(x,t,s) = p \circ \widetilde{H}(x,t)$. Then the diagram

$$
\begin{array}{ccc}
X \times \{0\} \cup A \times [0,1] & \xrightarrow{\widetilde{H} \cup K} & E \\
\downarrow & & \downarrow{\scriptstyle p} \\
X \times [0,1] & \xrightarrow{\quad H \quad} & B
\end{array}
$$

is commutative. By the covering homotopy extension property (Theorem 4.6.31), we obtain a homotopy $\widetilde{K} : X \times [0,1] \to E$ which is an extension of K and a lift of H.

Since

$$\widetilde{K}((S^{n-1} \times \{0\} \cup \{*\} \times [0,1]) \times [0,1])$$
$$= (\widetilde{H} \cup K)((S^{n-1} \times \{0\} \cup \{*\} \times [0,1]) \times [0,1])$$
$$= K((S^{n-1} \times \{0\} \cup \{*\} \times [0,1]) \times [0,1])$$
$$= G'((S^{n-1} \times \{0\} \cup \{*\} \times [0,1]) \times [0,1])$$
$$= \{*\},$$

it induces a map

$$\tilde{K} : D^n \times [0,1] = (S^{n-1} \times [0,1]/S^{n-1} \times \{0\} \cup \{*\} \times [0,1]) \times [0,1] \longrightarrow E.$$

Define $g = \tilde{K}|_{D^n \times \{1\}}$. Then

$$g(S^{n-1}) = \tilde{K}(S^{n-1} \times \{1\} \times \{1\}) = \{*\}$$

and we obtain an element $[g] \in \pi_n(E)$ with $p_*[g] = [f]$. This completes the proof. $\square$

By combining these propositions together, we obtain a long exact sequence.

Corollary 4.8.12. *For a fiber bundle $p : E \to B$ with fiber F, there exists a long exact sequence of homotopy groups.*

$$\cdots \longrightarrow \pi_n(F) \xrightarrow{i_*} \pi_n(E) \xrightarrow{p_*} \pi_n(B) \xrightarrow{\partial} \pi_{n-1}(F) \longrightarrow \cdots \xrightarrow{p_*} \pi_1(B).$$

It often happens that the last map in the sequence is surjective.

Lemma 4.8.13. *Let $E \to B$ be a fiber bundle with an arcwise connected fiber F. Then $p_* : \pi_1(E) \to \pi_1(B)$ is surjective.*

Proof. For an element $[\ell] \in \pi_1(B)$, we may assume that it is represented by a map $\ell : [0,1] \to B$ with $\ell(0) = \ell(1) = *$. Regard $[0,1] = \{*\} \times [0,1]$ so that ℓ a homotopy to the base space B. Take $x_0 \in p^{-1}(*)$ and let $\{*\} \times \{0\} \to E$ be the constant map to x_0. Then the square of the diagram

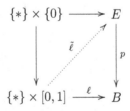

is commutative. By the covering homotopy theorem, we obtain a continuous map $\tilde{\ell} : [0,1] \to E$ making the diagram commutative. (See Figure 4.19.)

Note that $\tilde{\ell}(1)$ may be different from x_0, but there exists a path $\omega : [0,1] \to E$ which connects $\tilde{\ell}(0) = x_0$ and $\tilde{\ell}(1)$ in $F = p^{-1}(*)$ by the arcwise connectivity of F. Define

$$\hat{\ell}(t) = \begin{cases} \tilde{\ell}(2t), & 0 \leq t \leq \frac{1}{2} \\ \omega(2t-1), & \frac{1}{2} \leq t \leq 1 \end{cases}.$$

Then $[\hat{\ell}] \in \pi_1(E)$ and $p_*([\hat{\ell}]) = [\ell]$. $\square$

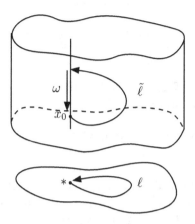

Fig. 4.19   a lift of loop

The surjectivity of a map can be expressed by the exactness of a sequence.

Exercise 4.8.14. Let $f : G \to H$ be a group homomorphism. Denote the group consisting only of the unit element by 0. Show that f is surjective if and only

$$G \xrightarrow{f} H \longrightarrow 0$$

is exact.

By Lemma 4.8.13, if the fiber is arcwise connected, we obtain a long exact sequence

$$\cdots \longrightarrow \pi_n(F) \xrightarrow{i_*} \pi_n(E) \xrightarrow{p_*} \pi_n(B) \xrightarrow{\partial} \pi_{n-1}(F) \longrightarrow \cdots \longrightarrow \pi_1(B) \longrightarrow 0. \tag{4.13}$$

Definition 4.8.15. The sequence (4.13) is called the *homotopy long exact sequence* of the fiber bundle $E \to B$.

Exercise 4.8.16. Let $p : E \to B$ be a principal G-bundle. Define a group structure on $\pi_0(G)$ by the group multiplication of G. Show that the sequence

$$\cdots \longrightarrow \pi_n(G) \xrightarrow{i_*} \pi_n(E) \xrightarrow{p_*} \pi_n(B) \xrightarrow{\partial} \pi_{n-1}(G) \longrightarrow \cdots \xrightarrow{p_*} \pi_1(B) \xrightarrow{\partial} \pi_0(G)$$

is exact. Furthermore, if E is arcwise connected, show that the sequence
$$\cdots \longrightarrow \pi_n(G) \xrightarrow{i_*} \pi_n(E) \xrightarrow{p_*} \pi_n(B) \xrightarrow{\partial} \pi_{n-1}(G) \longrightarrow \cdots$$
$$\xrightarrow{p_*} \pi_1(B) \xrightarrow{\partial} \pi_0(G) \longrightarrow 0$$
is also exact.

4.9   Construction of Universal Bundles: Steenrod's Approach

Recall that, given a topological group G, we want to construct a principal G-bundle $EG \to BG$ with the total space EG ∞-connected, i.e. all the homotopy groups of EG vanish. If we could find an ∞-connected space EG with G action, there would be a good chance that the projection $EG \to EG/G$ is a principal G-bundle, as we have discussed in §3.6 and §3.7, and we could define $BG = EG/G$. And this is what we are going to do.

We first need to find a space whose homotopy groups vanish. An obvious choice is $\{*\}$ but this space is too small to admit a nontrivial action of topological groups.

Let us consider spheres, since they the only spaces we know a part of their homotopy groups. A good news is that S^n is $(n-1)$-connected by Corollary 4.6.30. If we could take the "limit" $n \to \infty$, we would obtain an ∞-connected space S^∞. But how can we take the limit of topological spaces? Although there is a modern theory of limits (and colimits) for geometric and algebraic objects such as topological spaces and groups, let us construct an infinite dimensional sphere concretely without using such an abstract theory.

Definition 4.9.1. Define
$$S^\infty = \left\{ (x_0, x_1, \ldots) \,\middle|\, x_i \in \mathbb{R}, x_i \neq 0 \text{ only for a finite number of } i, \sum_{i=0}^{\infty} x_i^2 = 1 \right\}$$
as a set. In order to define a topology, we use an approximation by finite dimensional spheres. Regard
$$S^n = \left\{ (x_0, x_1, \ldots, x_n) \in \mathbb{R}^{n+1} \,\middle|\, x_0^2 + \cdots + x_n^2 = 1 \right\}$$
and define $i_n : S^n \to S^{n+1}$ by $i_n(x_0, \ldots, x_n) = (x_0, \ldots, x_n, 0)$. This is obviously an embedding of S^n into S^{n+1} and we have a sequence of inclusions
$$\cdots \subset S^n \subset S^{n+1} \subset \cdots \subset S^\infty.$$
This sequence is complete in the sense that
$$S^\infty = \bigcup_{n=1}^{\infty} S^n$$

(as sets). Define a topology of S^∞ by

$$A \subset S^\infty \text{ is closed} \iff A \cap S^n \text{ is closed in each } S^n.$$

The definition of this topology is similar to the condition (W) in the definition of CW complexes. In general, we have the following definition.

Definition 4.9.2. Let X be a topological space and $\{X_\lambda\}_{\lambda \in \Lambda}$ a covering of X by subspaces. We say X has the *weak topology* with respect to this covering if

$$A \subset X \text{ is closed} \iff A \cap X_\lambda \text{ is closed in each } X_\lambda.$$

Let X be a set and $\{X_\lambda\}_{\lambda \in \Lambda}$ a family of topological spaces such that

$$X = \bigcup_{\lambda \in \Lambda} X_\lambda$$

as sets. Then define a topology on X by

$$A \subset X \text{ is closed} \iff A \cap X_\lambda \text{ is closed in each } X_\lambda,$$

in which case we say X is *topologized by the weak topology* with respect to $\{X_\lambda\}_{\lambda \in \Lambda}$.

Exercise 4.9.3. Show that any topological space has a weak topology with respect to any open covering.

Exercise 4.9.4. Show that any topological space has a weak topology with respect to any finite closed covering.

In the terminology of Definition 4.9.2, S^∞ is topologized by the weak topology with respect to the covering

$$S^\infty = \bigcup_{n=1}^{\infty} S^n.$$

The following fact is essential when we show the homotopy groups of S^∞ vanish.

Lemma 4.9.5. *Let X be a Hausdorff space and $\{X_n\}_{n \in \mathbb{N}}$ a sequence of closed subspaces of X satisfying the following conditions:*

(1) $X_n \subset X_{n+1}$ for each n,

(2) $X = \bigcup_{n=1}^{\infty} X_n$, and

(3) X has the weak topology with respect to the covering $\{X_n\}_{n \in \mathbb{N}}$.

Then, for any compact subset $K \subset X$, there exists n such that $K \subset X_n$.

Proof. We prove by contradiction. Suppose that K is not contained in any one of X_n's. Then there exist an infinitely many integers n such that $K \cap (X_n \setminus X_{n-1}) \neq \emptyset$. Let $\{n_1, n_2, \ldots\}$ be the collection of such numbers. Take a point x_k from each $K \cap (X_{n_k} \setminus X_{n_k-1})$ and set $A = \{x_1, x_2, \ldots\}$. Then each $A \cap X_n$ is closed in X_n, since $A \cap X_n$ consists of a finite number of points and X is Hausdorff. By the definition of weak topology, A is closed in X. Since A is a closed subset of a compact set K, it is also compact.

Define $A_i = A \setminus \{x_i\}$ for each i. Then

$$A_i \cap X_n = \begin{cases} \{x_1, \ldots, x_{i-1}\}, & n = n_i \\ \phi, & n \notin \{n_1, n_2, \ldots\} \end{cases} .$$

By the same argument as above A_i is closed, which means that each $\{x_i\}$ is an open subset of A. Now A has an open covering $A = \bigcup\limits_{i=1}^{\infty} \{x_i\}$ consisting of infinitely many open sets. This contradicts the compactness of A. $\square$

Corollary 4.9.6. *For any i, we have $\pi_i(S^\infty) = 0$. Thus the infinite dimensional sphere S^∞ is ∞-connected.*

Proof. For a continuous based map $f(S^i)$ is a compact subset of S^∞. By Lemma 4.9.5, it is contained in some S^n. We may assume that $n > i$. By Corollary 4.6.30, f is homotopic to the constant map to the base point in S^n. Let $H : S^i \times [0,1] \to S^n$ be the homotopy. Then the composition

$$S^i \times [0,1] \xrightarrow{H} S^n \hookrightarrow S^\infty$$

is a homotopy between f and the constant map to the base point in S^∞. $\square$

Thanks to this corollary, if there exists a principal G-bundle whose total space is S^∞, then it is a universal G-bundle. When G is not so complicated, it is not difficult to find such a bundle.

Exercise 4.9.7. Recall that an action of the group C_2 of order 2 on S^n is defined in Example 3.6.16. The quotient space was denoted by $\mathbb{RP}^n = S^n/C_2$ and called the n-dimensional real projective space.

Show that the projection $S^n \to \mathbb{RP}^n$ is a principal C_2-bundle.

The inclusions $S^n \hookrightarrow S^{n+1}$ induce embeddings $\mathbb{RP}^n \hookrightarrow \mathbb{RP}^{n+1}$. Define

$$\mathbb{RP}^\infty = \bigcup_{n=1}^{\infty} \mathbb{RP}^n$$

and topologize it by the weak topology. This is called the *infinite dimensional real projective space.*

Proposition 4.9.8. *The projection $S^\infty \to \mathbb{R}P^\infty$ is a principal C_2-bundle, hence is a universal C_2-bundle.*

This is a special case of Theorem 4.9.20 and the proof is included in the proof of Theorem 4.9.20.

Corollary 4.9.9. *For any CW complex, we have a bijection $[X, \mathbb{R}P^\infty] \cong P_{C_2}(X)$.*

Remark 4.9.10 (For those who are familiar with cohomology). Let $H^*(X; \mathbb{Z}/2\mathbb{Z})$ be the singular cohomology of X with coefficients in $\mathbb{Z}/2\mathbb{Z}$. Then it is known that we have a bijection
$$H^1(X; \mathbb{Z}/2\mathbb{Z}) \cong [X, \mathbb{R}P^\infty].$$
In fact, the right hand side has a structure of an Abelian group and this is an isomorphism of Abelian groups. Combined with Corollary 4.9.9, we see that fiber bundles over a CW complex X with structure group C_2 are determined by degree 1 mod 2 cohomology classes of X.

For a principal C_2-bundle $\xi = (E \xrightarrow{p} X)$, the element of $H^1(X; \mathbb{Z}/2\mathbb{Z})$ obtained by the bijections
$$P_{C_2}(X) \cong [X, \mathbb{R}P^\infty] \cong H^1(X; \mathbb{Z}/2\mathbb{Z})$$
is denoted by $w_1(\xi)$ and called the *first Stiefel-Whitney class* of ξ. In other words, isomorphism classes of principal C_2-bundles are determined by the first Stiefel-Whitney class.

More generally, many kinds of cohomology classes of the base space have been introduced for a principal G-bundle, especially when G is a compact Lie group. Such cohomology classes are called *characteristic classes*. See [Milnor and Stasheff (1974)] or [Husemoller (1994)] for more details.

We have succeeded in classifying fiber bundles with structure group C_2. How can we generalize this result to larger topological groups? The problem is the infinite dimensional sphere S^∞ is not large enough for most topological groups. Although S^∞ is an infinite dimensional space, it is too simple to admit free actions of more complex groups. Let us try to enlarge S^∞ and make a space on which larger groups act.

Note that we have a homeomorphism
$$S^n \cong O(n+1)/O(n) \tag{4.14}$$
proved in Example 3.6.42. In order to enlarge S^n, an obvious idea is to enlarge the numerator of this quotient.

Definition 4.9.11. Define a map $O(n) \hookrightarrow O(n+k)$ by

$$A \longmapsto \begin{pmatrix} A & 0 \\ 0 & I_k \end{pmatrix}$$

and regard $O(n)$ as a subgroup of $O(n+k)$. The quotient space is denoted by $St_{n,k}(\mathbb{R}) = O(n+k)/O(n)$ and is called the *real Stiefel manifold*.

Theorem 4.9.12. *The Stiefel manifold* $St_{n,k}(\mathbb{R})$ *is* $(n-1)$*-connected.*

We need the following general fact to prove this theorem.

Proposition 4.9.13. *Let G be a topological group and H a closed subgroup such that the projection $G \to G/H$ admits local cross sections. Then for any closed subgroup K of H, the projection $G/K \to G/H$ is a fiber bundle with fiber H/K.*

Proof. By assumption, the projection

$$G \longrightarrow G/H \tag{4.15}$$

is a principal H-bundle. Define an action of H on H/K $\mu : H \times H/K \to H/K$ by $\mu(h, h'K) = hh'K$. The fiber bundle $G \times_H H/K \to G/H$ associated with the principal H-bundle (4.15) is isomorphic to $G/K \to G/H$ by Exercise 4.9.14 next. $\square$

Exercise 4.9.14. Define a map $G \times_H H/K \to G/K$ by $(g, hK) \longmapsto ghK$. Show that this is a homeomorphism and makes the diagram

$$
\begin{array}{ccc}
G \times_H H/K & \longrightarrow & G/K \\
\downarrow & & \downarrow \\
G/H & =\!=\!= & G/H
\end{array}
$$

commutative.

Proof of Theorem 4.9.12. We proceed by induction on k. When $k = 1$, $St_{n,1}(\mathbb{R}) = S^n$ and is $(n-1)$-connected.

Suppose we have proved that $St_{n,k}$ is $(n-1)$-connected for all n. Let us prove that $St_{n,k+1}(\mathbb{R})$ is $(n-1)$-connected. By Proposition 4.9.13, the projection

$$O(n+k+1)/O(n) \longrightarrow O(n+k+1)/O(n+1)$$

is a fiber bundle with fiber $O(n+1)/O(n) \cong S^n$. In other words, we have a fiber bundle $\text{St}_{n,k+1}(\mathbb{R}) \to \text{St}_{n+1,k}(\mathbb{R})$ with fiber S^n. Let

$$\cdots \longrightarrow \pi_i(S^n) \longrightarrow \pi_i(\text{St}_{n,k+1}(\mathbb{R})) \longrightarrow \pi_i(\text{St}_{n+1,k}(\mathbb{R})) \longrightarrow \cdots$$

be the long exact sequence of homotopy groups of this fiber bundle. By the inductive hypothesis, we have

$$\pi_i(S^n) = 0$$
$$\pi_i(\text{St}_{n+1,k}(\mathbb{R})) = 0$$

for $i < n$ and we obtain an exact sequence

$$0 \longrightarrow \pi_i(\text{St}_{n,k+1}(\mathbb{R})) \longrightarrow 0,$$

which implies that $\pi_i(\text{St}_{n,k+1}(\mathbb{R})) = 0$. □

As is the case of spheres, we want to take the "limit" as $n \to \infty$.

Definition 4.9.15. Define a map $O(n+k) \hookrightarrow O(n+k+1)$ by

$$A \longmapsto \begin{pmatrix} 1 & 0 \\ 0 & A \end{pmatrix}.$$

It induces an embedding.

$$\text{St}_{n,k}(\mathbb{R}) = O(n+k)/O(n) \hookrightarrow O(n+k+1)/O(n+1) = \text{St}_{n+1,k}(\mathbb{R}).$$

Regard $\text{St}_{n,k}(\mathbb{R})$ as a subspace of $\text{St}_{n+1,k}(\mathbb{R})$ under this embedding and define

$$EO(k) = \bigcup_{n=1}^{\infty} \text{St}_{n,k}(\mathbb{R}).$$

It is topologized by the weak topology.

The same argument as in the proof of Corollary 4.9.6 can be used to prove the next fact by using Lemma 4.9.5.

Corollary 4.9.16. *The space $EO(k)$ is ∞-connected.*

Since $C_2 = O(1)$ acts on $S^\infty = EO(1)$, it seems that $O(k)$ acts on $EO(k)$.

Definition 4.9.17. Regard $O(n) \times O(k)$ as a subgroup of $O(n+k)$ under the correspondence

$$(A, B) \longmapsto \begin{pmatrix} A & 0 \\ 0 & B \end{pmatrix}.$$

Denote the quotient space by $\text{Gr}_{n,k}(\mathbb{R}) = O(n+k)/O(n) \times O(k)$ and call it the *Grassmannian manifold*.

By Proposition 4.9.13, the projection

$$\mathrm{St}_{n,k}(\mathbb{R}) = O(n+k)/O(n) \longrightarrow O(n+k)/O(n) \times O(k) = \mathrm{Gr}_{n,k}(\mathbb{R})$$

is a fiber bundle with fiber $O(n) \times O(k)/O(n) = O(k)$.

Lemma 4.9.18. *The projection*

$$p_{n,k} : \mathrm{St}_{n,k}(\mathbb{R}) \longrightarrow \mathrm{Gr}_{n,k}(\mathbb{R}) \tag{4.16}$$

is a principal $O(k)$-bundle.

Proof. Regard $O(k)$ as a subgroup of $O(n+k)$ under the correspondence

$$A \longmapsto \begin{pmatrix} 1_n & 0 \\ 0 & A \end{pmatrix},$$

which defines an action $O(k) \times O(n+k) \to O(n+k)$ of $O(k)$ on $O(n+k)$. This is compatible with the action of $O(n)$ and induces a map $O(k) \times O(n+k)/O(n) \to O(n+k)/O(n)$. The projection (4.16) is nothing but a projection to the quotient by this action. Since (4.16) is a fiber bundle, this is a principal $O(k)$-bundle. □

Definition 4.9.19. As is the case of the Stiefel manifolds, the inclusions $O(n+k) \hookrightarrow O(n+k+1)$ induce embeddings $\mathrm{Gr}_{n,k}(\mathbb{R}) \hookrightarrow \mathrm{Gr}_{n+1,k}(\mathbb{R})$ with which we define

$$BO(k) = \bigcup_{n=1}^{\infty} \mathrm{Gr}_{n,k}(\mathbb{R})$$

with weak topology.

Theorem 4.9.20. *The projection*

$$p_k : EO(k) \longrightarrow BO(k) \tag{4.17}$$

is a principal $O(k)$-bundle. Hence this is a universal $O(k)$-bundle.

Remark 4.9.21. Proposition 4.9.8 is the case $k = 1$.

We need the following property of smooth manifolds.

Theorem 4.9.22. *Let M be a smooth manifold and N a closed submanifold. Then there exists an open neighborhood U of N and a map $\pi : U \to N$ satisfying the following conditions:*

(1) π fiber bundle with fiber $\mathrm{Int}\,D^{\dim M - \dim N}$. The inclusion $i : N \hookrightarrow U$ is a cross section.

(2) There exists a bundle map

$$
\begin{array}{ccc}
U \times [0,1] & \xrightarrow{\ h\ } & U \\
{\scriptstyle \pi \times 1_{[0,1]}} \downarrow & & \downarrow {\scriptstyle \pi} \\
N \times [0,1] & \xrightarrow[\mathrm{pr}_1]{} & N
\end{array}
$$

satisfying the following properties

(a) $h|_{U \times \{0\}} = 1_U$, *and*
(b) $h|_{U \times \{1\}} = i \circ \pi$.

In particular, N is a deformation retract of U.

This neighborhood U is called a *tubular neighborhood* of N. A proof of this fact can be found in most textbooks on manifolds. For example, see §6.2 of [Cannas da Silva (2001)] or §3.3 of [Oliva (2002)].

Proof of Theorem 4.9.20. For each $x \in BO(k)$, we construct a neighborhood of x and a local trivialization which in an extension of the local trivialization of Lemma 4.9.18.

By the definition of $BO(k)$, there exists n such that $x \in \mathrm{Gr}_{n,k}(\mathbb{R})$. Let

$$
\varphi_{x,n} : p_n^{-1}(V_{x,n}) \xrightarrow{\ \cong\ } V_{x,n} \times O(k)
$$

be the local trivialization of Lemma 4.9.18 around x. Starting from this local trivialization, we construct and open neighborhood $V_{n+\ell,x}$ of x in $\mathrm{Gr}_{n+\ell,k}(\mathbb{R})$ and a local trivialization

$$
\varphi_{x,n+\ell} : p_{n+\ell}^{-1}(V_{x,n+\ell}) \xrightarrow{\ \cong\ } V_{x,n+\ell} \times O(k)
$$

which is an extension of $\varphi_{x,n+\ell-1}$ for each ℓ.

By induction, suppose we have constructed $\varphi_{x,n+\ell}$. Let $U_{n+\ell}$ be a tubular neighborhood of $\mathrm{Gr}_{n+\ell,k}(\mathbb{R})$ in $\mathrm{Gr}_{n+\ell+1,k}(\mathbb{R})$ and $\pi_{n+\ell} : U_{n+\ell} \to \mathrm{Gr}_{n+\ell,k}(\mathbb{R})$ the projection in Theorem 4.9.22. Define

$$
V_{x,n+\ell+1} = \pi_{n+\ell}^{-1}(V_{x,n+\ell}).
$$

Since $\mathrm{Gr}_{n+\ell,k}(\mathbb{R})$ is a deformation retract of $U_{n+\ell}$ and $\pi_{n+\ell}$ the projection of a fiber bundle, $V_{x,n+\ell}$ is a deformation retract of $V_{x,n+\ell+1}$. By

Corollary 4.3.23, we obtain a bundle map $V_{x,n+\ell+1} \to V_{x,n+\ell}$ by restricting $\pi_{n+\ell}$ and we have a commutative diagram

$$
\begin{array}{ccccc}
p^{-1}_{n+\ell+1,k}(V_{x,n+\ell+1}) & \xrightarrow{\tilde{\pi}_{n+\ell}} & p^{-1}_{n+\ell+1,k}(V_{x,n+\ell}) & =\!\!=\!\!= & p^{-1}_{n+\ell,k}(V_{x,n+\ell}) \\
\downarrow{\scriptstyle p_{n+\ell+1,k}} & & \downarrow{\scriptstyle p_{n+\ell+1,k}} & & \downarrow{\scriptstyle p_{n+\ell,k}} \\
V_{x,n+\ell+1} & \xrightarrow{\pi_{n+\ell}} & V_{x,n+\ell} & =\!\!=\!\!= & V_{x,n+\ell}.
\end{array}
$$

The composition

$$
p^{-1}_{n+\ell+1,k}(V_{x,n+\ell+1}) \xrightarrow{\tilde{\pi}_{n+\ell}} p^{-1}_{n+\ell,k}(V_{x,n+\ell}) \cong V_{x,n+\ell} \times O(k) \xrightarrow{\mathrm{pr_2}} O(k)
$$

and $p_{n+\ell+1,k}$ give us a bundle map

$$
p^{-1}_{n+\ell+1,k}(V_{x,n+\ell+1}) \longrightarrow V_{x,n+\ell+1} \times O(k)
$$

and we obtain a local trivialization on $V_{x,n+\ell+1}$.

Now define $V_{x,n} = \bigcup\limits_{\ell=0}^{\infty} V_{x,n+\ell}$. Then, since $BO(k)$ has the weak topology, this is open in $BO(K)$. By the commutativity of the diagram

$$
\begin{array}{ccc}
p^{-1}_{n+\ell,k}(V_{x,n+\ell}) & \xrightarrow{\;\cong\;} & V_{x,n+\ell} \times O(k) \\
\cup\Big\uparrow & & \cup\Big\uparrow \\[2mm]
\Big\downarrow & & \Big\downarrow \\
p^{-1}_{n+\ell+1,k}(V_{x,n+\ell+1}) & \xrightarrow[\cong]{} & V_{x,n+\ell+1} \times O(k),
\end{array}
$$

we obtain a local trivialization

$$
p^{-1}_{k}(V_{x,n}) \xrightarrow{\;\cong\;} V_{x,n} \times O(k)
$$

on $V_{x,n}$. $\qquad\qquad\qquad\qquad\qquad\qquad\qquad\qquad\qquad\qquad\quad$ $\square$

Corollary 4.9.23. *For any CW complex X, we have a one-to-one correspondence* $[X, BO(k)] \cong P_{O(k)}(X)$.

Thus we obtain a classification theorem for fiber bundles having the orthogonal group $O(k)$ as a structure group. The orthogonal group $O(k)$ is becomes larger as k increases. For example, it is known that $O(k)$ has a structure of smooth manifold (in fact a Lie group) of dimension $\dim O(k) = \frac{1}{2}k(k-1)$. The following theorem says that $O(k)$ becomes large enough to contain most compact topological groups as k increases. A proof can be found in [Mimura and Toda (1991)].

Theorem 4.9.24. *Let G be a compact topological group. If the unit e of G has a neighborhood U such that the only subgroup contained in U is $\{e\}$, then G can be embedded in $O(k)$ for some k.*

For such topological groups, we obtain a universal bundle from $EO(k) \to BO(k)$.

Theorem 4.9.25. *For a closed subgroup G of $O(k)$, define*

$$BG = \bigcup_{n=1}^{\infty} O(n+k)/O(n) \times G = EO(k)/G.$$

Then the projection $EO(k) \to BG$ is a principal G-bundle, hence is a universal G-bundle.

Corollary 4.9.26. *Theorem 4.5.21 holds for any closed subgroup of $O(k)$. Namely, for any CW complex X, we have a bijection $[X, BG] \cong P_G(X)$.*

4.10 Construction of Universal Bundles: The Bar Construction

We have succeeded in constructing a universal bundle for a topological group G which can be embedded in the orthogonal group $O(k)$ as a closed subgroup in §4.9. Thus we obtain a classification theorem for most compact topological groups. How about noncompact groups?

Example 4.10.1. Regard the additive group $\mathbb{Z}$ of integers as a topological group by the discrete topology. This is not compact because it consists of infinitely many points. It cannot be embedded in an orthogonal group and the construction in Theorem 4.9.25 cannot be applied. However, we have already seen a universal $\mathbb{Z}$-bundle in this book. For example, it is not difficult to show that the map $\exp : \mathbb{R} \to S^1$ appeared in Exercise 2.1.6 and Example 3.3.25 is a principal $\mathbb{Z}$-bundle. Since $\mathbb{R}$ is contractible, it is ∞-connected and $\exp : \mathbb{R} \to S^1$ is a universal $\mathbb{Z}$-bundle. $\square$

Exercise 4.10.2. Verify that the map $\exp : \mathbb{R} \to S^1$ is a principal $\mathbb{Z}$-bundle.

Of course, it is not easy to extend Example 4.10.1 to other groups. But it suggests the existence of universal bundles for noncompact groups. In fact, J.W. Milnor [Milnor (1956)] found a construction of universal bundles for general topological groups. Milnor's construction was reformulated by Milgram [Milgram (1967)], which is the subject of this section.

Milgram focused on the properties of the base space BG of a universal bundle $EG \to BG$, i.e. the classifying space of G. It is known that the classifying space BG of G contains an important information on G. Constructions of universal bundles which give rise to good properties of BG are desired. In particular, the following two properties are useful.

(1) Any continuous homomorphism $f : G \to H$ between topological groups induces a natural[5] continuous map

$$Bf : BG \longrightarrow BH.$$

(2) There exists a natural homeomorphism $B(G \times H) \cong BG \times BH$.

Milnor's construction was remarkable in the sense that it has the first property, besides the fact that it can be applied to arbitrary topological groups. Unfortunately, however, it does not have the second condition. It was Milgram who found a way to improve Milnor's construction.

In order to explain Milgram's construction, let us first take a look at Milnor's idea. Milnor noticed that a construction on spaces, called join, has an important property concerning connectivities.

Definition 4.10.3. For topological spaces X and Y, define their *join* $X * Y$ by

$$X * Y = X \times [0,1] \times Y/_\sim,$$

where the relation $\sim$ is the equivalence relation generated by

$$(x, 0, y) \sim (x', 0, y)$$
$$(x, 1, y) \sim (x, 1, y')$$

for $x, x' \in X, y, y' \in Y$. (See Figure 4.20.)

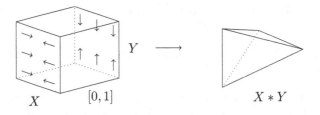

Fig. 4.20   join

---

[5] See Remark 4.6.21 for the meaning of "natural".

More generally, for a sequence of topological spaces $X_1, \ldots, X_n$, define

$$X_1 * X_2 * \cdots * X_n = (\cdots ((X_1 * X_2) * X_3) \cdots) * X_n.$$

The following is the property of join Milnor used.

Lemma 4.10.4. *For any nonempty topological space X, the $(n+1)$-fold join of X ks $(n-1)$-connected, i.e.*

$$\pi_i(\underbrace{X * \cdots * X}_{n+1}) = 0$$

for $i < n$.

One of the most popular ways to prove this fact is to show that the homology groups vanish up to the degree $n-1$ by using the Mayer-Vietoris sequence and then apply the Whitehead theorem to show the homotopy groups vanish. Unfortunately, we can not explain the proof here, since it take too many pages to prepare for such tools.

Milnor noticed that a space of high connectivity with G action can be obtained by taking iterated joins of a topological group G with itself and introduced the following construction.

Definition 4.10.5. For a topological group G and a nonnegative integer n, define

$$E_n^{\text{join}} G = \underbrace{G * \cdots * G}_{n+1}.$$

The group multiplication of G on each G defines a right action

$$\mu : E_n^{\text{join}} G \times G \longrightarrow E_n^{\text{join}} G$$

of G on $E_n^{\text{join}} G$. The quotient space is denoted by $B_n^{\text{join}} G = E_n^{\text{join}} G / G$ and the projection is denoted by $p_n : E_n^{\text{join}} G \to B_n^{\text{join}} G$.

Milnor modified topologies of $E_n^{\text{join}} G$ and $B_n^{\text{join}} G$ in an appropriate way and showed that the projection $p_n : E_n^{\text{join}} G \to B_n^{\text{join}} G$ is a principal G-bundle. By Lemma 4.10.4, $E_n^{\text{join}} G$ is $(n-1)$-connected, hence p_n is an n-universal. By taking the "limit" as $n \to \infty$, we obtain a universal G-bundle

$$p_\infty : E_\infty^{\text{join}} G \longrightarrow B_\infty^{\text{join}} G.$$

An important observation of Milgram is that Milnor's iterated join construction can be expressed by using simplices. Let us recall the definition of simplices, in order to understand Milgram's idea. Intuitively an n-simplex is an n-dimensional version of triangle. An n-dimensional extension of the right triangle is called the standard n-simplex.

Definition 4.10.6. For a nonnegative integer n, define

$$\Delta^n = \left\{ (t_0, t_1, \ldots, t_n) \in \mathbb{R}^{n+1} \,\middle|\, \sum_{i=0}^{n} t_i = 1, t_0 \geq 0, \ldots, t_n \geq 0 \right\}$$

as a subspace of $\mathbb{R}^{n+1}$. This is called the *standard n-simplex*.

Example 4.10.7. The standard 0-simplex $\Delta^0 = \{1\}$ is a single point and the standard 1-simplex

$$\Delta^1 = \left\{ (t_0, t_1 \in \mathbb{R}^2 \,\middle|\, t_0 + t_1 = 1, t_0 \geq 0, t_1 \geq 0 \right\} \cong [0, 1]$$

is homeomorphic to a closed interval. The standard 2-simplex is given by

$$\Delta^2 = \left\{ (t_0, t_1, t_2) \in \mathbb{R}^3 \,\middle|\, t_0 + t_1 + t_2 = 1, t_0 \geq 0, t_1 \geq 0, t_2 \geq 0 \right\}.$$

It is the region obtained from the plane $t_0 + t_1 + t_2 = 1$ by cutting it by the xy-plane, the yz-plane, and the zx-plane and is a right triangle.

The 3-simplex is the right tetrahedron in $\mathbb{R}^4$. $\qquad\square$

Remark 4.10.8. The $(n-1)$-simplex Δ^{n-1} is contained as a face in Δ^n. For example, the 2-simplex Δ^2 can be identified with four faces in Δ^3 defined by $t_0 = 0$, $t_1 = 0$, $t_2 = 0$, and $t_3 = 0$. More generally, the n-simplex Δ^n has $(n+1)$ faces defined by $t_0 = 0$, $t_1 = 0$, ..., $t_n = 0$. The union of all these faces is denoted by

$$\partial \Delta^n = \bigcup_{i=0}^{n} \{ (t_0, \ldots, t_n) \in \Delta^n \,|\, t_i = 0 \}$$

and is called the *boundary* of Δ^n.

Lemma 4.10.9. *For a sequence of topological spaces $X_0, X_1, \ldots, X_n$, we have a homeomorphism*

$$X_0 * X_1 * \cdots * X_n \cong (X_0 \times \cdots \times X_n \times \Delta^n)/\sim,$$

where the relation $\sim$ is the equivalence relation generated by

$$(x_0, \ldots, x_i, \ldots, x_n, t_0, \ldots, t_{i-1}, 0, t_{i+1}, \ldots, t_n) \sim$$
$$(x_0, \ldots, x_i', \ldots, x_n, t_0, \ldots, t_{i-1}, 0, t_{i+1}, \ldots, t_n).$$

Proof. Let us denote the right hand side by $J(X_0, \ldots, X_n)$. We construct a homeomorphism

$$\varphi_n : X_0 * \cdots * X_n \longrightarrow J(X_0, \ldots, X_n)$$

by induction on n. It is defined to be the identity map when $n = 0$.

Suppose φ_k is constructed. The join $X_0 * \cdots * X_k * X_{k+1}$ is a quotient of $(X_0 * \cdots * X_k) \times [0,1] \times X_{k+1}$. By using the homeomorphism φ_k, we regard it as a quotient of $J(X_0, \ldots, X_k) \times [0,1] \times X_{k+1}$. Name it can be regarded as a quotient of $X_0 \times \cdots \times X_k \times \Delta^k \times [0,1] \times X_{k+1}$. Define a map $\psi_{k+1} : J(X_0, \ldots, X_k) * X_{k+1} \to J(X_0, \ldots, X_{k+1})$ by

$$\psi_{k+1}([[x_0, \ldots, x_k, s_0, \ldots, s_k], t, x_{k+1}]) = [x_0, \ldots, x_{k+1}, ts_0, \ldots, ts_k, 1-t].$$

Then this is a homeomorphism. Now define φ_{k+1} to be the composition

$$X_0 * X_1 * \cdots * X_k * X_{k+1} \xrightarrow{\varphi_k * 1_{X_{k+1}}} J(X_0, \ldots, X_k) * X_{k+1} \xrightarrow{\psi_{k+1}} J(X_0, \ldots, X_{k+1}).$$

$\square$

Remark 4.10.10. Notice that the essential part of this proof is the identification $\Delta^k * \Delta^0 \cong \Delta^{k+1}$.

In particular, $\underbrace{X * \cdots * X}_{n+1}$ can be expressed as a quotient of $X^{n+1} \times \Delta^n$ under an equivalence relation involving faces of Δ^n. On the other hand, there is a well-known construction defined in terms of faces of simplices, i.e. the geometric realization of a simplicial complex. See Appendix A.3 for a summary. Comparing these two equivalence relations, we obtain the following construction.

Definition 4.10.11. Let n be a nonnegative integer or ∞. For a topological group G, define

$$E_n G = \left(\coprod_{k=0}^{n} G^{k+1} \times \Delta^k \right) \Big/ \sim,$$

where the relation $\sim$ is the equivalence relation generated by the following four relations: for $(g_1, g_2, \ldots, g_n, g_{n+1}; t_0, \ldots, t_n) \in G^{n+1} \times \Delta^n$,

(1) when $t_0 = 0$,

$$(g_1, g_2, \ldots, g_n, g_{n+1}; 0, t_1, \ldots, t_n) \sim$$
$$(g_2, g_3, \ldots, g_n, g_{n+1}; t_1, \ldots, t_n)$$

(2) when $t_k = 0$ for $1 \le k \le n-1$,

$$(g_1, \ldots, g_{n+1}; t_0, \ldots, t_{i-1}, 0, t_{i+1}, \ldots, t_n) \sim$$
$$(g_1, \ldots, g_{k-1}, g_k g_{k+1}, g_{k+2}, \ldots, g_n; t_0, \ldots, t_{k-1}, t_{k+1}, , \ldots, t_n)$$

(3) when $t_n = 0$,

$$(g_1, \ldots, g_{n-1}, g_n, g_{n+1}; t_0, \ldots, t_{n-1}, 0) \sim$$
$$(g_1, \ldots, g_{n-1}, g_n g_{n+1}; t_0, \ldots, t_{n-1})$$

(4) when $g_k = e$ for $1 \leq k \leq n$,

$$(g_1, \ldots, g_{k-1}, e, g_{k+1}, \ldots, g_{n+1}; t_0, \ldots, t_n) \sim$$
$$(g_1, \ldots, g_{k-1}, g_{k+1}, \ldots, g_{n+1}; t_0, \ldots, t_{k-2}, t_{k-1} + t_k, t_{k+1}, \ldots, t_n).$$

Define a right action $\mu : E_n G \times G \to E_n G$ of G on $E_n G$ by

$$\mu([g_1, \ldots, g_n, g_{n+1}; t_0, \ldots, t_n], g) = [g_1, \ldots, g_n, g_n g; t_0, \ldots, t_n].$$

Finally define

$$B_n G = \left(\coprod_{k=0}^{n} G^k \times \Delta^k \right) \Big/ _{\sim},$$

where $\sim$ is the equivalence relation generated by

(1) when $t_0 = 0$,

$$(g_1, g_2, \ldots, g_n; 0, t_1, \ldots, t_n) \sim (g_2, g_3, \ldots, g_n; t_1, \ldots, t_n)$$

(2) when $t_k = 0$ for $1 \leq k \leq n - 1$,

$$(g_1, \ldots, g_n; t_0, \ldots, t_{i-1}, 0, t_{i+1}, \ldots, t_n) \sim$$
$$(g_1, \ldots, g_{k-1}, g_k g_{k+1}, g_{k+2}, \ldots, g_n; t_0, \ldots, t_{k-1}, t_{k+1}, , \ldots, t_n)$$

(3) when $t_n = 0$,

$$(g_1, \ldots, g_{n-1}, g_n; t_0, \ldots, t_{n-1}, 0) \sim (g_1, \ldots, g_{n-1}; t_0, \ldots, t_{n-1})$$

(4) when $g_k = e$ for $1 \leq k \leq n$,

$$(g_1, \ldots, g_{k-1}, e, g_{k+1}, \ldots, g_n; t_0, \ldots, t_n) \sim$$
$$(g_1, \ldots, g_{k-1}, g_{k+1}, \ldots, g_n; t_0, \ldots, t_{k-2}, t_{k-1} + t_k, t_{k+1}, \ldots, t_n)$$

for $(g_1, g_2, \ldots, g_n; t_0, \ldots, t_n) \in G^n \times \Delta^n$.

Both $E_n G$ and $B_n G$ are topologized by the quotient topology.

Remark 4.10.12.   (1) Milgram defined $E_n G$ as a quotient of $\coprod_{k=0}^{n} G \times \Delta^k \times G^k$ and define the action of G on $E_n G$ from the left. Since we define the action of a structure group on the total space from the right, Milgram's definition is modified as above.

(2) Milgram used another family of "standard simplices" defined by

$$\Delta^n = \{(t_1, t_2, \ldots, t_n) \in \mathbb{R}^n \mid 0 \le t_1 \le \cdots \le t_n \le 1\}.$$

The equivalence relation in Definition 4.10.11 is the corresponding relation when we use the standard simplices in Definition 4.10.6.

We need to impose a condition on a neighborhood of the unit in G to prove that $p_n : E_n G \to B_n G$ is a principal G-bundle.

Definition 4.10.13. For a topological space X and a closed subspace A, we say the pair (X, A) is an *NDR pair* (Neighborhood Deformation Retract pair) if there exist continuous maps

$$u : X \longrightarrow [0,1]$$
$$h : X \times [0,1] \longrightarrow X$$

satisfying the following conditions:

(1) $u^{-1}(0) = A$.
(2) Denote $U = u^{-1}([0,1))$. Then $h : X \times [0,1] \to X$ is a deformation retract of U onto A in X, namely

    (a) for any $a \in A$, $h(a,t) = a$,
    (b) $h_{X \times \{0\}} = 1_X$, and
    (c) $h(U \times \{1\}) \subset A$.

The pair (h, u) is called an *NDR representation* of (X, A).

When (X, x_0) is an NDR pair for point $x_0 \in X$, the point x_0 is called a *nondegenerate base point*.

Remark 4.10.14. We refer the reader to §6 of Steenrod's exposition [Steenrod (1967)], the Appendix of May's monograph [May (1972)], and Chapter 6 of May's book [May (1999)].

There is a close relationship between NDR pairs and cofibrations, which will be investigated in §5.9.

An important example is the following.

Example 4.10.15. For a CW complex X its subcomplex A, the pair (X, A) is always an NDR pair as we will see in Corollary 5.9.10. In particular, the pair $(\Delta^n, \partial \Delta^n)$ is an NDR pair. $\qquad\qquad\square$

The following is one of basic properties of NDR pairs. A proof can be found in [Steenrod (1967)].

Lemma 4.10.16. *Let (X, A) and (Y, B) be NDR pairs. Then both $(X \times Y, A \times Y)$ and $(X \times Y, X \times B \cup A \times Y)$ are NDR pairs.*

Definition 4.10.17. For pairs of spaces (X, A) and (Y, B), denote

$$(X, A) \times (Y, B) = (X \times Y, X \times B \cup A \times Y)$$
$$(X, A) \times Y = (X \times Y, A \times Y).$$

With this definition, Lemma 4.10.16 says that the product of NDR pairs is again an NDR pair. This fact can be used to prove the following.

Proposition 4.10.18. *The pairs $(E_{n+1}G, E_nG)$ and $(B_{n+1}G, B_nG)$ are NDR pairs.*

Proof. Since any element in $E_{n+1}G$ can be represented by an element in $G^{n+2} \times \Delta^{n+1}$, we may regard $E_{n+1}G$ as a quotient of $G^{n+2} \times \Delta^{n+1}$. Then E_nG corresponds to the subspace

$$\left(\{e\} \times G^{n+1} \times \Delta^{n+1}\right) \cup \left(G \times \{e\} \times G^n \times \Delta^{n+1}\right) \cup \cdots$$
$$\cup \left(G^n \times \{e\} \times G \times \Delta^{n+1}\right) \cup \left(G^{n+2} \times \partial\Delta^{n+1}\right).$$

Let us denote this space by $\widetilde{E}_nG$. Then we have

$$(G^{n+2} \times \Delta^{n+1}, \widetilde{E}_nG) = \underbrace{(G, \{e\}) \times \cdots \times (G, \{e\})}_{n+1} \times G \times (\Delta^{n+1}, \partial\Delta^{n+1}).$$

This is an NDR pair by Lemma 4.10.16. The NDR representation of this pair is compatible with the defining relation of $E_{n+1}G$ and E_nG and induces an NDR representation of $(E_{n+1}G, E_nG)$.

It is also compatible with the action of G and the NDR representation of $(E_{n+1}G, E_nG)$ descends to that of $(B_{n+1}G, B_nG)$. $\square$

Theorem 4.10.19 (Milgram). *Let n be a nonnegative integer or ∞. For a topological group G whose unit is a nondegenerate base point, the projection*

$$p_n : E_nG \longrightarrow B_nG$$

is a principal G-bundle.

Remark 4.10.20. Many important classes of topological groups satisfy this condition. For example, if G has the discrete topology, the condition is obviously satisfied. When G is a Lie group, the condition is satisfied because G is a manifold.

Proof of Theorem 4.10.19. By induction on n.

When $n = 0$, we have

$$E_0 G = G$$
$$B_0 G = \{*\}$$

and p_0 is a trivial bundle over a single point.

Suppose we have shown that p_n is a principal G-bundle. In order to construct local trivializations of p_{n+1}, we need to find an appropriate open covering of $B_{n+1}G$. Since $B_n G$ is contained in $B_{n+1}G$ as a closed subset, the complement $B_{n+1}G \setminus B_n G$ is an open set. Define a map

$$\varphi_{n+1} : p_{n+1}^{-1}(B_{n+1}G \setminus B_n G) = E_{n+1}G \setminus E_n G \longrightarrow (B_{n+1}G \setminus B_n G) \times G \quad (4.18)$$

by

$$\varphi_{n+1}([g_1, \ldots, g_{n+1}, g_{n+2}; t_0, \ldots, t_{n+1}]) = ([g_1, \cdots, g_{n+1}; t_0, \cdots, t_{n+1}], g_{n+2}).$$

Then it is easy to verify that this is a homeomorphism. Thus it remains to find a family of open subsets $\{U_\alpha\}$ in $B_{n+1}G$ such that

$$\bigcup_\alpha U_\alpha \supset B_n G$$

and each restriction

$$p_{n+1} : p_{n+1}^{-1}(U_\alpha) \longrightarrow U_\alpha$$

can be trivialized.

By the inductive hypothesis, there exists an open covering

$$B_n G = \bigcup_\alpha V_\alpha$$

of $B_n G$ and trivializations

$$\psi_\alpha : p_n^{-1}(V_\alpha) \xrightarrow{\cong} V_\alpha \times G$$

of

$$p_n = p_{n+1}|_{E_n G} : E_n G \longrightarrow B_n G.$$

Although these subsets V_α are not open in $B_{n+1}G$, we may use the NDR representation of (G, e) to enlarge V_α to an open subset of $B_{n+1}G$.

By Proposition 4.10.18, there exist NDR representations $(\tilde{h}, \tilde{u})$ and (h, u) of $(B_{n+1}G, B_n G)$ and $(E_{n+1}G, E_n G)$, respectively, which make the following diagram commutative

$$
\begin{array}{ccc}
E_{n+1}G \times [0,1] & \xrightarrow{\tilde{h}} & E_{n+1}G \\
\downarrow{\scriptstyle p_{n+1}} & & \downarrow{\scriptstyle p_{n+1}} \\
B_{n+1}G \times [0,1] & \xrightarrow{h} & B_{n+1}G
\end{array}
\qquad
\begin{array}{ccc}
E_{n+1}G & \xrightarrow{\tilde{u}} & [0,1] \\
\downarrow{\scriptstyle p_{n+1}} & & \| \\
B_{n+1}G & \xrightarrow{u} & [0,1]
\end{array}
$$

$$G \times E_{n+1}G \times [0,1] \xrightarrow{1_G \times \tilde{h}} G \times E_{n+1}G \qquad\qquad G \times E_{n+1}G \xrightarrow{\text{pr}_2} E_{n+1}G$$

$$\mu \times 1_{[0,1]} \Big\downarrow \qquad\qquad \Big\downarrow \mu \qquad\qquad \mu \Big\downarrow \qquad\qquad \Big\downarrow \tilde{u}$$

$$E_{n+1}G \times [0,1] \xrightarrow{\tilde{h}} E_{n+1}G. \qquad\qquad E_{n+1}G \xrightarrow{\tilde{u}} [0,1].$$

Define $r : B_{n+1}G \longrightarrow B_{n+1}G$ by

$$r(x) = h(x,1).$$

By the definition of NDR pairs, r is a deformation retract of $u^{-1}([0,1))$ onto B_nG and we obtain a map

$$r : u^{-1}([0,1)) \longrightarrow B_nG.$$

Define

$$U_\alpha = r^{-1}(V_\alpha).$$

Since $u^{-1}([0,1))$ is open in $B_{n+1}G$, U_α is an open subset of $B_{n+1}G$ and we obtain a family of open sets satisfying

$$\bigcup_\alpha U_\alpha \supset B_nG.$$

The next task is to find a local trivialization

$$\varphi_\alpha : p_{n+1}^{-1}(U_\alpha) \longrightarrow U_\alpha \times G$$

for each α. Define

$$\tilde{r} : E_{n+1}G \longrightarrow E_{n+1}G$$

by $\tilde{r}(x) = \tilde{h}(x,1)$. Then the diagram

$$p_{n+1}^{-1}(U_\alpha) \xrightarrow{\tilde{r}} p_n^{-1}(V_\alpha)$$

$$p_{n+1} \Big\downarrow \qquad\qquad \Big\downarrow p_n$$

$$U_\alpha \xrightarrow{r} V_\alpha$$

is commutative. Now define

$$\varphi_\alpha : p_{n+1}^{-1}(U_\alpha) \longrightarrow U_\alpha \times G$$

by

$$\varphi_\alpha(x) = (p_{n+1}(x), \text{pr}_2(\psi_\alpha(\tilde{r}(x)))).$$

Then the diagram

$$p_{n+1}^{-1}(U_\alpha) \xrightarrow{\ \varphi_\alpha\ } U_\alpha \times G$$

is commutative.

In order to define an inverse to φ_α

$$\varphi_\alpha^{-1} : U_\alpha \times G \longrightarrow p_{n+1}^{-1}(U_\alpha),$$

we first define

$$\gamma_\alpha : p_{n+1}^{-1}(U_\alpha) \times G \longrightarrow p_{n+1}^{-1}(U_\alpha)$$

by

$$\gamma_\alpha(x, g) = g(\mathrm{pr}_2 \circ \psi_\alpha \circ \tilde{r})(x)^{-1}x.$$

This is continuous and, for $g' \in G$,

$$\begin{aligned}
\gamma_\alpha(g'x, g) &= g(\mathrm{pr}_2 \circ \psi_\alpha \circ \tilde{r})(g'x)^{-1}g'x \\
&= g(g'(\mathrm{pr}_2 \circ \psi_\alpha \circ \tilde{r})(x))^{-1}g'x \\
&= g(\mathrm{pr}_2 \circ \psi_\alpha \circ \tilde{r})(x)^{-1}g'^{-1}g'x \\
&= g(\mathrm{pr}_2 \circ \psi_\alpha \circ \tilde{r})(x)^{-1}x \\
&= \gamma_\alpha(x, g).
\end{aligned}$$

Thus it induces a continuous map

$$\bar{\gamma}_\alpha : U_\alpha \times G \longrightarrow p_{n+1}^{-1}(U_\alpha).$$

It is left to the reader to verify that this in an inverse to φ_α.

The case of $n = \infty$ is analogous to the proof of Theorem 4.9.20 and is omitted. □

Now we have a principal G-bundle $p_\infty : E_\infty G \to B_\infty G$. The next step is to show that all the homotopy groups of the total space $E_\infty G$ vanish. The following fact is useful when we want to show the triviality of homotopy groups of filtered spaces such as $E_\infty G$.

Lemma 4.10.21. *Let X be a based Hausdorff space. Suppose it has a sequence of closed subspaces $\{X_n\}_{i=1,2,...}$ satisfying the following conditions:*

(1) X_1 contains the base point.
(2) $X_n \subset X_{n+1}$.

(3) $X = \bigcup_{n=1}^{\infty} X_n$.

(4) X has the weak topology with respect to the covering $\{X_n\}$.

(5) For each n, the inclusion $X_n \hookrightarrow X_{n+1}$ is homotopic to the constant map to the base point.

Then we have $\pi_i(X) = 0$ for all i.

Proof. An element $[f] \in \pi_i(X)$ is represented by a based continuous map $f : S^i \to X$. The first three conditions guarantee that Lemma 4.9.5 can be applied. Since S^i is compact, there exists an integer n such that $f(S^i) \subset X_n$. In other words, f factors as

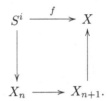

By the last condition, the inclusion $X_n \hookrightarrow X_{n+1}$ is homotopic to the constant map. Since all maps in the diagram preserve base points, f is homotopic to the constant map to the base point and hence $[f] = 0 \in \pi_i(X)$. $\square$

Exercise 4.10.22. Verify that the filtration $\{E_n G\}_{n=1,2,\dots}$ on $E_\infty G$ satisfy the first three conditions of Lemma 4.9.5.

By Lemma 4.10.21 and Exercise 4.10.22, in order to show that $E_\infty G$ is ∞-connected, it suffices to show that each inclusion map $E_n G \hookrightarrow E_{n+1} G$ is homotopic to the constant map. It is not difficult to find such homotopies.

Proposition 4.10.23. *For each n, the inclusion map $E_n G \hookrightarrow E_{n+1} G$ is homotopic to the constant map to the base point.*

Proof. As we have done in the proof of Proposition 4.10.18, any element of $E_n G$ can be represented by an element $(g_1, \dots, g_n, g_{n+1}; t_0, \dots, t_n) \in G^{n+1} \times \Delta^n$ and elements in $E_{n-1} G$ are represented by elements with $t_i = 0$ for some i or $g_j = e$ for some j. By the relation

$$(g_1, \dots, g_{j-1}, e, g_{j+1}, g_n, g_{n+1}; t_0, \dots, t_n)$$

$$\sim (g_1, \dots, g_{j-1}, g_{j+1}, g_n, g_{n+1}; t_0, \dots, t_{j-2}, t_{j-1} + t_j, t_{j+1}, \dots, t_n)$$

$$\sim (g_1, \dots, g_{j-1}, g_{j+1}, g_n, g_{n+1}, e; t_0, \dots, t_{j-2}, t_{j-1} + t_j, t_{j+1}, \dots, t_n, 0),$$

we see that elements of $E_{n-1}G$ are represented by elements of $G^{n+1} \times \partial \Delta^n$. Thus if we construct a homotopy which moves t_n to 0, we can deform E_nG into $E_{n-1}G$.

Such a homotopy $H : E_nG \times [0,1] \to E_{n+1}G$ can be defined by

$$H([g_1, \ldots, g_n, g_{n+1}; t_0, \ldots, t_n], s) = [g_1, \ldots, g_{n+1}, e; (1-s)t_0, \ldots, (1-s)t_n, s].$$

Since

$$
\begin{aligned}
H([g_1, \ldots, g_n, g_{n+1}; t_0, \ldots, t_n], 0) &= [g_1, \ldots, g_n, g_{n+1}, e; t_0, \ldots, t_n, 0] \\
&= [g_1, \ldots, g_n, g_{n+1}e; t_0, \ldots, t_n] \\
&= [g_1, \ldots, g_n, g_{n+1}; t_0, \ldots, t_n] \\
H([g_1, \ldots, g_n, g_{n+1}; t_0, \ldots, t_n], 1) &= [g_1, \ldots, g_n, g_{n+1}; \underbrace{0, 0, \ldots, 0}_{n+1}, 1] \\
&= [g_2, \ldots, g_n, g_{n+1}; \underbrace{0, 0, \ldots, 0}_{n}, 1] \\
&\quad \vdots \\
&= [g_{n+1}, e; 0, 1] \\
&= [e, 1],
\end{aligned}
$$

H is a homotopy we wanted. $\qquad\square$

Remark 4.10.24. Note that if we try to move t_n to 0, when $t_0 = \cdots = t_{n-1} = 0$, the relation $t_0 + \cdots + t_n = 1$ in Δ^n cannot be satisfied. This is the reason we need to use $E_{n+1}G$ as the range of our homotopy instead of E_nG.

Actually, the space $E_\infty G$ is a contractible space. This stronger result is proved in Appendix A.3 for simplicial spaces. Note, however, that Lemma 4.10.21 itself provides us with an important technique for proving that homotopy groups vanish.

Corollary 4.10.25. *For a topological group G satisfying Theorem 4.10.19, $p_\infty : E_\infty G \to B_\infty G$ is a universal G-bundle.*

We choose Milgram's construction as our standard construction of universal bundles and denote $EG = E_\infty G$ and $BG = B_\infty G$. The projection is denoted by $p_G = p_\infty$.

Let us verify that Milgram's construction satisfies the desiderata stated at the beginning of this section.

Theorem 4.10.26. *For any continuous homomorphism $f : G \to H$ of topological groups, there exist map Ef and Bf making the following diagram commutative*

$$
\begin{array}{ccc}
EG & \xrightarrow{\ Ef\ } & EH \\
\Big\downarrow{\scriptstyle p_G} & & \Big\downarrow{\scriptstyle p_H} \\
BG & \xrightarrow[\ Bf\ ]{} & BH.
\end{array}
$$

For another homomorphism $H \to K$ of topological groups, we have $E(g \circ f) = Eg \circ Ef$ and $B(g \circ f) = Bg \circ Bf$. For the identity map $1_G : G \to G$, we have $E(1_G) = 1_{EG}$ and $B(1_G) = 1_{BG}$.

Proof. Since $EG = \bigcup_{n=0}^{\infty} E_n G$, it suffices to construct a map $E_n f : E_n G \to E_n H$ for each n that are compatible with inclusions. For example, such a map is defined by

$$E_n f([g_1, \ldots, g_n, g_{n+1}; t_0, \ldots, t_n]) = [f(g_1), \ldots, f(g_n), f(g_{n+1}); t_0, \ldots, t_n]$$

for $[g_1, \ldots, g_n, g_{n+1}; t_0, \ldots, t_n] \in E_n G$. □

Remark 4.10.27. In other words, the pair of E and B define functors from the category of topological groups and continuous homomorphisms to the category of fiber bundles and fiber-preserving maps.

Theorem 4.10.28. *For topological groups G and H, we have a natural homomorphism*

$$B(G \times H) \cong BG \times BH.$$

This is a consequence of a general properties of the geometric realization of simplicial spaces. A proof is given in Appendix A.3 after a summary of basic properties of simplicial spaces.

Recall from Definition 4.5.23 that the space BG is called the classifying space of G. These two properties imply the following important property of classifying spaces.

Corollary 4.10.29. *When G is a commutative topological group, the classifying space BG has a structure of commutative topological group again.*

Proof. Since G is commutative, the group multiplication $\mu : G \times G \to G$ is a homomorphism. By Theorem 4.10.26, we obtain a continuous map

$B\mu : B(G \times G) \to BG$. Composed with the homeomorphism in Theorem 4.10.28, we obtain a map

$$BG \times BG \cong B(G \times G) \xrightarrow{B\mu} BG.$$

This multiplication is associative, since the homeomorphism in Theorem 4.10.28 is associative. The unique element in B_0G is the unit. Furthermore, since G is commutative, the map $\nu : G \to G$ which maps $g \in G$ to g^{-1} is a homeomorphism and defines a continuous map $B\nu : BG \to BG$, which define the inverse in BG. Thus BG becomes a topological group. The commutativity of μ implies that of $B\mu$. □

A remarkable fact is that we may iterate taking the classifying space construction and obtain a sequence of commutative topological groups $\{B^nG\}_n$.

Definition 4.10.30. Let G be a commutative group equipped with the discrete topology. For a nonnegative integer n, we denote

$$K(G, n) = B^nG.$$

This is called the *Eilenberg-Mac Lane space* of type (G, n).

We have seen two constructions of universal bundles in §4.9 and in this section. Although Steenrod's construction in §4.9 is limited to closed subgroups of the orthogonal groups, it has an advantage that the total space and the base space can be approximated by smooth manifolds.[6] Such an approach are useful when we study smooth manifolds. In fact, we may use differential forms on the Grassmannian manifolds to represent important characteristic classes of vector bundles and related bundles.

On the other hand, Milgram's construction provides an elementary and functorial construction of Eilenberg-Mac Lane spaces, which played an important role, for example, in the work of Ravenel and Wilson [Ravenel and Wilson (1980)].

4.11 Covering Spaces Revisited

We started a real discussion on fiber bundles in Chapter 2, where covering spaces are studied as a toy model of fiber bundles. In fact, as is discussed after Definition 2.1.1, covering spaces can be regarded as fiber bundles.

---

[6]Smooth structures on Milgram's universal bundles are studied by Gajer in [Gajer (1997)].

Proposition 4.11.1. *A continuous map $p : E \to B$ is a covering space in the sense of Definition 2.1.1 if and only if it is a fiber bundle whose fiber has the discrete topology.*

Based on this viewpoint, we apply theorems on fiber bundles proved in Chapter 3 and Chapter 4 to covering spaces and obtain properties of covering spaces in this section.

We first need to know structure groups of covering spaces.

Proposition 4.11.2. *Let $p : E \to B$ be a covering space over an arcwise connected, locally arcwise connected, and locally simply connected space. Then p is a fiber bundle with structure group $\pi_1(B)$, where the action of $\pi_1(B)$ on fiber is given by the map (2.1).*

Proof. Although local trivializations are already included in the definition of covering space, we need a description of it in terms of lifts of paths in order to relate structure group and $\pi_1(B)$.

For $x \in B$, let U_x be an open neighborhood of x which is simply connected. By Theorem 2.3.7, we obtain an isomorphism of covering spaces

$$\varphi_x : p^{-1}(U_x) \longrightarrow U_x \times p^{-1}(x).$$

By the proof of the theorem, this map is given as follows: For $e \in p^{-1}(U_x)$, let ℓ be a path in U_x from $p(e)$ to x. Let $\tilde{\ell}$ be the lift of of ℓ starting from e. Then φ_x is defined by $\varphi_x(e) = (p(e), \tilde{\ell}(1))$. We use this map as a local trivialization around x.

In order to describe coordinate transformations, take $x' \in B$ and its simply-connected open neighborhood $U_{x'}$ and suppose that $U_x \cap U_{x'} \neq \emptyset$. Then the composition

$$(U_x \cap U_{x'}) \times p^{-1}(x) \xrightarrow{\varphi_x^{-1}} p^{-1}(U_x \cap U_{x'}) \xrightarrow{\varphi_{x'}} (U_x \cap U_{x'}) \times p^{-1}(x')$$

is given by the following procedure: For $(a, y) \in (U_x \cap U_{x'}) \times p^{-1}(x)$,

(1) take a path ℓ in U_x from x to a,
(2) take a path m in $U_{x'}$ from a to x',
(3) let $\widetilde{\ell * m}$ the lift of $\ell * m$ with initial point y,
(4) and then assign the end point of $\widetilde{\ell * m}$.

Take a path from x' to x and compose with the induced bijection $p^{-1}(x') \cong p^{-1}(x)$ to obtain a map

$$(U_x \cap U_{x'}) \times p^{-1}(x) \xrightarrow{\cong} (U_x \cap U_{x'}) \times p^{-1}(x)$$

which is given by the action of $\pi_1(B, x)$. $\qquad\square$

One of the most fundamental properties of covering spaces is Theorem 2.2.3. This is an immediate consequence of the covering homotopy theorem (Corollary 4.3.16). In fact, let $p : E \to B$ be a covering space. Take X to be a single point $*$, $H : X \times [0,1] \to B$ to be a path ℓ under the identification $X \times [0,1] \cong [0,1] \cong [a,b]$. Define maps $f : X \to B$ and $\tilde{f} : X \to E$ by $f(*) = \ell(a)$ and $\tilde{f}(*) = e$, respectively. By Corollary 4.3.16, we obtain a map $\tilde{H} : X \times [0,1] \to E$ which is a lift H of H satisfying $\tilde{H}(*,0) = \tilde{f}(*) = e$. And we obtain a lift of ℓ in Theorem 2.2.3. Since fibers have discrete topology, the uniqueness of lift follows.

Let us consider homotopy groups. By applying Corollary 4.8.12 to a covering space $p : E \to B$, we obtain an exact sequence

$$\cdots \longrightarrow \pi_n(F) \xrightarrow{i_*} \pi_n(E) \xrightarrow{p_*} \pi_n(B) \xrightarrow{\partial} \pi_{n-1}(F) \longrightarrow \cdots \longrightarrow \pi_1(B).$$

Since F has the discrete topology, $\pi_n(F) = 0$ for $n \geq 1$.

Corollary 4.11.3. *Let $p : E \to B$ be a covering space. For $x_0 \in B$ and $e_0 \in E$ with $p(e_0) = x_0$, the induced map $p_* : \pi_n(E, e_0) \to \pi_n(B, x_0)$ is an isomorphism for $n \geq 2$. When $n = 1$, $p_* : \pi_1(E, e_0) \to \pi_1(B, x_0)$ is injective.*

Now suppose that $p : E \to B$ is a principal G-bundle for a group G and E is arcwise connected. Then by Exercise 4.8.16, we obtain an exact sequence

$$1 \longrightarrow \pi_1(E, e_0) \xrightarrow{p_*} \pi_1(B, x_0) \longrightarrow G \longrightarrow 1.$$

Here we used 1 instead of 0 for the trivial group, since the fundamental groups may not be commutative. This is equivalent to saying that $\pi_1(E, e_0)$ can be regarded as a normal subgroup of $\pi_1(B, x_0)$ under the map p_*, and that the quotient group is isomorphic to G. Then when is a covering space a principal bundle? As we have seen in §3.7, the group G should act on the total space E. On the other hand, we have the notion of covering transformation groups. Recall from Definition 2.3.5 that an isomorphism $f : E \to E$ of covering spaces is called a covering transformation of p.

Definition 4.11.4. For a covering space $p : E \to B$, define

$$\mathrm{Aut}_B(E) = \{ f : E \to E \,|\, \text{covering transformation} \} .$$

This is called the *covering transformation group* of p.

Covering transformations are compatible with the action of the fundamental group.

Lemma 4.11.5. *Let* $p : E \to B$ *be a covering space and* $f \in \mathrm{Aut}_B(E)$ *a covering transformation. Then for a path* $\ell : [0,1] \to B$ *in* B *and* $e \in E$, *we have*

$$\varphi_\ell(f(e)) = f(\varphi_\ell(e)).$$

In particular, f *commutes with the action of the fundamental group of* B *given by (2.1).*

We have the following important corollary.

Corollary 4.11.6. *Let* $p : E \to B$ *be a covering space with both* B *and* E *arcwise connected and* f, g *covering transformations of* p. *If there exists* $e \in E$ *with* $f(e) = g(e)$, *then* $f = g$.

Proof. Let f be a covering transformation of p. For $e' \in E$, take a path $\tilde{\ell}$ in E from e to e' by the arcwise connectivity of E and define $\ell = p \circ \tilde{\ell}$. Then

$$f(e') = f(\varphi_\ell(e)) = \varphi_\ell(f(e)),$$

which means that $f(e')$ is determined by $f(e)$. In other words, f is determined by the value at a single point. □

Corollary 4.11.7. *Let* $p : E \to B$ *be a covering space with* E *arcwise connected. For* $x_0 \in B$, *the map* $\mathrm{Aut}_B(E) \to \mathrm{Aut}(p^{-1}(x_0))$ *obtained by restricting covering transformations to the fiber* $p^{-1}(x_0)$ *over* x_0 *is an isomorphism of groups.*

Proposition 4.11.8. *Let* $p : E \to B$ *be a covering space. Suppose that* B *is arcwise connected, locally arcwise connected, and locally simply connected and that* E *is arcwise connected. Let* $e_0 \in E$ *and* $x_0 \in B$ *satisfy* $p(e_0) = x_0$. *If* $p_*(\pi_1(E, e_0))$ *is a normal subgroup of* $\pi_1(B, x_0)$, *then we have an isomorphism* $\mathrm{Aut}_B(E) \cong \pi_1(B, x_0)/p_*(\pi_1(E, e_0))$ *and that* p *is a principal* $\mathrm{Aut}_B(E)$-*bundle.*

Proof. By Proposition 4.11.2, p is a fiber bundle with structure group $\pi_1(B, x_0)$. The result follows once the following two facts are proved:

(1) The action of $\pi_1(B, x_0)$ on the fiber $p^{-1}(x_0)$ is transitive and its isotropy subgroup[7] is $p_*(\pi_1(E, e_0))$.

(2) $\mathrm{Aut}_B(E) \cong \pi_1(B, x_0)/p_*(\pi_1(E, e_0))$.

---
[7]Definition 3.6.37.

The transitivity of the action of $\pi_1(B, x_0)$ on $p^{-1}(x_0)$ follows from the assumption that E is arcwise connected. In order to determine the isotropy subgroup, suppose $\varphi_\ell(e_0) = e_0$ for $[\ell] \in \pi_1(B, x_0)$. Then we can take a loop based on e_0 as a lift of ℓ starting from e_0. Hence $[\ell] \in p_*(\pi_1(E, e_0))$. Thus the isotropy subgroup is $\pi_*(E, e_0)$ under the identification by p_* and we obtain an identification of the fiber over x_0 with $\pi_1(B, x_0)/p_*(\pi_1(E, e_0))$ including the action of $\pi_1(B, x_0)$. See Remark 3.6.36. This means that, if $p_*(\pi_1(E, e_0))$ is a normal subgroup of $\pi_1(B, x_0)$, then p is a principal bundle.

Let

$$\varphi : \pi_1(B, x_0) \longrightarrow \mathrm{Aut}(p^{-1}(x_0)) \cong \mathrm{Aut}_B(E)$$

be the homomorphism obtained by the action of $\pi_1(B, x_0)$. This is surjective, since, for $f \in \mathrm{Aut}_B(E)$, choose a path m in E connecting e_0 and $f(e_0)$ and define $\ell = p \circ m$, then $\varphi_\ell(e_0) = f(e_0)$ and by Corollary 4.11.6, we have $f = \varphi_\ell = \varphi([\ell])$. It is also easy to see that $\mathrm{Ker}\,\varphi = p_*(\pi_1(E, e_0))$ and we obtain an isomorphism $\mathrm{Aut}_B(E) \cong \pi_1(B, x_0)/p_*(\pi_1(E, e_0))$. $\square$

Let us compare universal bundles and universal coverings next. In the definition of universal covering, we only required that $\pi_0(E)$ and $\pi_1(E)$ are trivial, while all $\pi_n(E)$ have to be trivial in the case of universal bundle. Because of this strong requirement, universal G-bundles exist only on spaces of the form BG. On the other hand, Theorem 2.4.9 states that we can always construct a universal covering over a space B as long as B satisfies a set of reasonable conditions.

Proof of Theorem 2.4.9. Fix $x_0 \in B$ and consider the set

$$P(B, x_0) = \{\ell : [0, 1] \to B \mid \ell(0) = x_0\}.$$

As we have done in the definition of the fundamental group, define as set $\widetilde{B}$ by

$$\widetilde{B} = P(B, x_0)/_{\simeq},$$

where $\simeq$ is the based homotopy relation. The map which sends a path to its end point is denoted by $p : \widetilde{B} \to B$. Define a topology on $\widetilde{B}$ as follows. By assumption, there exists an arcwise connected open subset U of B satisfying the condition of semilocally-simply-connectivity. Let $\mathcal{U}$ be the collection of such open sets. For $[\ell] \in \widetilde{B}$ and $U \in \mathcal{U}$ with $\ell(1) \in U$, define

$$U_{[\ell]} = \{[\ell * m] \mid m : [0, 1] \to U, \ell(1) = m(0)\}.$$

Use

$$\{U_{[\ell]} \mid U \in \mathcal{U}, \ell \in P(B, x_0), \ell(1) \in U\} \tag{4.19}$$

as an open base to define a topology on $\widetilde{B}$.

We claim that $p : \widetilde{B} \to B$ is a universal covering over B. We need to verify the following:

(1) p is a covering space.
(2) $\widetilde{B}$ is simply connected.

In order to show that p is a covering space, take $U \in \mathcal{U}$ with $\ell(1) \in U$ for $[\ell] \in \widetilde{B}$. Then the restriction $p : U_{[\ell]} \to U$ of p is a homeomorphism, since it is a bijective open continuous map. The fact that p is open and continuous follows from the definition. The arcwise connectivity of U implies that surjectivity. The injectivity follows from the fact that any loop in U is homotopic to the trivial loop in X.

If $U_{[\ell]} \cap V_{[\ell']} \neq \emptyset$, take $[m] \in U_{[\ell]} \cap V_{[\ell']}$. Then there exist paths n, n' with

$$\ell * n \simeq m \simeq \ell' * n'.$$

Since

$$\ell \simeq \ell' * n' * \nu(n),$$

$[\ell] \in V_{[\ell']}$. Similarly we have $[\ell'] \in U_{[\ell]}$ and thus $U_{[\ell]} = V_{[\ell']}$. This completes the proof of the fact that $p : \widetilde{B} \to B$ is a covering space.

The space $\widetilde{B}$ is arcwise connected, since any $[\ell] \in \widetilde{B}$ can be connected to the constant path at c_{x_0} by shrinking ℓ with x_0 fixed. It remains to show that the fundamental group of $\widetilde{B}$ based at $[c_{x_0}]$ is trivial. Let $\ell : [0, 1] \to \widetilde{B}$ be a loop based at $[c_{x_0}]$. Then $p \circ \ell$ is a loop in B based at x_0. Define

$$L(t)(s) = \begin{cases} (p \circ \ell)(s), & 0 \leq s \leq t \\ (p \circ \ell)(t), & t \leq s \leq 1. \end{cases}$$

Then $L(t) \in P(X, x_0)$ for any t. Define $\widetilde{\ell}(t) = [L(t)]$. Then we obtain a path $\widetilde{\ell} : [0, 1] \to \widetilde{B}$ on $\widetilde{B}$. We have

$$(p \circ \widetilde{\ell})(s) = \widetilde{\ell}(1)(s) = L(1, s) = (p \circ \ell)(s)$$

and thus $\widetilde{\ell}$ is a lift of $p \circ \ell$. By the uniqueness of lifts, we have $\ell = \widetilde{\ell}$. Therefore

$$[c_{x_0}] = \ell(1) = \widetilde{\ell}(1) = [L(1)] = [p \circ \ell]$$

and $p \circ \ell$ is the trivial loop in B. By the existence of lifts of homotopies, ℓ is homotopic to the trivial loop in $\widetilde{B}$. $\square$

Exercise 4.11.9. Verify that the family of sets $\widetilde{B}$ (4.19) used in the above proof satisfies the condition for an open base. In other words, for any $U_{[\ell]}$, $V_{[m]}$ in (4.19) and $x \in U_{[\ell]} \cap V_{[m]}$, show that there exist $W \in \mathcal{U}$ and ω with $x \in W_{[\omega]} \subset U_{[\ell]} \cap V_{[m]}$.

The construction of $\widetilde{B}$ allows us to describe the action of the fundamental group explicitly. The proof is omitted.

Proposition 4.11.10. *Let B be the space satisfying the assumption of Theorem 2.4.9. Define a map*

$$\mu_B : \widetilde{B} \times \pi_1(B, x_0) \longrightarrow \widetilde{B}$$

by $\mu_B([\ell], [\omega]) = [\ell * \omega]$ for $[\ell] \in \widetilde{B}$ and $[\omega] \in \pi_1(B, x_0)$. Then this is a right action of $\pi_1(B, x_0)$.

In the case of fiber bundles, given an action of G on F, any fiber bundle with fiber F and structure group G can be obtained from a principal G-bundle as the associated bundle. In the case of covering spaces, we have asked the existence of the corresponding construction as Problem 2.4.5.

Proposition 4.11.11. *Suppose B has a universal covering $p : \widetilde{B} \to B$. Given a set F regarded as a topological space with discrete topology and a group homomorphism $f : \pi_1(B, x_0) \to \mathrm{Aut}(F)$, define a right action of $\pi_1(B, x_0)$ on $\widetilde{B} \times F$ by*

$$(e, y) \cdot [\ell] = (\mu_B(e, [\ell]), \varphi^{\mathrm{op}}([\ell], y)).$$

Define

$$E_f = E \times F \big/ \pi_1(B, x_0)$$

and $p_f : E_f \to B$ by $p_f([e, y]) = p(x)$. Then this is a covering space with fiber F and the action of $\pi_1(B, x_0)$ on the fiber agrees with f.

Proof. Since $p : \widetilde{B} \to B$, is a universal covering, it satisfies the condition of Proposition 4.11.8. It is in particular a principal $\pi_1(B, x_0)$-bundle. The construction in this proposition is, in fact, the construction of the associated bundle with fiber F. □

Remark 4.11.12. In most textbooks on topology, the classification of covering spaces are stated in terms of subgroups of $\pi_1(B, x_0)$. This is analogous to the Galois theory. In fact, it was Grothendieck [Grothendieck (2003)] who notice this similarity and developed the theory of Galois categories which encompasses these two. We recommend Lestra's lecture note [Lenstra (2008)] for those who are interested in this topic.

Chapter 5

Fibrations

●

5.1 Why Further Generalizations of Fiber Bundles?

In this chapter, we study generalizations of fiber bundles. Why do we generalize fiber bundles? There are two reasons.

(1) The definition of fiber bundles, especially of fiber bundles with structure groups is quite complicated. Given a map $p : E \to B$, it is difficult to tell if this is a fiber bundle or not from the definition. It would be useful, if we could simplify the definition without loosing basic properties.

(2) There are maps that are not fiber bundles but behave analogously to fiber bundles. We would like to develop a theory including such maps.

What is, then, a generalization or an abstraction? In mathematics, we often use the term generalization in the following sense:

generalization or abstraction =
to extract most important (essential) properties.

We have already seen generalizations in this sense. For example,

(1) $\mathbb{Z}, \mathbb{Z}/p\mathbb{Z}, D_{2n}, \mathrm{GL}_n(\mathbb{R})$ and $O(n)$ have associative binary operations with unit and inverses. The notion of groups was introduced by extracting this property.

(2) When we study spaces such as $\mathbb{R}^n, D^n$ and S^n and maps between them, open sets play an essential role. The notion of topological space was introduced by extracting properties of open sets.

(3) The notion of fiber bundles is also obtained by extracting the local triviality from concrete examples such as the Möbius band, the

199

Hopf bundle $S^3 \to S^2$, and the projection $G \to G/H$ for a Lie group G and its closed subgroup H.

In order to find a further generalization of fiber bundles, we need to find an important and essential property of fiber bundles, which is the subject of §5.2.

5.2 Serre Fibrations and Hurewicz Fibrations

Let us recall important properties of fiber bundles proved in Chapter 4.

The main theorem in Chapter 4 is, of course, Theorem 4.5.21, whose proof consists of two parts. In the first half, which is §4.7, the homotopy invariance, i.e. Theorem 4.3.25, is essential, while the long exact sequence of homotopy groups, i.e. Corollary 4.8.12 is used in the construction of universal bundle in §4.9. And it is the covering homotopy theorem (Corollary 4.3.16) that plays an essential role the proofs of all of these theorems.

Definition 5.2.1. Let $p : E \to B$ be a continuous map and X a topological space. We say p has the *covering homotopy property* with respect to X if, given a commutative square

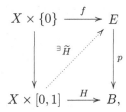

there exists a dotted map $\widetilde{H}$ which makes two triangles commutative.

We often abbreviate covering homotopy property as *CHP*.

Definition 5.2.2. A continuous map $p : E \to B$ is called a *Serre fibration* if p has CHP for any CW complex. It is called a *Hurewicz fibration* if it has CHP for any topological space.

In both cases, spaces B and E are called the *base space* and the *total space*. The map p is called the *projection*. For $x \in B$, $p^{-1}(x)$ is called the *fiber* over x.

Remark 5.2.3. Fibrations are called fiber spaces in the old literature such as [Cartan (1951); Serre (1951); Hurewicz (1955)]. We choose to use the term "fibration" since it is much more popular now. They are already called fibrations in the famous textbook by George Whitehead [Whitehead

(1978)]. It is also confusing to use the term "space" for a map. The term fibration is now quite popular because of the development of model categories and ∞-categories.

We have already seen an important example of Serre fibrations.

Corollary 5.2.4. *Any fiber bundle is a Serre fibration.*

Proof. By Corollary 4.3.16, any fiber bundle has CHP with respect to paracompact Hausdorff spaces and it is a well-known fact that CW complexes are paracompact Hausdorff. □

Remark 5.2.5. We refer the reader to the book [Lundell and Weingram (1969)] for a proof of the fact that a CW complex is paracompact.

Here is a simple example of a Hurewicz fibration, hence a Serre fibration, which is not a fiber bundle.

Example 5.2.6. Let $E = \left\{(x, y) \in \mathbb{R}^2 \,\middle|\, 0 \leq y \leq x \leq 1\right\}$ and $B = [0, 1]$. Define a map $p : E \to B$ by the projection as is shown in Figure 5.1.

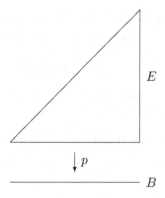

Fig. 5.1   a fibration which is not a fiber bundle

Let us verify that this is a Hurewicz fibration. Suppose we have a commutative diagram

$$X \times \{0\} \xrightarrow{\ f\ } E$$

$$X \times [0,1] \xrightarrow{\ H\ } B$$

for a topological space X. In order to construct a map

$$\widetilde{H} : X \times [0,1] \longrightarrow E,$$

write $f(x) = (f_1(x), f_2(x))$ for $x \in X$ and define

$$\widetilde{H}(x,t) = (H(x,t), \min\{H(x,t), f_2(x)\})$$

$(x,t) \in X \times [0,1]$. Then $\widetilde{H}$ is continuous and makes the above diagram commutative. And thus $p : E \to B$ is a Hurewicz.

This is not a fiber bundle, since the fibers on 0 and 1 are

$$p^{-1}(0) = \{(0,0)\}$$
$$p^{-1}(1) = \{1\} \times [0,1]$$

and not homeomorphic. □

The set of paths $P(X, x_0)$ used in the construction of universal covering in the proof of Theorem 2.4.9 also plays an important role when we construct fibrations. Here we regard it as a topological space.

Definition 5.2.7. For a based space (X, x_0), regard $P(X, x_0)$ as a topological space as a subspace of $\mathrm{Map}([0,1], X)$, which is topologized by the compact-open topology.[1] It is called the *path space* of (X, x_0). When the base point is obvious from the context, it is simply denoted by PX.

Proposition 5.2.8. *Define a map $p : PX \to X$ by $p(\ell) = \ell(1)$. Then this is continuous and is a Hurewicz fibration.*

Definition 5.2.9. This fibration is called the *path-loop fibration* on X.

In order to prove the continuity of p, we need the following corollary to Theorem A.1.11.

Corollary 5.2.10. *If X is locally compact Hausdorff, for any space Y and $x \in X$, the map*

$$\mathrm{ev}_x : \mathrm{Map}(X, Y) \longrightarrow Y$$

defined by $\mathrm{ev}_x(f) = f(x)$ is continuous.

---

[1]Definition 3.4.4.

Proof. Let $i_x : \{x\} \hookrightarrow X$ be the inclusion map. Then $\mathrm{ev}_x = \mathrm{ev} \circ i_x$ and ev is continuous by Theorem A.1.11. □

Proof of Proposition 5.2.8. The continuity of p follows from Corollary 5.2.10.

In order to show that p has CHP for any topological space Y, suppose a commutative diagram

$$
\begin{array}{ccc}
Y \times \{0\} & \xrightarrow{\ f\ } & PX \\
\Big\downarrow & & \Big\downarrow{\scriptstyle p} \\
Y \times [0,1] & \xrightarrow{\ H\ } & X
\end{array}
\tag{5.1}
$$

is given.

For each $y \in Y$, $f(y) \in PX$ is a path in X and the adjoint to $H(y, -)$

$$\mathrm{ad}(H)(y) : [0,1] \longrightarrow X$$

is also a path in X. By the commutativity of the diagram (5.1), we have

$$\mathrm{ad}(H)(y)(0) = H(y, 0) = p(f(y)) = f(y)(1).$$

Thus we may concatenate these two paths to obtain a path $(f(y) * \mathrm{ad}(H))(y)$. Recall that

$$
(f(y) * \mathrm{ad}(H)(y))(t) = \begin{cases} f(y)(2t) & 0 \le t \le \frac{1}{2} \\ H(y, 2t - 1) & \frac{1}{2} \le t \le 1. \end{cases}
$$

See Figure 5.2.

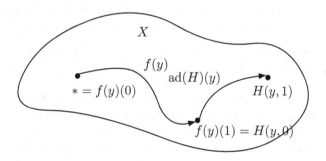

Fig. 5.2   concatenation of paths

The conditions that a lift
$$\widetilde{H} : Y \times [0,1] \longrightarrow PX$$
of H has to satisfy are
$$p(\widetilde{H}(y,t)) = H(y,t),$$
$$\widetilde{H}(y,0) = f(y).$$
In other words, $\widetilde{H}(y,t)$ is a path in X with end point $H(y,t)$ and that $\widetilde{H}(y,0) = f(y)$. Such a path can be constructed by concatenating $f(y)$ and the path obtained from $\mathrm{ad}(H)(y)$ by cutting it at time t. A precise definition is given by
$$\widetilde{H}(y,t)(s) = \begin{cases} f(y)(s(1+t)), & 0 \le s \le \frac{1}{1+t}, \\ H(y, s(1+t) - 1), & \frac{1}{1+t} \le s \le 1. \end{cases}$$
It is easy to verify that this map makes the diagram

$$
\begin{array}{ccc}
Y \times \{0\} & \xrightarrow{\ f\ } & PX \\
\Big\downarrow & \nearrow_{\widetilde{H}} & \Big\downarrow{\scriptstyle p} \\
Y \times [0,1] & \xrightarrow{\ H\ } & X
\end{array}
$$

commutative. $\square$

Remark 5.2.11. The path-loop fibration $p : PX \to X$ is not a fiber bundle in general, either. For example, the fiber over the base point is
$$p^{-1}(*) = \{\omega : [0,1] \to X \mid \omega(0) = \omega(1) = *\},$$
while, for other point x, the fiber is given by
$$p^{-1}(x) = \{\omega : [0,1] \to X \mid \omega(0) = *, \ \omega(1) = x\}.$$
As is shown in Figure 5.3, they look different.

Notice that the fiber $p^{-1}(*)$ agrees with the set $\Omega(X, x_0)$ of loops in Definition 2.3.1. Here we topologize it as a subspace of $P(X, x_0)$.

Definition 5.2.12. For a based space (X, x_0), equip $\Omega(X, x_0)$ with the compact-open topology. This is called the *loop space* of (X, x_0).

In the rest of this book, we do not distinguish Serre fibrations and Hurewicz fibrations and simply call them fibrations, as long as no confusion arises. One of the reasons is that most spaces used in topology are CW complexes. There are many exotic topological spaces but we may safely ignore them. Another reason is that many of naturally occurring fibrations such as the path-loop fibration are Hurewicz fibrations. Thus the reader may assume that fibrations are Hurewicz fibrations.

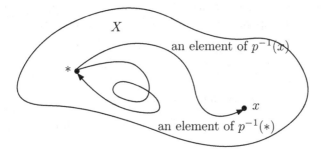

Fig. 5.3 difference of fibers

5.3 Loop Spaces

The path-loop fibration appeared in Proposition 5.2.8 is an important example of fibrations. We have also seen that the fiber over the base space is the loop space ΩX. It turns out that loop spaces and related spaces of paths play central roles in this chapter. In this section, we study basic properties of loop spaces.

Recall that, for a based space (X, x_0), the loop space is defined by

$$\Omega X = \{\ell : [0,1] \to X \,|\, \text{continuous}, \ell(0) = \ell(1) = x_0\}$$
$$= \{\ell : [0,1] \to X \,|\, \text{continuous}, \ell(\{0,1\}) = x_0\}$$
$$= \{\ell : [0,1] \to X \,|\, \text{continuous}, \ell(\partial[0,1]) = x_0\}.$$

As we have discussed in §4.6, this set can be identified with

$$\Omega X = \{\ell : [0,1]/\partial[0,1] \to X \,|\, \text{continuous}, \ell(*) = x_0\}$$
$$\cong \{\ell : S^1 \to X \,|\, \text{continuous}, \ell(*) = x_0\}.$$

See Example 4.6.4 and Lemma 4.6.20. Here $*$ is the base point of $[0,1]/\partial[0,1] \cong S^1$.

Thus the loop space of X is the space of all based continuous maps from S^1 to X. Recall from Definition 4.6.3 that, for based spaces (X, x_0) and (Y, y_0), the set of based maps from X to Y is denoted by

$$\text{Map}_*(X, Y) = \{f : X \to Y \,|\, \text{continuous}, f(x_0) = y_0\}.$$

Since $\text{Map}_*(X, Y)$ is a subset of $\text{Map}(X, Y)$, we topologize it by the relative topology.

With this notation, we may write

$$\Omega X = \text{Map}_*(S^1, X).$$

Recall also that, in §2.3, we defined the fundamental group $\pi_1(X)$ from $\Omega(X, x_0)$. The group operation was defined by the concatenation[2] of loops

$$\mu : \Omega X \times \Omega X \longrightarrow \Omega X.$$

This map is called the *loop product*.

Lemma 5.3.1. *The loop product μ is continuous.*

Proof. Let us first recall the compact-open topology on $\mathrm{Map}([0,1], X)$. We used

$$\{W(K, U) \mid K \subset [0, 1] : \text{compact}, U \subset X : \text{open}\}$$

as a subbasis and generated a topology on $\mathrm{Map}([0, 1], X)$, where

$$W(A, B) = \{\ell : [0, 1] \to X \mid \ell(A) \subset B\}$$

for $A \subset [0, 1]$ and $B \subset X$. Denote

$$\overline{W}(K, U) = W(K, U) \cap \Omega X.$$

Then

$$\left\{\overline{W}(K, U) \,\middle|\, K \subset [0, 1] : \text{compact}, U \subset X : \text{open}\right\}$$

is a subbasis of the topology on ΩX. Note that any compact subset K of $[0, 1]$ is a disjoint union of finite number of closed intervals

$$K = \coprod_{i=1}^{n} [t_i, t_i'].$$

Since

$$W\left(\coprod_{i=1}^{n} [t_i, t_i'], U\right) = \bigcap_{i=1}^{n} W([t_i, t_i'], U),$$

the topology of ΩX is generated by

$$\left\{\overline{W}([s, t], U) \,\middle|\, 0 \le s \le t \le 1, U \subset X : \text{open}\right\}.$$

Thus in order to prove the continuity of μ, it suffices to show that, for any open set $U \subset X$,

$$\mu^{-1}\left(\overline{W}([s, t], U)\right) = \{(\ell_1, \ell_2) \in \Omega X \times \Omega X \mid \mu(\ell_1, \ell_2)([s, t]) \subset U\}$$

is open in $\Omega X \times \Omega X$. Take (ℓ_1, ℓ_2) from $\mu^{-1}(\overline{W}([s, t], U))$. We would like to find an open neighborhood of this point in $\mu^{-1}(\overline{W}([s, t], U))$. The proof splits into three cases as is shown in Figure 5.4.

---

[2]Definition 2.3.10.

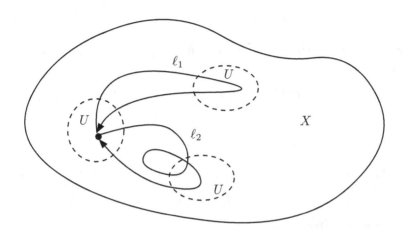

Fig. 5.4   continuity of loop product

(1) When $\frac{1}{2} \in [s, t]$, we have $\mu(\ell_1, \ell_2)([s, t]) \subset U$, which means that

$$\ell_1([2s, 1]) \subset U$$
$$\ell_2([0, 2t - 1]) \subset U.$$

Thus $\overline{W}([2s, 1], U) \times \overline{W}([0, 2t - 1], U)$ is an open neighborhood of (ℓ_1, ℓ_2) in $\mu^{-1}(\overline{W}([s, t], U))$.

(2) When $t < \frac{1}{2}$, by $\mu(\ell_1, \ell_2)([s, t]) \subset U$, we have

$$\ell_1([2s, 2t]) \subset U.$$

Thus $\overline{W}([2s, 2t], U) \times \Omega X$ is an open neighborhood of (ℓ_1, ℓ_2) in $\mu^{-1}(\overline{W}([s, t], U))$.

(3) When $\frac{1}{2} < s$, as is the case of $t < \frac{1}{2}$, $\Omega X \times \overline{W}([2s - 1, 2t - 1])$ is an open neighborhood of (ℓ_1, ℓ_2) in $\mu^{-1}(\overline{W}([s, t], U))$.

This completes the proof of the continuity of μ. □

Unfortunately, the loop product does not make ΩX into a topological monoid, since it is not associative. On the other hand, Proposition 2.3.12 says that it satisfies the condition obtained from the associativity by replacing $=$ by $\simeq$. Furthermore the homotopy can be taken to be independent of paths ℓ_1, ℓ_2, ℓ_3.

Proposition 5.3.2. *For the loop product μ on ΩX, we have*

$$\mu \circ (1_{\Omega X} \times \mu) \simeq \mu \circ (\mu \times 1_{\Omega X}).$$

Proof. Let H be the homotopy used in the proof of Proposition 2.3.12. Define a map

$$\widetilde{H} : \Omega X \times \Omega X \times \Omega X \times [0,1] \longrightarrow \Omega X$$

by

$$\widetilde{H}(\ell_1, \ell_2, \ell_3; s)(t) = H(\ell_1, \ell_2, \ell_3)(s, t).$$

It suffices to show that $\widetilde{H}$ is continuous. By Lemma 3.4.6, we show that

$$\mathrm{ad}^{-1}(\widetilde{H}) : \Omega X \times \Omega X \times \Omega X \times [0,1] \times [0,1] \longrightarrow X$$

is continuous. Define a map $h : [0,1] \times [0,1] \to [0,1]$ by

$$h(t,s) = \begin{cases} \frac{t}{s+1}, & 0 \le t \le \frac{s+1}{4} \\ t - \frac{s}{4}, & \frac{s+1}{4} \le t \le \frac{s+2}{4} \\ \frac{2t-s}{2-s}, & \frac{s+2}{4} \le t \le 1. \end{cases}$$

Then this is continuous and $\mathrm{ad}^{-1}(\widetilde{H})$ is given by the composition

$$\Omega X \times \Omega X \times \Omega X \times [0,1] \times [0,1] \xrightarrow{(\mu \circ (\mu \times 1_{\Omega X})) \times h} \Omega X \times [0,1] \xrightarrow{\mathrm{ev}} X$$

and is continuous. $\qquad\square$

Let us consider the inverse and the unit next. Define

$$\nu : \Omega X \longrightarrow \Omega X$$

by $\nu(\ell)(t) = \ell(1 - t)$ and define $e \in \Omega X$ by $e(t) = x_0$.

Proposition 5.3.3. *The map ν is continuous and we have*

$$\mu \circ (1_{\Omega X} \times \nu) \simeq c_{x_0} \tag{5.2}$$
$$\mu \circ (\nu \times 1_{\Omega X}) \simeq c_{x_0}$$
$$\mu \circ (1_{\Omega X} \times e) \simeq 1_{\Omega X}$$
$$\mu \circ (e \times 1_{\Omega X}) \simeq 1_{\Omega X}.$$

Proof. The continuity of ν is left to the reader, since it is much easier than the proof of μ in Lemma 5.3.1. We construct a homotopy for (5.2) only. Homotopies for the other three can be constructed analogously.

We use the homotopy in Proposition 2.3.16 for the proof of (5.2). Since this homotopy depends on ℓ, let us denote it by $H(\ell)$. Recall that it is defined by

$$H(\ell)(t, s) = \begin{cases} \ell(2ts), & 0 \leq t \leq \frac{1}{2} \\ \ell((2 - 2t)s), & \frac{1}{2} \leq t \leq 1. \end{cases}$$

Then H can be regarded as a map $H : \Omega X \to \mathrm{Map}([0, 1] \times [0, 1], X)$ and we obtain a map $\mathrm{ad}^{-1}(H) : \Omega X \times [0, 1] \times [0, 1] \to X$. Now define a map $\widetilde{H} : \Omega X \times [0, 1] \to \Omega X$ by

$$\widetilde{H}(\ell, s)(t) = H(\ell, t, s).$$

Then this is a homotopy for (5.2) if it is continuous.

In order to prove the continuity, define a map

$$k : \mathrm{Map}([0, 1], X) \times [0, 1] \to \mathrm{Map}([0, 1], X)$$

by

$$k(\ell, t)(s) = \ell(ts).$$

Then we have a commutative diagram

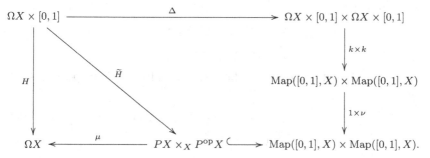

Here the space $P^{\mathrm{op}}X$ in the middle of the bottom row is defined by

$$P^{\mathrm{op}}X = \{\ell : [0, 1] \to X \mid \ell(1) = x_0\}$$

and

$$PX \times_X P^{\mathrm{op}}X = \{(\ell, \ell') \in PX \times P^{\mathrm{op}}X \mid \ell(1) = \ell'(0)\}.$$

And the map Δ in the top row is the diagonal map and the left map in the bottom row μ is given by the concatenation of paths. Since the image of the composition $(1 \times \nu) \circ (k \times k) \circ \Delta$ is contained in $PX \times_X P^{\mathrm{op}}X$, we have a continuous map $\widetilde{H}$ in the diagram. The proof of the continuity of the loop product can be used without a change to show that μ is continuous. Thus H is continuous. $\qquad\square$

Exercise 5.3.4. Prove the continuity of ν and construct the remaining three homotopies in the above proof.

Remark 5.3.5. Recall that we used commutative diagrams in the definition of groups in Definition 2.3.11 instead of equations such as $\mu \circ (1 \times \mu) = \mu \circ (\mu \times 1)$. In general, commutative diagrams are useful to visualize relations among maps. Diagrammatic descriptions are also useful for homotopy relations among continuous maps. For example, when

$$\mu \circ (1_{\Omega X} \times \mu) \simeq \mu \circ (\mu \times 1_{\Omega X})$$

we say the diagram

$$
\begin{array}{ccc}
\Omega X \times \Omega X \times \Omega X & \xrightarrow{\mu \times 1_{\Omega X}} & \Omega X \times \Omega X \\
\downarrow{\scriptstyle 1_{\Omega X} \times \mu} & & \downarrow{\scriptstyle \mu} \\
\Omega X \times \Omega X & \xrightarrow{\quad \mu \quad} & \Omega X
\end{array}
$$

is *homotopy commutative*.

Definition 5.3.6. Let X be a based space. When it is equipped with continuous maps

$$\mu : X \times X \longrightarrow X$$
$$\nu : X \longrightarrow X$$

satisfying the following five conditions, it is called a *Hopf space* or an *H-space*:

$$\mu \circ (1_X \times \mu) \simeq \mu \circ (\mu \times 1_X) \tag{5.3}$$
$$\mu \circ (1_X \times \nu) \simeq * \tag{5.4}$$
$$\mu \circ (\nu \times 1_X) \simeq * \tag{5.5}$$
$$\mu \circ (1_X \times *) \simeq 1_X \tag{5.6}$$
$$\mu \circ (* \times 1_X) \simeq 1_X. \tag{5.7}$$

Remark 5.3.7. In most books and papers written by experts of Hopf spaces, the conditions (5.4) and (5.5) are not included in the definition of Hopf spaces. Furthermore the condition (5.3) is often excluded, too. The condition (5.3) is called the *homotopy associativity* and is a key property in the study of Hopf spaces. There is a serious gap between (5.3) and the strict associativity as is described in the famous work of Stasheff [Stasheff (1963)]. One of the most famous spaces which do not satisfy (5.3) is the 7-dimensional S^7. The multiplications in the real numbers $\mathbb{R}$, the complex numbers $\mathbb{C}$, the quaternions $\mathbb{H}$, and the Cayley numbers $\mathbb{O}$ induce multiplications on their unit spheres S^0, S^1, S^3 and S^7, respectively. By the non-associativity of the multiplication in $\mathbb{O}$, the multiplication in S^7 is not associative. We want S^7 to be a friend of S^1 and S^3 and thus the condition (5.3) is not required.

By the way, it is a famous result of Adams [Adams (1960)] that the only spheres having a continuous multiplication with a unit up to homotopy are the above four.

The reason we use this non-standard terminology is that the corresponding algebraic structure is called *Hopf algebra* whose definition requires the associativity and the existence of inverse. When X is a Hopf space in our sense, the homology $H_*(X; k)$ of X with coefficients in a field k becomes a Hopf algebra. If we do not require (5.3), the multiplication of $H_*(X; k)$ may not be associative and $H_*(X; k)$ cannot be called a Hopf algebra.

Hopf algebras are now quite popular algebraic objects, playing important roles in many fields including mathematical physics. The author thinks it is better to follow the terminology in the theory of Hopf algebras. See Sweedler's book [Sweedler (1969)] or Abe's book [Abe (1980)] for more details on Hopf algebra. For Hopf algebras used in algebraic topology, the paper [Milnor and Moore (1965)] by Milnor and Moore is fundamental. Note, however, that the terminologies Milnor and Moore used are not popular nowadays. For example, they do not require the existence of a map, called an *antipode*, which corresponds to an inverse, and an antipode is called a canonical anti-homomorphism.

The reason that we do not need to require (5.4) and (5.5) is the following.

Theorem 5.3.8. *Let X be an arcwise connected CW complex. If a continuous map $\mu : X \times X \to X$ satisfies (5.3), (5.6), and (5.7), then there exists a continuous map $\nu : X \to X$ satisfying (5.4) and (5.5).*

With the terminology in Definition 5.3.6, ΩX is a Hopf space, which means that it has a structure analogous to a topological group. On the other hand, it is not easy to deal with maps if we allow continuous deformations. It would be nice if we could make ΩX into a topological group. Although it is not possible to make it a genuine topological group, there is a model introduced by J.C. Moore whose multiplication is strictly associative.

Definition 5.3.9. For a based space (X, x_0), define

$$\Omega_M(X, x_0) = \{(\ell, L) \in \mathrm{Map}([0, \infty), X) \times [0, \infty) \mid \ell(0) = x_0, \ell([L, \infty)) = x_0\}.$$

When the base point is obvious from the context, it is simply denoted by $\Omega_M X$. Define a multiplication

$$\mu_M : \Omega_M X \times \Omega_M X \longrightarrow \Omega_M X$$

by

$$\mu_M(\ell_1, L_1; \ell_2, L_2) = (\bar{\mu}(\ell_1, \ell_2), L_1 + L_2)$$

for $(\ell_1, L_1), (\ell_2, L_2) \in \Omega_M X$, where

$$\bar{\mu}(\ell_1, \ell_2)(t) = \begin{cases} \ell_1(t) & 0 \le t \le L_1 \\ \ell_2(t - L_1) & L_1 \le t. \end{cases}$$

This is called the *Moore loop space* of X.

Proposition 5.3.10. *The multiplication μ_M is associative. If e denote the constant loop at the base point, $(e, 0)$ is a unit. In other words, $\Omega_M X$ becomes a topological monoid.*[3]

Exercise 5.3.11. Prove this fact.

The following fact allows us to use the Moore loop space $\Omega_M X$ and the standard loop space ΩX interchangeably.

Proposition 5.3.12. *For a based space X, define a map $r : \Omega_M X \to \Omega X$ by $r(\ell, L)(t) = \ell(Lt)$. Then it preserves multiplications. Define another map $i : \Omega X \to \Omega_M X$ by $i(\ell) = (\ell, 1)$. Then we have*

$$r \circ i = 1_{\Omega X}$$

$$i \circ r \simeq 1_{\Omega_M X}.$$

Proof. It is easy to verify that r preserves multiplications, i.e. the diagram

$$
\begin{array}{ccc}
\Omega_M X \times \Omega_M X & \xrightarrow{\ \mu_M\ } & \Omega_M X \\
{\scriptstyle r \times r}\downarrow & & \downarrow{\scriptstyle r} \\
\Omega X \times \Omega X & \xrightarrow{\ \mu\ } & \Omega X
\end{array}
$$

is commutative. Since

$$r \circ i(\ell) = r(\ell, 1) = \ell = 1_{\Omega X}(\ell),$$

it remains to show that $i \circ r \simeq 1_{\Omega_M X}$. Such a homotopy $H : \Omega_M X \times [0, 1] \to \Omega_M X$ can be defined by

$$H(\ell, L; s) = \left(h(\ell, L; s), \frac{L}{\ell(1 - s) + s} \right)$$

$$h(\ell, L; s)(t) = \omega((\ell(1 - s) + s)t).$$

$\square$

---

[3]Definition 3.4.20.

If ΩX is a subspace of $\Omega_M X$ and i is the inclusion map, we can say that ΩX is a deformation retract[4] of $\Omega_M X$. More generally, we use the following terminology.

Definition 5.3.13. Topological spaces X and Y are said to be *homotopy equivalent* if there exist continuous maps $f : X \to Y$ and $g : Y \to X$ such that

$$g \circ f \simeq 1_X$$
$$f \circ g \simeq 1_Y.$$

In this case we denote $X \simeq Y$.

These maps f and g are called *homotopy equivalence* and g is called a *homotopy inverse* to f.

Exercise 5.3.14. Verify that $\sim$ is an equivalence relation.

With this terminology, Proposition 5.3.12 says that the map $r : \Omega_M X \to \Omega X$ is a homotopy equivalence which preserves multiplications.

5.4 Comparing Fiber Bundles and Fibrations

Let us go back to the discussion on fibrations. Since fibrations are introduced as a generalizations, a natural question is how far theorems on fiber bundles can be extended to fibrations. As we have discussed at the beginning of §5.2, fundamental properties of fiber bundles include

- the long exact sequence of homotopy groups,
- homotopy invariance, and
- the classification theorem.

In this section, we first study the homotopy invariance of fibrations. The definition of pullback[5] of fiber bundles can used without a change for fibrations. In fact, we have observed right after Definition 4.2.10 that pullbacks can be defined for any continuous maps. Let us recall the definition.

Definition 5.4.1. For continuous map $f : X \to B$ and $g : Y \to B$, define

$$X \times_B Y = \{(x, y) \in X \times Y \mid f(x) = g(y)\}$$

---

[4]Definition 4.3.21.
[5]Definition 4.2.1.

Fiber Bundles and Homotopy

as a subspace of $X \times Y$. This is called the *fiber product* of X and Y over B. And the diagram

$$
\begin{array}{ccc}
X \times_B Y & \xrightarrow{\; g^*(f) \;} & Y \\
{\scriptstyle f^*(g)} \big\downarrow & & \big\downarrow {\scriptstyle g} \\
X & \xrightarrow{\quad f \quad} & B
\end{array}
$$

is called a *pullback diagram*.

In particular, when g is a fibration, the left map $f^*(g)$ is called the *pullback* of g along f.

As is the case of fiber bundles, pullbacks of fibrations are always fibrations.

Proposition 5.4.2. *For any fibration $p : E \to B$ and any continuous map $f : X \to B$, the pullback*

$$
f^*(p) : f^*(E) \longrightarrow X
$$

is a fibration.

Proof. Following the definition of fibration, we show that $f^*(p)$ has CHP for any space. Suppose we have a commutative diagram

$$
\begin{array}{ccc}
Z \times \{0\} & \xrightarrow{\; h \;} & f^*(E) \\
\big\downarrow & & \big\downarrow {\scriptstyle f^*(p)} \\
Z \times [0,1] & \xrightarrow{\; H \;} & X.
\end{array}
$$

By attaching the pullback diagram to the left, we obtain a commutative diagram

$$
\begin{array}{ccccc}
Z \times \{0\} & \xrightarrow{\; h \;} & f^*(E) & \xrightarrow{\; \bar{f} \;} & E \\
\big\downarrow & & \big\downarrow {\scriptstyle f^*(p)} & & \big\downarrow {\scriptstyle p} \\
Z \times [0,1] & \xrightarrow{\; H \;} & X & \xrightarrow{\; f \;} & B.
\end{array}
$$

In the outside rectangle

$$
\begin{array}{ccc}
Z \times \{0\} & \xrightarrow{\; \bar{f} \circ h \;} & E \\
\big\downarrow & & \big\downarrow {\scriptstyle p} \\
Z \times [0,1] & \xrightarrow{\; f \circ H \;} & B,
\end{array}
$$

there exists a homotopy $G : Z \times [0,1] \to E$ satisfying

$$G|_{Z \times \{0\}} = \bar{f} \circ h$$
$$p \circ G = f \circ H,$$

since p is a fibration. Now define

$$\widetilde{H}(z,t) = (H(z,t), G(z,t))$$

for $(z,t) \in Z \times [0,1]$. $\widetilde{H}$ takes values in $f^*(E)$, since $p \circ G = f \circ H$. Thus we obtain a map

$$\widetilde{H} : Z \times [0,1] \longrightarrow f^*(E)$$

making the diagram

$$
\begin{array}{ccc}
Z \times \{0\} & \xrightarrow{\ h\ } & f^*(E) \\
\downarrow & \overset{\widetilde{H}}{\nearrow} & \downarrow{\scriptstyle f^*(p)} \\
Z \times [0,1] & \xrightarrow{\ H\ } & X
\end{array}
$$

commutative. □

Now we are ready to study the homotopy invariance of fibrations. Let $p : E \to B$ be a fibration and $f_0, f_1 : X \to B$ homotopic maps. We would like to know the relation between $f_0^*(E)$ and $f_1^*(E)$. When p is a fiber bundle, these two bundles are isomorphic, and hence $f_0^*(E)$ and $f_1^*(E)$ are homeomorphic as spaces. However, this is too much to expect for fibrations.

Example 5.4.3. Recall Example 5.2.6 (Figure 5.5). The total space and the base space are given by

$$E = \left\{ (x,y) \in \mathbb{R}^2 \,\middle|\, 0 \leq y \leq x \leq 1 \right\}$$
$$B = [0,1]$$

and the projection $p : E \to B$ is given $p(x,y) = x$.

Let X be a single point space $\{*\}$ and define

$$f_0 : X \longrightarrow B$$
$$f_1 : X \longrightarrow B$$

by $f_0(*) = 0$ and $f_1(*) = 1$. Then f_0 and f_1 are homotopic. For example, a homotopy $H : X \times [0,1] \to B$ is given by $H(*,t) = t$. On the other hand,

$$f_0^*(E) = \{(0,0)\}$$
$$f_1^*(E) = \{1\} \times [0,1]$$

and they are not homeomorphic. □

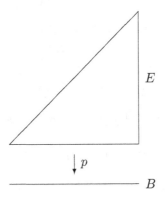

Fig. 5.5   the fibration in Example 5.2.6

The reader might suspect from this example that $f_0^*(E)$ and $f_1^*(E)$ are homotopy equivalent. In fact, they are related by a relation called fiber homotopy equivalence. In order to state and prove the relation, we first need a precise definition of maps between fibrations.

Definition 5.4.4. Let $p : E \to B$ and $p' : E' \to B'$ be fibrations.

(1) a *fiber-preserving map* from p to p' or a map of fibrations is a pair of maps

$$\tilde{f} : E \longrightarrow E'$$
$$f : B \longrightarrow B'$$

making the diagram

$$
\begin{array}{ccc}
E & \xrightarrow{\ \tilde{f}\ } & E' \\
\downarrow{\scriptstyle p} & & \downarrow{\scriptstyle p'} \\
B & \xrightarrow{\ f\ } & B'
\end{array}
$$

commutative.

(2) Let $(f, \tilde{f})$ and $(g, \tilde{g})$ be maps of fibrations from p to p'. A *fiber homotopy* or *fiberwise homotopy* from $(f, \tilde{f})$ to $(g, \tilde{g})$ is a pair $(H, \widetilde{H})$ of a homotopy $H : B \times [0, 1] \to B'$ from f to g and a homotopy

$\tilde{H} : E \times [0,1] \to E'$ from $\tilde{f}$ to $\tilde{g}$ making the diagram

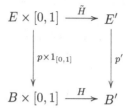

commutative.

(3) If such a fiber homotopy exists, we say $(f, \tilde{f})$ and $(g, \tilde{g})$ are *fiber homotopic* or *fiberwise homotopic*).

(4) When $B = B'$ and $f = g = 1_B$, if $(1_B, \tilde{f})$ and $(1_B, \tilde{g})$ are fiber homotopic by a fiber homotopy $(H, \tilde{H})$ with $H(x,t) = x$, then we denote

$$\tilde{f} \underset{B}{\simeq} \tilde{g}.$$

(5) Finally fibrations $p : E \to B$ and $p' : E \to B$ are said to be *fiber homotopy equivalent* or *fiberwise homotopy equivalent*, if there exist maps of fibrations

$$(1_B, f) : (B, E) \longrightarrow (B, E')$$
$$(1_B, g) : (B, E') \longrightarrow (B, E)$$

such that $g \circ f \underset{B}{\simeq} 1_E$ and $f \circ g \underset{B}{\simeq} 1_{E'}$. In this case we write $(B, E) \underset{B}{\simeq} (B, E')$ or $E \underset{B}{\simeq} E'$.

Remark 5.4.5. If $E \to B$ and $E' \to B$ are fiber homotopy equivalent $E \underset{B}{\simeq} E'$, then E and E' are homotopy equivalent $E \simeq E'$.

Remark 5.4.6. Notice that in the definition of fiber homotopy and fiber homotopy equivalence, we did not used the assumption that maps p and p' are fibrations.

Now we are ready to state the homotopy invariance of fibrations.

Theorem 5.4.7. *Let $p : E \to B$ be a Hurewicz fibration. If continuous maps $f, g : X \to B$ are homotopic $f \simeq g$, then the pullbacks along f and g are fiber homotopy equivalent*

$$f^*(E) \underset{X}{\simeq} g^*(E)$$

to each other.

Proof. An obvious idea is to lift a homotopy between f and g to obtain a fiber homotopy between $f^*(E)$ and $g^*(E)$ and this is in fact what we are going to do.

Let $H : X \times [0,1] \to B$ be a homotopy from f to g. Denote the inclusions of X to $X \times \{0\}$ and $X \times \{1\}$ by

$$i_0 : X \hookrightarrow X \times [0,1]$$
$$i_1 : X \hookrightarrow X \times [0,1],$$

respectively. Then

$$f = H \circ i_0$$
$$g = H \circ i_1$$

and we obtain a commutative diagram

$$
\begin{array}{ccc}
f^*(E) & \xrightarrow{\ \bar{f}\ } & E \\
\downarrow & & \downarrow{\scriptstyle p} \\
X & \xrightarrow{\ f\ } & B \\
\downarrow{\scriptstyle i_0} & & \| \\
X \times [0,1] & \xrightarrow{\ H\ } & B.
\end{array}
$$

We also have another commutative diagram

$$
\begin{array}{ccc}
f^*(E) & \xrightarrow{\ =\ } & f^*(E) \\
\downarrow & & \downarrow{\scriptstyle p} \\
{\scriptstyle i_0} & & X \\
& & \downarrow{\scriptstyle i_0} \\
f^*(E) \times [0,1] & \xrightarrow{p \times 1_{[0,1]}} & X \times [0,1].
\end{array}
$$

By combining these two diagrams, we obtain

$$
\begin{array}{ccc}
f^*(E) & \xrightarrow{\hspace{4cm}} & E \\
\downarrow{\scriptstyle i_0} & & \downarrow{\scriptstyle p} \\
f^*(E) \times [0,1] & \xrightarrow[p \times 1_{[0,1]}]{} X \times [0,1] \xrightarrow{\ H\ } & B.
\end{array}
$$

By CHP, we obtain a homotopy $\widetilde{H}$ making the diagram

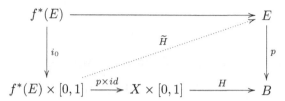

commutative. The bottom triangle can be rewritten as

$$
\begin{array}{ccc}
f^*(E) \times [0,1] & \xrightarrow{\widetilde{H}} & E \\
\downarrow{\scriptstyle p \times 1_{[0,1]}} & & \downarrow{\scriptstyle p} \\
X \times [0,1] & \xrightarrow{H} & B
\end{array}
\tag{5.8}
$$

which implies that $(H, \widetilde{H})$ is a map of fibrations.

We construct a fiber homotopy equivalence between $f^*(E)$ and $g^*(E)$ by using $\widetilde{H}$. Define a map $h : f^*(E) \to X \times E$ by

$$h(x, e) = (x, \widetilde{H}(x, e, 1)).$$

By the commutativity of (5.8), we have

$$p(\widetilde{H}(x, e, 1)) = H(x, 1) = g(x)$$

and thus $h(x, e) \in g^*(E)$. And we obtain a map of fibrations

$$
\begin{array}{ccc}
f^*(E) & \xrightarrow{h} & g^*(E) \\
\downarrow & & \downarrow \\
X & \xrightarrow{=} & X.
\end{array}
$$

Define $H'(x, t) = H(x, 1 - t)$ and use H' in place of H' to obtain a commutative diagram

$$
\begin{array}{ccc}
g^*(E) & \xrightarrow{h'} & f^*(E) \\
\downarrow & & \downarrow \\
X & \xrightarrow{=} & X,
\end{array}
$$

where $h'(x, e) = (x, \widetilde{H}'(x, e, 1))$ and $\widetilde{H}' : g^*(E) \times [0,1] \to E$ is a lift of H'.

It remains to show that

$$h \circ h' \underset{X}{\simeq} 1_{g^*(E)}$$

$$h' \circ h \underset{X}{\simeq} 1_{f^*(E)}.$$

These fiber homotopies can be obtained by CHP. Denote

$$K = \partial[0,1] \times [0,1] \cup [0,1] \times \{0\}$$

and define $F : f^*(E) \times K \to E$ by

$$F(x,e,0,t) = \widetilde{H}(x,e,1-t)$$

$$F(x,e,1,t) = \widetilde{H}'(h(x,e),t)$$

$$F(x,e,s,0) = \widetilde{H}(x,e,1)$$

for $x \in X, e \in E, s,t \in [0,1]$. Then we have

$$\widetilde{H}(x,e,1-0) = \widetilde{H}(x,e,1)$$

$$\widetilde{H}'(h(x,e),0) = \bar{f}(h(x,e))$$

$$= \widetilde{H}(x,e,1)$$

and F is well-defined and continuous. And the diagram

$$
\begin{array}{ccc}
f^*(E) \times K & \xrightarrow{\quad F \quad} & E \\
\downarrow & & \downarrow{\scriptstyle p} \\
f^*(E) \times [0,1] \times [0,1] & \xrightarrow{\quad G \quad} & B
\end{array}
$$

is also commutative, where

$$G(x,e,s,t) = H'(x,t) = H(x,1-t)$$

and the left vertical map is given by the inclusion of K into $[0,1] \times [0,1]$.

Furthermore we have a homeomorphism of pairs of spaces

$$([0,1] \times [0,1], K) \cong ([0,1] \times [0,1], [0,1] \times \{0\}) \tag{5.9}$$

as is shown in Figure 5.6 and we have a commutative diagram

$$
\begin{array}{ccccc}
f^*(E) \times [0,1] \times \{0\} & \xrightarrow{\quad\cong\quad} & f^*(E) \times K & \xrightarrow{\ F\ } & E \\
\downarrow & & \downarrow & & \downarrow{\scriptstyle p} \\
f^*(E) \times [0,1] \times [0,1] & \xrightarrow{\quad\cong\quad} & f^*(E) \times [0,1] \times [0,1] & \xrightarrow{\ G\ } & B.
\end{array}
$$

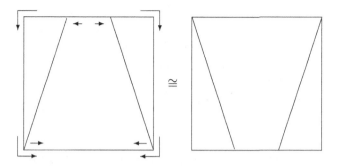

Fig. 5.6 a homeomorphism of pairs

The outside rectangle in the above diagram admits CHP. Thanks to the homeomorphism of pairs (5.9), we may apply CHP to the right square. Thus we obtain a homotopy $\widetilde{F}$ making the diagram

$$
\begin{array}{ccc}
f^*(E) \times K & \xrightarrow{\quad F \quad} & E \\
\downarrow & \overset{\widetilde{F}}{\nearrow} & \downarrow p \\
f^*(E) \times [0,1] \times [0,1] & \xrightarrow{\quad G \quad} & B
\end{array}
$$

commutative. By the bottom triangle, we obtain

$$
\begin{array}{ccc}
f^*(E) \times [0,1] \times [0,1] & \xrightarrow{\hspace{4cm} \widetilde{F} \hspace{4cm}} & E \\
\downarrow & & \downarrow p \\
X \times [0,1] \times [0,1] & \xrightarrow{\;\mathrm{pr}\;} X \times [0,1] \xrightarrow{\;H'\;} & B,
\end{array}
$$

where pr the map which removes the middle $[0,1]$. On the other hand, by the top triangle, we obtain

$$
\widetilde{F}(x,e,0,t) = F(x,e,0,t) = \widetilde{H}(x,e,1-t)
$$
$$
\widetilde{F}(x,e,1,t) = F(x,e,1,t) = \widetilde{H}'(h(x,e),t).
$$

Now define a map $\varphi : f^*(E) \times [0,1] \to f^*(E)$ by

$$
\varphi(x,e,t) = (x, \widetilde{F}(x,e,t,1)),
$$

then we have

$$\varphi(x, e, 0) = (x, \widetilde{F}(x, e, 0, 1))$$
$$= (x, \widetilde{H}(x, e, 0))$$
$$= (x, e)$$
$$\varphi(x, e, 1) = (x, \widetilde{F}(x, e, 1, 1))$$
$$= (x, \widetilde{H}'(h(x, e), 1))$$
$$= h'(h(x, e)),$$

which implies that φ is a fiber homotopy between $1_{f^*(E)}$ and $h' \circ h$. We obtain

$$h \circ h' \underset{X}{\simeq} 1_{g^*(E)}$$

analogously. $\qquad\square$

Remark 5.4.8. Note that the above proof works for Serre fibrations, if all spaces appearing in the proof are CW complexes.

Corollary 5.4.9. *Let $p : E \to B$ be a Hurewicz fibration over an arcwise connected space B. For any $x, x' \in B$, we have a homotopy equivalence*

$$p^{-1}(x) \simeq p^{-1}(x').$$

Proof. Let $X = \{*\}$, $f(*) = x$, and $g(*) = x'$ in Theorem 5.4.7. Then $f, g : X \to B$ are continuous map. Since B is arcwise connected, there exists a path $\ell : [0, 1] \to B$ connecting x and x'. We regard it as a homotopy between f and g. By Theorem 5.4.7, we have

$$f^*(E) \underset{X}{\simeq} g^*(E).$$

Since

$$f^*(E) = p^{-1}(x)$$
$$g^*(E) = p^{-1}(x'),$$

we have $p^{-1}(x) \simeq p^{-1}(x')$. $\qquad\square$

Remark 5.4.10. When $p : E \to B$ is a fiber bundle, we have a homeomorphism $p^{-1}(x) \cong p^{-1}(x')$ for any $x, x' \in B$. This corollary says that an analogous result holds for Hurewicz fibrations if we replace $\cong$ by $\simeq$.

As we will see in §5.8, when p is a Serre fibration, we have an isomorphism of homotopy groups $\pi_n(p^{-1}(x)) \cong \pi_n(p^{-1}(x'))$ for all n.

We also have the following analogous but slightly different result. The proof is left to the reader.

Corollary 5.4.11. *If Hurewicz fibrations $p : E \to B$ and $p' : E' \to B$ are fiber homotopy equivalent $E \underset{B}{\simeq} E'$, then for any $x \in B$, the fibers over x are homotopy equivalent to each other, i.e. $p^{-1}(x) \simeq {p'}^{-1}(x)$.*

An analogue of Theorem 4.4.3 also follows from the homotopy invariance.

Corollary 5.4.12. *Let $p : E \to B$ be a Hurewicz fibration. When B is contractible to a point $b \in B$, let us denote $F = p^{-1}(b)$. Then we have a fiber homotopy equivalence*

$$B \times F \underset{B}{\simeq} E.$$

Proof. Let $H : B \times [0,1] \to B$ be a contraction to b. It is a homotopy with $H|_{B \times \{0\}} = 1_B$ and $H|_{B \times \{1\}}$ is the constant map c_b to b. By Theorem 5.4.7, we have a fiber homotopy equivalence

$$1_B^*(E) \underset{B}{\simeq} c_b^*(E).$$

Since $1_B^*(E) = E$ and

$$\begin{aligned} c_b^*(E) &= \{(x, e) \in B \times E \mid c_b(x) = p(e)\} \\ &= \{(x, e) \in B \times E \mid b = p(e)\} \\ &= B \times F, \end{aligned}$$

we obtain a fiber homotopy equivalence $E \underset{B}{\simeq} B \times F$. □

Definition 5.4.13. A fibration $p : E \to B$ is said to be *trivial* if it is fiber homotopy equivalent to a fibration of the form $\mathrm{pr}_1 : B \times F \to B$, where pr_1 is the projection onto the first coordinate.

Recall that, in the proof of the covering homotopy theorem (Corollary 4.3.16), the existence of partition of unity[6] played an essential role. Dold [Dold (1963)] showed that the existence of a numerable covering can be also used to construct fiber homotopy equivalences. See Theorem 3.3 in Dold's paper [Dold (1963)].

---

[6]Definition 4.3.14.

Theorem 5.4.14. *Given a commutative diagram*

$$
\begin{array}{ccc}
E & \xrightarrow{\;\;f\;\;} & E' \\
& {}_{p}\searrow \quad \swarrow{}_{p'} & \\
& B, &
\end{array}
$$

if B has a numerable covering $\{U_\lambda\}_{\lambda \in \Lambda}$ and f is a fiber homotopy equivalence on each U_λ, then f is a fiber homotopy equivalence.

Remark 5.4.15. Note that maps p and p' do not have to be fibrations.

5.5   Deforming Continuous Maps into Fibrations

As we have seen, many properties of fiber bundles can be translated to corresponding properties of fibrations by replacing $=$ or $\cong$ by $\simeq$ or $\underset{B}{\simeq}$. In this section, we prove a very important property of fibrations which does not have an analogue for fiber bundles; any continuous map can be deformed continuously into a fibration. Because of this property, fibrations are much more useful than fiber bundles when continuous deformations are allowed. In order to give a precise statement, we need the following construction.

Definition 5.5.1. For a continuous map $f : X \to Y$, define

$$
E_f = \{(x, \ell) \in X \times \mathrm{Map}([0,1], Y) \mid f(x) = \ell(0)\}
$$

and define maps

$$
\begin{aligned}
p_f &: E_f \longrightarrow Y \\
i_f &: X \longrightarrow E_f \\
r_f &: E_f \longrightarrow X
\end{aligned}
$$

by

$$
\begin{aligned}
p_f(x, \ell) &= \ell(1) \\
i_f(x) &= \big(x, c_{f(x)}\big) \\
r_f(x, \ell) &= x,
\end{aligned}
$$

where $c_{f(x)}$ is the constant map to $f(x)$. When f is obvious from the context, these maps are abbreviated by p, i, and r, respectively.

This is called the *mapping track* of f.

Theorem 5.5.2. *For any continuous map $f : X \to Y$, the following hold.*

(1) The map $p_f : E_f \to Y$ is a Hurewicz fibration.

(2) The map f factors as $f = p_f \circ i_f$

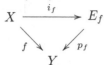

and we have $f \circ r_f \simeq p_f$.

(3) The maps i_f and r_f are homotopy inverse to each other and homotopies H and H' for $i_f \circ r_f \simeq 1_{E_f}$ and $r_f \circ i_f \simeq 1_X$ can be chosen to make the diagram

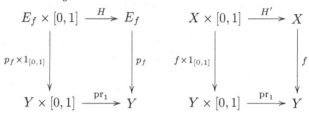

commutative.

Remark 5.5.3. We can say that i_f a fiber preserving map and H' is a fiber homotopy equivalence. If r_f is a fiber preserving map and H is a fiber homotopy, the third part can be stated as "f and p_f" are fiber homotopy equivalence. Unfortunately, this is not the case.

Definition 5.5.4. We say the map $p_f : E_f \to Y$ is obtained by *replacing* $f : X \to Y$ *by a fibration.* For $y \in Y$, we write

$$\mathrm{hofib}_y(f) = p_f^{-1}(y).$$

This is called the *homotopy fiber* of f over y. When Y is arcwise connected, it is sometimes denoted by $\mathrm{hofib}(f)$ since $\mathrm{hofib}_y(f)$ and $\mathrm{hofib}_{y'}(f)$ are homotopy equivalent for any $y, y' \in Y$.

Proof of Theorem 5.5.2. The proof of the fact that $p_f : E_f \to Y$ in (1) is a Hurewicz fibration is basically the same as the case of the path-loop fibration $PX \to X$ (Proposition 5.2.8). We sketch a proof.

Suppose a commutative diagram

$$
\begin{CD}
Z \times \{0\} @>g>> E_f \\
@VVV @VV{p_f}V \\
Z \times [0,1] @>G>> Y
\end{CD}
$$

is given. In order to construct a map
$$\widetilde{G} : Z \times [0,1] \longrightarrow E_f,$$
write $g(z) = (g_1(z), g_2(z))$ and define
$$G_1(z,t) = g_1(z)$$
$$G_2(z,t) = \begin{cases} g_2(z)(s(1+t)), & 0 \leq s \leq \frac{1}{1+t} \\ G(z, s(1+t)-1), & \frac{1}{1+t} \leq s \leq 1. \end{cases}$$
Then the map $\widetilde{G}(z) = (G_1(z), G_2(z))$ takes values in E_f and makes the diagram commutative.

The second part is easy to prove and is left to the reader.

In order to prove the third part, we need to find homotopies. A homotopy $H : E_f \times [0,1] \to E_f$ between $i \circ r$ and 1_{E_f} is given by $H(x, \ell, t) = (x, h(\ell, t))$ where h is the map defined by $h(\ell, t)(s) = \ell(st)$. It is immediate to verify that the diagram

$$\begin{array}{ccc} E_f \times [0,1] & \xrightarrow{\ H\ } & E_f \\ {\scriptstyle p_f \times 1_{[0,1]}} \downarrow & & \downarrow {\scriptstyle p_f} \\ Y \times [0,1] & \xrightarrow{\ \mathrm{pr}_1\ } & Y \end{array}$$

is homotopy commutative. The case of $r \circ i$ is much easier, since $r \circ i = 1_X$. Let $H' : X \times [0,1] \to X$ to be the projection onto the first coordinate, then the diagram

$$\begin{array}{ccc} X \times [0,1] & \xrightarrow{\ H'\ } & X \\ {\scriptstyle f \times 1_{[0,1]}} \downarrow & & \downarrow {\scriptstyle f} \\ Y \times [0,1] & \xrightarrow{\ \mathrm{pr}_1\ } & Y \end{array}$$

is commutative. $\qquad\square$

Example 5.5.5. The reader might have notice that the path-loop fibration can be obtained as a special case. Namely, when (X, x_0) is a based space, the path-loop fibration $PX \to X$ is obtained by replacing the inclusion map $i : \{x_0\} \hookrightarrow X$ by a fibration. In fact,
$$\begin{aligned} E_i &= \{(x, \ell) \in \{x_0\} \times \mathrm{Map}([0,1], X) \mid i(x) = \ell(0)\} \\ &= \{\ell \in \mathrm{Map}([0,1], X) \mid \ell(0) = x_0\} \\ &= PX \end{aligned}$$
and the $p : E_i \to X$ coincides with the projection $PX \to X$.

We also obtain the evaluation map

$$\text{ev}_1 : \text{Map}([0,1], X) \longrightarrow X$$

by replacing the identity map $1_X : X \to X$ by a fibration. Thus ev_1 is a fibration. □

Exercise 5.5.6. Show that the homotopy fiber of a continuous map $f : X \to Y$ over $y \in Y$ can be obtained as the following pullback

$$
\begin{array}{ccc}
\text{hofib}_y(f) & \longrightarrow & X \\
\downarrow & & \downarrow f \\
P(Y,y) & \xrightarrow{\text{ev}_1} & Y.
\end{array}
$$

Show also that

$$
\begin{array}{ccc}
E_f & \xrightarrow{p_f} & X \\
\downarrow & & \downarrow f \\
\text{Map}([0,1], Y) & \xrightarrow{\text{ev}_1} & Y
\end{array}
$$

is a pullback diagram.

Fibrations can be also used to make a homotopy commutative diagram into a strictly commutative diagram.

Proposition 5.5.7. *Let*

$$
\begin{array}{ccc}
X & \xrightarrow{h} & E \\
f \downarrow & & \downarrow p \\
Y & \xrightarrow{g} & B
\end{array}
$$

be a homotopy commutative diagram in which p is a fibration. Then there exists a map $\tilde{h} : X \to E$ which is homotopic to h and makes the diagram commutative.

Proof. Let $H : X \times [0,1] \to B$ be a homotopy from $p \circ h$ to $g \circ f$. Since $H(x,0) = p \circ h$, the outside square of the diagram

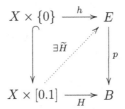

is commutative. Since p is a fibration, there exists $\widetilde{H}$ making the diagram commutative. Define $\tilde{h}(x) = \widetilde{H}(x,1)$. Then $h \simeq \tilde{h}$ and $(p \circ \tilde{h})(x) = (p \circ \widetilde{H})(x,1) = H(x,1) = g \circ f$. Thus the diagram

$$
\begin{array}{ccc}
X & \xrightarrow{\tilde{h}} & E \\
f \downarrow & & \downarrow p \\
Y & \xrightarrow{g} & B
\end{array}
$$

is commutative. $\qquad\square$

We may also replace homotopy by fiber homotopy.

Proposition 5.5.8. *Let $p : E \to B$ and $p' : E' \to B'$ be fibrations and $(f,\tilde{f})$ and $(g,\tilde{g})$ maps of fibrations from p to p'. Suppose there exist homotopies $H : B \times [0,1] \to B'$ and $\widetilde{H} : E \times [0,1] \to E'$ from f to g and from $\tilde{f}$ to $\tilde{g}$, respectively. If p' is a Serre fibration, we also assume that B and E are CW complexes.*

If the diagram

$$
\begin{array}{ccc}
E \times [0,1] & \xrightarrow{\widetilde{H}} & E' \\
p \times 1 \downarrow & & \downarrow p' \\
B \times [0,1] & \xrightarrow{H} & B'
\end{array}
$$

is homotopy commutative and the homotopy $G : E \times [0,1] \times [0,1] \to B'$ can be taken to satisfy

$$
G(e,0,t) = p' \circ \tilde{f}
$$
$$
G(e,1,t) = p' \circ \tilde{g}.
$$

Then $(f,\tilde{f})$ and $(g,\tilde{g})$ are fiber homotopic.

Proof. The idea of the proof is similar to the construction of fiber homotopy in the latter half of the proof of Theorem 5.4.7. Denote

$$K = \partial[0,1] \times [0,1] \cup [0,1] \times \{0\}$$

and define $F : E \times K \to E'$ by

$$F(e,s,t) = \begin{cases} \tilde{f}(e), & s = 0 \\ \tilde{g}(e), & s = 1 \\ \tilde{H}(e,s), & t = 0. \end{cases}$$

Then the outside square of the diagram

$$
\begin{array}{ccc}
E \times K & \xrightarrow{\quad F \quad} & E' \\
\big\uparrow & \overset{\tilde{F}}{\nearrow} & \big\downarrow{\scriptstyle p'} \\
E \times [0,1] \times [0,1] & \xrightarrow[\quad G \quad]{} & B'
\end{array}
$$

is commutative, where G is a homotopy from $p' \circ \tilde{H}$ to $H \circ (p \times 1)$. By the same argument as we used in the proof of Theorem 5.4.7, we obtain a map $\tilde{F}$ making the diagram commutative. Define

$$\tilde{G}(e,s) = \tilde{F}(e,s,1).$$

Then we have

$$\begin{aligned}
(p' \circ \tilde{G})(e,s) &= (p' \circ \tilde{F})(e,s,1) \\
&= G(e,s,1) \\
&= (p \circ (H \times 1))(e,s),
\end{aligned}$$

which implies that the pair $(H, \tilde{G})$ is a fiber homotopy. On the other hand, we have

$$\tilde{G}(e,0) = \tilde{F}(e,0,1) = F(e,0,1) = \tilde{f}(e)$$
$$\tilde{G}(e,1) = \tilde{F}(e,1,1) = F(e,1,1) = \tilde{g}(e).$$

Thus the pair $(H, \tilde{G})$ is a fiber homotopy from $(f, \tilde{f})$ to $(g, \tilde{g})$. $\qquad\square$

We have seen in Theorem 5.5.2 that any continuous map can be made into a fibration up to a homotopy. What happens if the map is already a fibration? As expected, the resulting fibration is fiber homotopy equivalent to the original fibration. This fact follows from the above two propositions.

Corollary 5.5.9. *If $p : E \to B$ is a Hurewicz fibration, the mapping track $E_p \to B$ of p is fiber homotopy equivalent to p.*

Exercise 5.5.10. Prove this fact by using Proposition 5.5.7 and Proposition 5.5.8.

By Corollary 5.4.11, we obtain the following.

Corollary 5.5.11. *Let $p : E \to B$ be a Hurewicz fibration. For any point $x \in B$, the fiber and the homotopy fiber over x are homotopy equivalent*

$$p^{-1}(x) \simeq \mathrm{hofib}_x(p).$$

Furthermore a homotopy equivalence is given by the map

$$i : p^{-1}(x) \longrightarrow \mathrm{hofib}_x(p)$$

defined by $i(y) = (y, c_x)$, where c_x is the constant loop at x.

Remark 5.5.12. When p is a Serre fibration, the map $i : p^{-1}(x) \to \mathrm{hofib}_x(p)$ may not be a homotopy equivalence but induces an isomorphism of homotopy groups $i_* : \pi_n(p^{-1}(x)) \to \pi_n(\mathrm{hofib}_x(p))$ for all n, as we will see in §5.8.

5.6 Homotopy Fiber Sequences

We have seen in the previous section that any continuous map $f : X \to Y$ can be deformed into a fibration $p : E_f \to Y$. For $y_0 \in Y$, let

$$j_f : \mathrm{hofib}_{y_0}(f) \hookrightarrow E_f$$

be the inclusion of the homotopy fiber $\mathrm{hofib}_{y_0}(f) = p^{-1}(y_0)$. Since this is a continuous map, we may replace it by a fibration. We obtain an infinite sequence of fibrations by iterating this construction. The resulting sequence is called the homotopy fiber sequence of f and is closely related to the long exact sequence of homotopy groups. The aim of this section is to give a construction of the homotopy fiber sequence and prove its basic properties. The relation to the long exact sequence is proved in §5.7 and §5.8 as an application of the result of this section.

What we are going to do in this section is to extend the sequence

$$\mathrm{hofib}_{y_0}(f) \xrightarrow{j_f} E_f \xrightarrow{p_f} Y$$

to the left by taking the homotopy fiber of j_f. Since E_f is homotopy equivalent to X, let us consider the homotopy fiber of the composition

$$r_f \circ j_f : \mathrm{hofib}_{y_0}(f) \xrightarrow{j_f} E_f \xrightarrow{r_f} X.$$

An interesting fact is that we do not need to replace $r_f \circ j_f$.

Proposition 5.6.1. *Denote* $\pi_f = r_f \circ j_f$. *Then the map* $\pi_f : \mathrm{hofib}_{y_0}(f) \to X$ *is a Hurewicz fibration for any* $y_0 \in Y$ *and the fiber over* x_0 *with* $f(x_0) = y_0$ *is homeomorphic to the loop space* $\Omega(Y, y_0)$.

Proof. By Exercise 5.5.6, we have a pullback diagram of the form

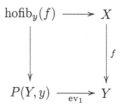

and the top map is π_f. Thus as a pullback of the path-loop fibration, π_f is a Hurewicz fibration. $\qquad\square$

Now we have a sequence

$$\Omega(Y, y_0) \xrightarrow{q_f} \mathrm{hofib}_{y_0}(f) \xrightarrow{\pi_f} X \xrightarrow{f} Y \qquad (5.10)$$

of continuous maps and topological spaces such that

- $\mathrm{hofib}_{y_0}(f)$ is the fiber of a map obtained from f by replacing it by a fibration, and
- $\Omega(Y, y_0)$ is the fiber of the map π_f.

Here the map q_f is the inclusion of the fiber over x_0 with $f(x_0) = y_0$ and is given by $q_f(\ell) = (x_0, \ell)$.

The next step is to extend the sequence (5.10) to the left. Namely we construct a fibration by replacing $\Omega(Y, y_0)$ or $\mathrm{hofib}_{y_0}(f)$ by a space homotopy equivalent to it and take the fiber. Instead, we replace π_f by a fibration $p_{\pi_f} : E_{\pi_f} \to X$. Since π_f is a fibration, by Corollary 5.5.9, π_f and p_{π_f} are fiber homotopy equivalent and the fiber homotopy equivalence is given by the map $i_{\pi_f} : \mathrm{hofib}_{y_0}(f) \hookrightarrow E_{\pi_f}$. By Corollary 5.4.11, the fiber $\mathrm{hofib}_{x_0}(\pi_f)$ of p_{π_f} and the fiber of π_f, i.e. $\Omega(Y, y_0)$ are homotopy equivalent by the restriction of i_{π_f}. On the other hand, the composition

$$\mathrm{hofib}_{x_0}(r \circ j) \xrightarrow{j_{\pi_f}} E_{\pi_f} \xrightarrow{r_{\pi_f}} \mathrm{hofib}_{y_0}(f)$$

is a fibration and the fiber over x_0 is $\Omega(X, x_0)$. Thus we obtain a commutative diagram

$$
\begin{array}{ccccc}
\Omega(Y, y_0) & \xrightarrow{\ q_f\ } & \mathrm{hofib}_{y_0}(f) & \xrightarrow{\ \pi_f\ } & X \\
\ \downarrow{\scriptstyle i_{\pi_f}} & & \| & & \| \\
\Omega(X, x_0) & \xrightarrow[\ q_{\pi_f}\ ]{} & \mathrm{hofib}_{x_0}(\pi_f) & \xrightarrow[r_{\pi_f} \circ j_{\pi_f}]{} & \mathrm{hofib}_{y_0}(f) & \xrightarrow{\ \pi_f\ } & X
\end{array}
\tag{5.11}
$$

in which the vertical maps are homotopy equivalences. Although $\Omega(X, x_0)$ and $\Omega(Y, y_0)$ are not directly connected by a map, there is a naturally defined by between loop spaces, which is already used in the definition of the induced homomorphism of homotopy groups (Definition 4.8.1).

Definition 5.6.2. Let (X, x_0) and (Y, y_0) be based spaces. For a based map $f : X \to Y$, define a map $\Omega f : \Omega(X, x_0) \to \Omega(Y, y_0)$ by $(\Omega f)(\ell) = f \circ \ell$ for $\ell \in \Omega(X, x_0)$.

This is easily seen to be a continuous map. It would be nice if it makes the diagram (5.11) homotopy commutative, but unfortunately that is not the case. We need to modify it by using the map $\nu : \Omega(X, x_0) \to \Omega(X, x_0)$.

Lemma 5.6.3. *We have* $q_{\pi_f} \simeq i_{\pi_f} \circ \Omega(f) \circ \nu$.

Proof. We first need explicit descriptions of maps E_{π_f} and $\mathrm{hofib}_{x_0}(\pi_f)$ to find a homotopy. We have

$$
E_{\pi_f} = \{((x, \ell), \ell') \in \mathrm{hofib}_{y_0}(f) \times \mathrm{Map}([0,1], X) \mid \pi_f(x, \ell) = \ell'(0)\}
$$

$$
= \left\{ (x, \ell, \ell') \in X \times \mathrm{Map}([0,1], Y) \times \mathrm{Map}([0,1], X) \;\middle|\; \begin{array}{l} f(x) = \ell(0), \\ \ell(1) = x_0, \\ x \quad = \ell'(0) \end{array} \right\}
$$

$$
\mathrm{hofib}_{x_0}(\pi_f) = \left\{ (x, \ell, \ell') \in X \times \mathrm{Map}([0,1], Y) \times \mathrm{Map}([0,1], X) \;\middle|\; \begin{array}{l} f(x) = \ell(0), \\ \ell(1) = y_0, \\ x \quad = \ell'(0), \\ \ell'(1) = x_0 \end{array} \right\}
$$

And maps $q_{\pi_f} : \Omega(X, x_0) \to \mathrm{hofib}_{x_0}(\pi_f)$ and $i_{\pi_f} \circ \Omega(f) : \Omega(X, x_0) \to \mathrm{hofib}_{x_0}(\pi_f)$ are given by

$$
q_{\pi_f}(\ell') = (x_0, c_{f(x_0)}, \ell')
$$
$$
(i_{\pi_f} \circ \Omega(f))(\ell') = (x_0, f \circ \ell', c_{x_0}).
$$

Now define a homotopy

$$
H : \Omega(X, x_0) \times [0,1] \longrightarrow \mathrm{hofib}_{x_0}(\pi_f)
$$

by $H(\ell', s) = (x_0, h_1(\ell', s), h_2(\ell', s))$, where

$$h_1(\ell', s)(t) = \begin{cases} y_0, & s \le t \le 1 \\ f(\ell'(s - t)), & 0 \le t \le s \end{cases}$$

$$h_2(\ell', s)(t) = \begin{cases} x_0, & s + t \ge 1 \\ \ell'(s + t), & s + t \le 1. \end{cases}$$

$\square$

Thus we have obtained a sequence of topological spaces and continuous maps

$$\Omega(X, x_0) \xrightarrow{-\Omega f} \Omega(Y, y_0) \xrightarrow{q} \mathrm{hofib}_{y_0}(f) \xrightarrow{\pi_f} X \xrightarrow{f} Y \qquad (5.12)$$

which has the following properties:

- $\mathrm{hofib}_{y_0}(f)$ is the homotopy fiber of f,
- $\Omega(Y, y_0)$ is the fiber of π_f, and
- in the commutative diagram

$$
\begin{array}{ccccc}
\Omega(X, x_0) & \xrightarrow{-\Omega(f)} & \Omega(Y, y_0) & \xrightarrow{q_f} & \mathrm{hofib}_{y_0}(f) \\
\Big\| & & \Big\downarrow{\scriptstyle i_{\pi_f}} & & \Big\| \\
\Omega(X, x_0) & \xrightarrow[q_{\pi_f}]{} & \mathrm{hofib}_{x_0}(\pi_f) & \xrightarrow[r_{\pi_f} \circ j_{\pi_f}]{} & \mathrm{hofib}_{y_0}(f)
\end{array}
$$

vertical maps are homotopy equivalences and the fiber of the composition $r_{\pi_f \circ j_{\pi_f}}$ in the bottom row is $\Omega(X, x_0)$.

Here we denote $(\Omega f) \circ \nu$ by $-\Omega f$ for simplicity.

As a framework for handling these sequences, we introduce the notion of homotopy fiber sequence. We need the notion of based fibration for this purpose.

Definition 5.6.4. Let E and B be based spaces. A based map $p : E \to B$ is said to have the *based covering homotopy property* (based CHP) with respect to a based space X if for any based homotopy $H : X \times [0, 1] \to B$ and a based map $f : X \to E$ making the diagram

$$
\begin{array}{ccc}
X \times \{0\} & \xrightarrow{f} & E \\
\Big\downarrow & & \Big\downarrow{\scriptstyle p} \\
X \times [0, 1] & \xrightarrow{H} & B
\end{array}
$$

commutative, there exists a based homotopy $\widetilde{H} : X \times [0,1] \to E$ which makes the diagram

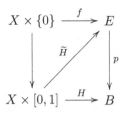

commutative.

Definition 5.6.5. Let E and B be based spaces. A based map $p : E \to B$ is called a *based Hurewicz fibration* if p has the based CHP with respect to any based space X. If p has the based CHP for any based CW complexes, it is called a *based Serre fibration*.

As is the unbased case, we do not distinguish based Serre fibrations and based Hurewicz fibrations and call them simply based fibrations.

Example 5.6.6. For a based map $f : (X, x_0) \to (Y, y_0)$, the map $p_f : E_f \to Y$ defined in Definition 5.5.1 is a based fibration. Here (x_0, c_{y_0}) is the base point of E_f.

Note that based versions of 2 and 3 of Theorem 5.5.2 also hold. □

It is straightforward to modify the definitions of map of fibrations, fiber homotopy, and fiber homotopy equivalence and define their based versions.

The following facts are fundamental and can proved in the same way as unbased cases. The proofs are omitted.

Proposition 5.6.7. *The pullback of a based fibration along a based map is again a based fibration.*

Theorem 5.6.8. *Let $p : E \to B$ be a based fibration. If based maps $f, g : X \to B$ are homotopic to each other, then the pullbacks of p along f and g are based fiber homotopy equivalent to each other.*

Example 5.6.9. For a based map $f : (X, x_0) \to (Y, y_0)$, the map $\pi_f : \mathrm{hofib}_{y_0}(f) \to X$ is a based fibration. In fact, the path-loop fibration $P(Y, y_0) \to Y$ is a based fibration and π_f is the pull back of the path-loop fibration along f. □

Now we are ready to introduce the notion of homotopy fiber sequence.

Definition 5.6.10. A sequence of based spaces and based maps

$$W \xrightarrow{g} X \xrightarrow{f} Y$$

is called a *homotopy fiber sequence* if there exist a based fibration $p : E \to B$ and based maps $h_1 : W \to F$, $h_2 : X \to E$, and $h_3 : Y \to B$ such that the diagram

$$
\begin{array}{ccccc}
W & \xrightarrow{g} & X & \xrightarrow{f} & Y \\
\downarrow{h_1} & & \downarrow{h_2} & & \downarrow{h_3} \\
F & \xrightarrow{i} & E & \xrightarrow{p} & B
\end{array}
$$

is based homotopy commutative, where y_0 is the base point of Y, $F = p^{-1}(y_0)$, and i is the inclusion map.

With this terminology the sequence (5.12) is a sequence in which any three consecutive sequence is a homotopy fiber sequence. In order to extend the sequence (5.12) further to the left, we can iterate the process by using q in place of f. It turns out, however, it is not necessary by the following fact, whose proof is postponed to §5.7.

Proposition 5.6.11. *For a homotopy fiber sequence*

$$W \xrightarrow{g} X \xrightarrow{f} Y,$$

the sequence

$$\Omega W \xrightarrow{\Omega g} \Omega X \xrightarrow{\Omega f} \Omega Y$$

is again a homotopy fiber sequence.

Thus we obtain a homotopy fiber sequence

$$\Omega \mathrm{hofib}_{y_0}(f) \xrightarrow{-\Omega \pi_f} \Omega(X, x_0) \xrightarrow{-\Omega f} \Omega(Y, y_0).$$

There is another homotopy fiber sequence

$$\Omega\Omega(Y, y_0) \xrightarrow{-\Omega q} \Omega \mathrm{hofib}_{y_0}(f) \xrightarrow{-\Omega \pi_f} \Omega(X, x_0).$$

By concatenating these sequences with (5.12), we obtain

$$\Omega\Omega(Y, y_0) \xrightarrow{-\Omega q} \Omega \mathrm{hofib}_{y_0}(f) \xrightarrow{-\Omega \pi_f} \Omega(X, x_0)$$

$$\xrightarrow{-\Omega f} \Omega(Y, y_0) \xrightarrow{q} \mathrm{hofib}_{y_0}(f) \xrightarrow{\pi_f} X \xrightarrow{f} Y.$$

Here $\Omega\Omega(Y, y_0)$ is the loop space on the loop space of Y. Precisely speaking, we should write $\Omega(\Omega(Y, y_0), c_{y_0})$ by using the constant loop c_{y_0} as a base point of $\Omega(Y, y_0)$. For simplicity, we used the following notation.

Definition 5.6.12. For a based space (X, x_0), the space obtained by taking loop spaces n times on X

$$\underbrace{\Omega\Omega\cdots\Omega}_{n \text{ times}}(X, x_0)$$

is denoted by $\Omega^n(X, x_0)$ and is called the *n-fold loop space* of X. We also denote it by $\Omega^n X$ when the base point is obvious from the context.

Theorem 5.6.13. *For a based map* $f : (X, x_0) \to (Y, y_0)$ *we obtain a sequence of based maps*

$$\cdots \longrightarrow \Omega^n\mathrm{hofib}(f) \xrightarrow{(-1)^n\Omega^n\pi_f} \Omega^n X \xrightarrow{(-1)^n\Omega^n f} \Omega^n Y$$

$$\xrightarrow{(-1)^{n-1}\Omega^{n-1}q} \Omega^{n-1}\mathrm{hofib}(f) \xrightarrow{(-1)^{n-1}\Omega^{n-1}\pi_f} \cdots$$

$$\cdots \longrightarrow \Omega\mathrm{hofib}(f) \xrightarrow{-\Omega\pi_f} \Omega X \xrightarrow{-\Omega f} \Omega Y \xrightarrow{q} \mathrm{hofib}(f) \xrightarrow{\pi_f} X \xrightarrow{f} Y$$

in which any consecutive three term is a homotopy fiber sequence.

Definition 5.6.14. This sequence is called the *Puppe sequence* associated to f.

5.7 Iterated Loop Spaces

Let us digress and study basic properties of iterated loop spaces appeared in the construction of the Puppe sequence.

Recall that, for a based space X, the n-fold loop space of X is defined by

$$\Omega^n X = \underbrace{\Omega\Omega\cdots\Omega}_{n \text{ times}} X,$$

which can be written as

$$\Omega^n X = \underbrace{\mathrm{Map}_*(S^1, \mathrm{Map}_*(S^1, \ldots, \mathrm{Map}_*(S^1, X)\cdots)}_{n \text{ times}}.$$

Fortunately, it can be simplified as follows.

Theorem 5.7.1. *For a based space X, we have a homeomorphism*

$$\Omega^n X \cong \mathrm{Map}_*(S^n, X).$$

In order to prove this fact, we need a couple of definitions.

Definition 5.7.2. For based spaces X and Y, define

$$X \wedge Y = X \times Y/X \times \{*\} \cup \{*\} \times Y = X \times Y/X \vee Y.$$

This is called *smash product* of X and Y.

The image of $(x, y) \in X \times Y$ under the projection

$$X \times Y \longrightarrow X \wedge Y$$

is denoted by $x \wedge y$. The base point of $X \wedge Y$ is also denoted by $*$. Note that we have

$$x \wedge * = * \wedge y = *$$

for any $x \in X$ and $y \in Y$.

Remark 5.7.3. Recall that $X \vee Y$ is called the wedge sum of X and Y. This might be confusing, since the symbol $\wedge$ is encoded by \wedge in the famous software TeX, which is used by most people who write books and papers on mathematics.

Example 5.7.4. The smash product $S^1 \wedge S^1$ is homeomorphic to S^2. In fact, $S^1 \times S^1$ is a torus and we obtain S^2 by collapsing $S^1 \vee S^1$ to a point, as is shown in Figure 5.7.

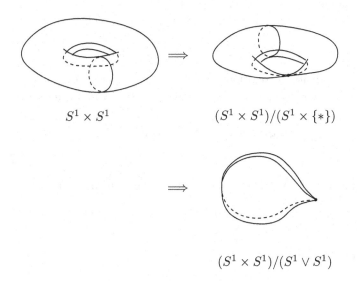

$$S^1 \times S^1 \qquad (S^1 \times S^1)/(S^1 \times \{*\})$$

$$(S^1 \times S^1)/(S^1 \vee S^1)$$

Fig. 5.7 $S^1 \wedge S^1$

A more precise argument is as follows. Regard the torus $S^1 \times S^1$ as a square $[0,1]^2$ with parallel edges identified, as in Figure 5.8. Or we can write

$$S^1 \times S^1 = [0,1]^2 / \sim$$

where the relation $\sim$ is the equivalence relation generated by

$$(s,0) \sim (s,1) \text{ for } s \in [0,1]$$
$$(0,t) \sim (1,t) \text{ for } t \in [0,1].$$

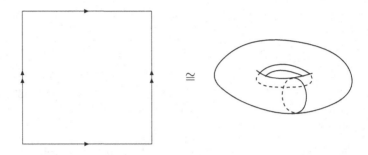

Fig. 5.8   torus from square

We may assume the four corners represent the base point so that the wedge sum $S^1 \vee S^1$ corresponds to the boundary $\partial [0,1]^2$. Thus we have

$$S^1 \wedge S^1 = (S^1 \times S^1)/(S^1 \vee S^1) = \frac{[0,1]^2 / \sim}{\partial [0,1]^2 / \sim}.$$

Since $\sim$ is the equivalence relation which identify any pair of points in $\partial [0,1]^2$, we may ignore $\sim$ if we collapse $\partial [0,1]^2$ first. And we have and identification

$$S^1 \wedge S^1 = [0,1]^2 / \partial [0,1]^2 \cong S^2.$$

□

This argument works without a change for higher dimensional spheres.

Lemma 5.7.5. *For any positive integers m, n, we have a homeomorphism $S^m \wedge S^n \cong S^{m+n}$.*

Proof. We regard spheres as

$$S^m = [0,1]^m / \partial [0,1]^m$$
$$S^n = [0,1]^n / \partial [0,1]^n.$$

Then

$$S^m \times S^n = ([0,1]^m/\partial[0,1]^m) \times ([0,1]^n/\partial[0,1]^n).$$

The right hand side is obtained from $[0,1]^m \times [0,1]^n$ by the identification

$$S^m \times S^n = [0,1]^m \times [0,1]^n/\sim$$

by the relation

$$(x,y) \sim (x,z) \text{ for } x \in [0,1]^m, \; y,z \in \partial[0,1]^n$$
$$(x,z) \sim (y,z) \text{ for } x,y \in \partial[0,1]^m, \; z \in [0,1]^n.$$

Since $S^m \vee S^n$ corresponds to $\partial([0,1]^m \times [0,1]^n)$, we have

$$\begin{aligned}
S^m \wedge S^n &= (S^m \times S^n)/(S^m \vee S^n) \\
&= ([0,1]^m \times [0,1]^n)/\partial([0,1]^m \times [0,1]^n) \\
&= [0,1]^{m+n}/\partial[0,1]^{m+n} \\
&\cong S^{m+n}.
\end{aligned}$$

$\square$

Now Theorem 5.7.1 can be obtained as a corollary to the following general fact.

Theorem 5.7.6. *Let X, Y, and Z be based spaces. If Y is locally compact Hausdorff, then we have a natural bijection*

$$\mathrm{Map}_*(X \wedge Y, Z) \cong \mathrm{Map}_*(X, \mathrm{Map}_*(Y, Z))$$

as sets, which is a homeomorphism when X and Y are compact Hausdorff.

Remark 5.7.7. In order for this theorem to hold for any space, a popular technique is to replace the topologies on the smash products and mapping spaces by the compactly generated topology. See §V.3 of Hu's book [Hu (1964)] or Steenrod's article [Steenrod (1967)].

Proof of Theorem 5.7.6. We define a map

$$\mathrm{ad} : \mathrm{Map}_*(X \wedge Y, Z) \longrightarrow \mathrm{Map}_*(X, \mathrm{Map}_*(Y, Z))$$

by mimicking Definition 3.4.3. For $f \in \mathrm{Map}_*(X \wedge Y, Z)$, $x \in X$, and $y \in Y$, define $\mathrm{ad}(f)(x)(y) = f(x \wedge y)$. Then $\mathrm{ad}(f)(x)$ is a map from Y to Z. As is the unbased case, this is continuous and preserves base points, since $\mathrm{ad}(f)(x)(*) = f(x \wedge *) = f(*) = *$ and we obtain and element

$\mathrm{ad}(f)(x) \in \mathrm{Map}_*(Y, Z)$. Since $\mathrm{ad}(f)(*)(y) = f(* \wedge y) = f(*) = *$, $\mathrm{ad}(f):$ $X \to \mathrm{Map}_*(Y, Z)$ preserves base points and we obtain a map

$$\mathrm{ad} : \mathrm{Map}_*(X \wedge Y, Z) \longrightarrow \mathrm{Map}_*(X, \mathrm{Map}_*(Y, Z)).$$

Define a map

$$\mathrm{ad}^{-1} : \mathrm{Map}_*(X, \mathrm{Map}_*(Y, Z)) \longrightarrow \mathrm{Map}_*(X \wedge Y, Z)$$

$\mathrm{ad}^{-1}(g)(x \wedge y) = g(x)(y)$ for $g \in \mathrm{Map}_*(X, \mathrm{Map}_*(Y, Z))$ and $x \wedge y$. This map takes values in $\mathrm{Map}_*(X \wedge Y, Z)$, since $g(x)(*) = g(*)(y) = *$. It also implies that $\mathrm{ad}^{-1}(g)$ is a based map. This is continuous since Y is locally compact.

It can be easily verified that ad^{-1} is inverse to ad. For example, we have $\varphi \circ \psi = 1$ by

$$\begin{aligned}
(\varphi \circ \psi)(g)(x)(y) &= \varphi(\psi(g))(x)(y) \\
&= \psi(g)(x \wedge y) \\
&= g(x)(y).
\end{aligned}$$

The rest is left to the reader. □

Definition 5.7.8. We also call $\mathrm{ad}(f)$ the *adjoint* to f.

Proof of Theorem 5.7.1. Apply the above theorem n times to obtain

$$\Omega^n X = \underbrace{\mathrm{Map}_*(S^1, \mathrm{Map}_*(S^1, \ldots, \mathrm{Map}_*(S^1, X)}_{n \text{ times}} \cdots)$$

$$= \mathrm{Map}_*(S^1 \wedge S^1, \underbrace{\mathrm{Map}_*(S^1, \mathrm{Map}_*(S^1, \ldots, \mathrm{Map}_*(S^1, X)}_{n-1 \text{ times}} \cdots)$$

$$= \mathrm{Map}_*(\underbrace{S^1 \wedge \cdots \wedge S^1}_{n \text{ times}}, X).$$

By Lemma 5.7.5, we have a homeomorphism

$$\underbrace{S^1 \wedge \cdots \wedge S^1}_{n \text{ times}} \cong S^n$$

and Theorem 5.7.1 is proved. □

The above proof tells us that we have another expression of $\Omega^n X$.

Corollary 5.7.9. *For a based space X, we have a homeomorphism*

$$\Omega^n X \cong \{f \in \mathrm{Map}([0,1]^n, X) \mid f(\partial[0,1]^n) = \{*\}\}.$$

By comparing with the description of homotopy groups in Lemma 4.6.20, we obtain the following relation to homotopy groups.

Corollary 5.7.10. *For a based space X, we have a bijection $\pi_0(\Omega^n X) \cong \pi_n(X)$.*

Proof. By Lemma 4.6.20, we have a natural bijection

$$\pi_n(X, x_0) \cong [((0, 1]^n, \partial[0, 1]^n), (X, x_0)],$$

while

$$\pi_0(\Omega^n X) = \{\text{arcwise connected components of } \Omega^n X\}.$$

By definition, two elements $f, g \in \Omega^n X$ belong to the same arcwise connected component if and only if they can be connected by a path in $\Omega^n X$ or there exists a map $h : [0, 1] \to \Omega^n X$ such that $h(0) = f$ and $h(1) = g$. Under the identification of Corollary 5.7.9, we regard f and g as maps of pairs $([0, 1]^n, \partial[0, 1]^n) \to (X, x_0)$. Define a homotopy $H : [0, 1]^n \times [0, 1] \to X$ by $H(x, t) = h(x)(t)$. Then this is a relative homotopy between f and g. Conversely a relative homotopy between maps of pairs $([0, 1]^n, \partial[0, 1]^n) \to (X, x_0)$ can be regarded as a path in $\Omega^n X$ and we obtain a bijection

$$\pi_0(\Omega^n X) \cong [((0, 1]^n, \partial[0, 1]^n), (X, x_0)] \cong \pi_n(X).$$

$\square$

As we have seen in §5.3, loop spaces have multiplications. More precisely, it is a Hopf space.[7] Since the n-fold loop space $\Omega^n X$ is a loop space on the $(n - 1)$-fold loop space $\Omega^{n-1} X$, it has a structure of Hopf space. Under the identification of Corollary 5.7.9, its multiplication can be described as follows. Regard $f, g \in \Omega^n X$ as maps $f, g : [0, 1]^n \to X$ with $f(\partial[0, 1]^n) = g(\partial[0, 1]^n) = *$. Under the identification $\Omega^n X = \Omega\Omega^{n-1} X$, the multiplication $f * g$ of f and g in $\Omega^n X$ is given by

$$(f * g)(t_1, t_2, \ldots, t_n) = \begin{cases} f(2t_1, t_2, \ldots, t_n), & 0 \le t_1 \le \frac{1}{2} \\ g(2t_1 - 1, t_2, \ldots, t_n), & \frac{1}{2} \le t_1 \le 1. \end{cases}$$

See Figure 5.9.

This is exactly the group operation in the homotopy group $\pi_n(X)$ in Lemma 4.6.22.

---

[7]Definition 5.3.6.

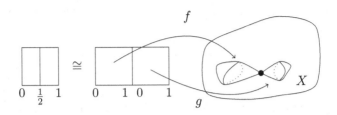

Fig. 5.9   concatenation of n-fold loops

Proposition 5.7.11. *The bijection $\pi_0(\Omega^n X) \cong \pi_n(X)$ in Corollary 5.7.10 is an isomorphism of groups for $n \geq 1$, where the multiplication in $\pi_0(\Omega^n X)$ is the one induced from the Hopf space structure on $\Omega^n X$.*

Corollary 5.7.12. *We have a natural isomorphism $\pi_m(\Omega^n X) \cong \pi_{m+n}(X)$. In particular, we have $\pi_n(\Omega X) \cong \pi_{n+1}(X)$.*

The discussion in §4.6 implies the following description of the multiplication in $\Omega^n X$.

Corollary 5.7.13. *Under the identification $\Omega^n X = \mathrm{Map}_*(S^n, X)$, the multiplication of $f, g \in \Omega^n X$ is given by the composition*

$$S^n \xrightarrow{\text{pinch}} S^n \vee S^n \xrightarrow{f \vee g} X \vee X \xrightarrow{\text{fold}} X.$$

Recall from Lemma 4.6.25 that the homotopy group $\pi_n(X)$ is commutative if $n \geq 2$, which implies that the multiplication in $\Omega^n X$ is also commutative "up to homotopy" if $n \geq 2$.

Corollary 5.7.14. *Let $\mu : \Omega^n X \times \Omega^n X \to \Omega^n X$ be the multiplication in $\Omega^n X$ and $T : \Omega^n X \times \Omega^n X \to \Omega^n X \times \Omega^n X$ the switching map. Then, for $n \geq 2$, we have a homotopy $\mu \circ T \simeq \mu$.*

Remark 5.7.15. The homotopy commutativity characterizes the difference between single loop spaces ΩX and double loop spaces $\Omega^2 Y$. It turns out that the difference between double loop spaces and triple loop spaces can be also described in terms of "higher homotopy commutativity." The study of higher homotopy commutativities was initiated by Kudo and Araki in [Kudo and Araki (1956b,a)] and developed by Boardman and Vogt [Boardman and Vogt (1968, 1973)] and then May [May (1972)]. The algebraic structure called *operad* introduced in May's work now plays very important roles in many fields.

We conclude this section by the proof of Proposition 5.6.11.

Proof of Proposition 5.6.11. It suffices to show that, if $p : E \to B$ is a based fibration with F the fiber over the base point, $\Omega p : \Omega E \to \Omega B$ is a based fibration with ΩF the fiber over the base point.

Let

$$
\begin{array}{ccc}
X \times \{0\} & \xrightarrow{\ f\ } & \Omega E \\
\big\downarrow & & \big\downarrow{\scriptstyle \Omega p} \\
X \times [0,1] & \xrightarrow[\ H\ ]{} & \Omega B
\end{array}
$$

be a commutative diagram of based spaces and based maps. Under the correspondence in Theorem 5.7.6, we obtain $\mathrm{ad}^{-1}(f) : X \wedge S^1 \xrightarrow{E}$. For each $t \in [0,1]$, take ad^{-1} of $h_t = H(-,t) : X \to \Omega B$ to obtain a homotopy

$$G : (X \wedge S^1) \times [0,1] \longrightarrow B.$$

These are a based map and a based homotopy making the diagram

$$
\begin{array}{ccc}
(X \wedge S^1) \times \{0\} & \xrightarrow{\ \mathrm{ad}(f)\ } & E \\
\big\downarrow & & \big\downarrow{\scriptstyle p} \\
(X \wedge S^1) \times [0,1] & \xrightarrow[\ G\ ]{} & B
\end{array}
$$

commutative. Since p is a based fibration, we obtain a lift $\widetilde{G} : (X \wedge S^1) \times [0,1] \to E$. Then, by taking ad for each $t \in [0,1]$, we obtain a map $\widetilde{H} : X \times [0,1] \to \Omega E$, which is a lift of H.

The fiber of Ωp over the base point can be easily seen to be ΩF. $\qquad\square$

5.8 Fibrations and Homotopy Groups

In this section prove the existence of a long exact sequence of homotopy groups for fibrations. Since the n-th homotopy group $\pi_n(X)$ is given by the based homotopy set $[S^n, X]_*$, we first investigate properties of based homotopy sets.

In order to discuss the exactness of a sequence consisting of based homotopy sets, we need to impose certain conditions on spaces under which based homotopy sets become groups.

Proposition 5.8.1. *If X is a Hopf space, then the multiplication of X induces a group structure on $[Y, X]_*$ for any based space X.*

We use the following construction in the proof of this fact.

Definition 5.8.2. For a based space X and a based map $f : Y \to Z$, define a map $f_* : [X, Y]_* \to [X, Z]_*$ by $f_*([g]) = [f \circ g]$ for $[g] \in [X, Y]_*$. This map f_* is called the *map induced from* f.

Define another map $f^* : [Z, X]_* \to [Y, X]_*$ by $f^*([g]) = [g \circ f]$. This is also called the map induced from f.

Note that if $f \simeq g$ for another based map $g : Y \to Z$, then $f_* = g_*$ as maps from $[X, Y]_*$ to $[X, Z]_*$.

Proof of Proposition 5.8.1. We first need to define a multiplication on $[Y, X]_*$. Let

$$\varphi : [Y, X]_* \times [Y, X]_* \longrightarrow [Y, X \times X]_*$$

be the map defined by $\varphi([f], [g]) = [f \times g]$, where $(f \times g)(y) = (f(y), g(y))$. Denote the multiplication of X by $\mu : X \times X \to X$ and define a product on $[Y, X]_*$ by the composition

$$\bar{\mu} = \mu_* \circ \varphi : [Y, X]_* \times [Y, X]_* \longrightarrow [Y, X \times X]_* \longrightarrow [Y, X]_*.$$

Let us verify that this multiplication satisfies the associativity. By the definition of Hopf space, we have a homotopy

$$\mu \circ (\mu \times 1) \simeq \mu \circ (1 \times \mu)$$

and we have

$$\mu_* \circ (\mu \times 1)_* = \mu_* \circ (1 \times \mu)_*$$

as maps from $[Y, X \times X \times X]_*$ to $[Y, X]_*$. This equation can be expressed as a commutative diagram

$$
\begin{array}{ccc}
[Y, X \times X \times X]_* & \xrightarrow{(1 \times \mu)_*} & [Y, X \times X]_* \\
\downarrow{\scriptstyle (\mu \times 1)_*} & & \downarrow{\scriptstyle \mu_*} \\
[Y, X \times X]_* & \xrightarrow{\quad \mu_* \quad} & [Y, X]_*.
\end{array}
$$

By composing with φ, we obtain a commutative diagram

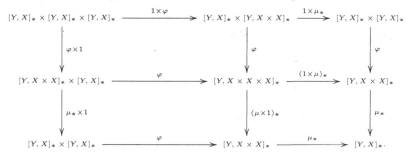

And we obtain a commutative diagram

$$[Y,X]_* \times [Y,X]_* \times [Y,X]_* \xrightarrow{1 \times \bar{\mu}} [Y,X]_* \times [Y,X]_*$$

$$\downarrow{\bar{\mu} \times 1} \qquad\qquad\qquad \downarrow{\bar{\mu}}$$

$$[Y,X]_* \times [Y,X]_* \xrightarrow{\qquad\bar{\mu}\qquad} [Y,X]_*$$

which implies the associativity of $\bar{\mu}$

$$\bar{\mu} \circ (\bar{\mu} \times 1) = \bar{\mu} \circ (1 \times \bar{\mu}).$$

The existences of unit and inverses also follow from the corresponding structures in Hopf spaces. □

Corollary 5.8.3. *For based spaces X and Y, the based homotopy set $[Y, \Omega X]_*$ becomes a group.*

Since $\pi_0(X) = [S^0, X]_*$, we have the following.

Corollary 5.8.4. *For a Hopf space X, $\pi_0(X)$ is a group.*

In order to obtain a group homomorphism $f_* : [Z,X]_* \to [Z,Y]_*$ from a based map $f : X \to Y$ between Hopf spaces, we need to require f to preserve multiplications, up to homotopy.

Definition 5.8.5. Let X and Y be Hopf spaces. Denote their multiplications by

$$\mu_X : X \times X \longrightarrow X$$
$$\mu_Y : Y \times Y \longrightarrow Y.$$

A based map $f : X \to Y$ is called a *map of Hopf spaces* if

$$\mu_Y \circ (f \times f) \simeq f \circ \mu.$$

Lemma 5.8.6. *If* $f : X \to Y$ *is a map of Hopf spaces, the induced map* $f_* : [Z, X]_* \to [Z, Y]_*$ *is a group homomorphism for any based space* Z.

Lemma 5.8.7. *Let* $f : X \to Y$ *be a based map. Then*

$$\Omega f : \Omega X \longrightarrow \Omega Y$$

is a map of Hopf spaces.

Proof. By Lemma 5.7.13 the multiplication of $\omega, \omega' \in \Omega X$ is given by the composition

$$S^1 \overset{\text{pinch}}{\longrightarrow} S^1 \vee S^1 \overset{\omega \vee \omega'}{\longrightarrow} X \vee X \overset{\text{fold}}{\longrightarrow} X.$$

By composing f, we obtain a commutative diagram

$$
\begin{array}{ccccccc}
S^1 & \overset{\text{pinch}}{\longrightarrow} & S^1 \vee S^1 & \overset{\omega \vee \omega'}{\longrightarrow} & X \vee X & \overset{\text{fold}}{\longrightarrow} & X \\
\Big\| & & \Big\| & & \Big\downarrow {\scriptstyle f \vee f} & & \Big\downarrow {\scriptstyle f} \\
S^1 & \overset{\text{pinch}}{\longrightarrow} & S^1 \vee S^1 & \overset{f \circ \omega \vee f \circ \omega'}{\longrightarrow} & Y \vee Y & \overset{\text{fold}}{\longrightarrow} & Y
\end{array}
$$

which is equivalent to saying

$$\Omega f(\omega * \omega') = \Omega f(\omega) * \Omega f(\omega').$$

$\square$

Corollary 5.8.8. *For based spaces* X, Y, *and* Z *and a based map* $f : X \to Y$, *the induced map*

$$(\Omega f)_* : [Z, \Omega X]_* \longrightarrow [Z, \Omega Y]_*$$

is a group homomorphism.

Here is the main theorem of this section.

Theorem 5.8.9. *Let* $p : E \to B$ *be a based Hurewicz fibration with fiber over the base point* F. *For any based space* X *and integer* $n \geq 1$, *the sequence*

$$[X, \Omega^n F]_* \overset{(\Omega^n i)_*}{\longrightarrow} [X, \Omega^n E]_* \overset{(\Omega^n p)_*}{\longrightarrow} [X, \Omega^n B]_*$$

is exact, where $i : F \hookrightarrow E$ *is the inclusion of the fiber.*

When p *is a Serre fibration, this sequence is exact for any CW complex* X.

Proof. Suppose p is a Hurewicz fibration. We need to show the following:

(1) For $[f] \in [X, \Omega^n F]_*$, $(\Omega^n i)_* \circ (\Omega^n p)_*([f]) = [*]$.

(2) For $[g] \in [X, \Omega^n E]_*$, if $(\Omega^n p)_*([g]) = [*]$, then there exists $[f] \in [X, \Omega^n F]_*$ with $(\Omega^n i)_*([f]) = [g]$.

Here $*$ is the constant map to the base point.

It is easy to show (1). We have

$$\Omega^n p \circ \Omega^n i \circ f = \Omega^n (p \circ i) \circ f.$$

Since $F = p^{-1}(*)$, $p \circ i = *$. Thus

$$\Omega^n p \circ \Omega^n i \circ f = *.$$

To show (2), take $[g] \in [X, \Omega^n E]_*$ and suppose $\Omega^n p(g) \simeq *$. Let

$$H : X \times [0, 1] \longrightarrow \Omega^n B$$

be a homotopy between $\Omega^n p(g)$ and $*$. Then, for $x \in X, t \in [0, 1], y \in S^n$, we have

$$H(x, 0)(y) = (\Omega^n p)(g)(x)(y) = p(g(x)(y))$$
$$H(x, 1) = *$$
$$H(*, t) = *.$$

Define

$$\widetilde{H} : X \times S^n \times [0, 1] \longrightarrow B$$

by $\widetilde{H}(x, y, t) = H(x, t)(y)$. Then the first condition for H implies that the diagram

$$
\begin{array}{ccc}
X \times S^n \times \{0\} & \xrightarrow{\ \tilde{g}\ } & E \\
\downarrow & & \downarrow{\scriptstyle p} \\
X \times S^n \times [0, 1] & \xrightarrow{\ \widetilde{H}\ } & B
\end{array}
$$

is commutative, where $\tilde{g}(x, y) = g(x)(y)$. Since both $\tilde{g}$ and $\widetilde{H}$ preserve base points, by the based CHP for p, we obtain a based homotopy

$$H' : X \times S^n \times [0, 1] \longrightarrow E$$

making the diagram

$$
\begin{array}{ccc}
X \times S^n \times \{0\} & \xrightarrow{\ \tilde{g}\ } & E \\
\downarrow & \nearrow{\scriptstyle H'} & \downarrow{\scriptstyle p} \\
X \times S^n \times [0, 1] & \xrightarrow{\ \widetilde{H}\ } & B
\end{array}
$$

commutative. By the bottom triangle, we have

$$p \circ H'(x, y, 1) = \widetilde{H}(x, y, 1) = *$$

for $(x, y) \in X \times S^n$. Define $f(x)(y) = H'(x, y, 1)$. Since $p(f(x)(y)) = *$, we obtain a map

$$f : X \longrightarrow \Omega^n F.$$

Since H' is a homotopy between $\Omega^n i(f)$ and g, we have $(\Omega^n i)_*([f]) = [g]$ and (2) is proved.

Suppose p is a Serre fibration. If X is a CW complex, by Theorem 4.7.5, $X \times S^n \times [0,1]$ is also a CW complex. Thus the above argument works without a change. □

Corollary 5.8.10. *Let* $f : X \to Y$ *be a based map. Then for any based space* Z, *the sequence induced from the Puppe sequence of* f

$$\cdots \longrightarrow [Z, \Omega^n \mathrm{hofib}(f)]_* \longrightarrow [Z, \Omega^n X]_* \overset{(\Omega^n f)_*}{\longrightarrow} [Z, \Omega^n Y]_*$$
$$\longrightarrow [Z, \Omega^{n-1} \mathrm{hofib}(f)]_* \longrightarrow \cdots$$
$$\cdots \longrightarrow [Z, \Omega \mathrm{hofib}(f)]_* \longrightarrow [Z, \Omega X]_* \overset{(\Omega f)_*}{\longrightarrow} [Z, \Omega Y]_*$$

is an exact sequence of groups.

Proof. By Theorem 5.5.2, $p_f : E_f \to Y$ is a Hurewicz fibration and we obtain a long exact sequence

$$\cdots \longrightarrow [Z, \Omega^n \mathrm{hofib}(f)]_* \overset{(\Omega^n j_f)}{\longrightarrow} [Z, \Omega^n E_f]_* \overset{(\Omega^n p_f)_*}{\longrightarrow} [Z, \Omega^n Y]_*$$
$$\longrightarrow [Z, \Omega^{n-1} \mathrm{hofib}(f)]_* \overset{(\Omega^{n-1} j_f)_*}{\longrightarrow} \cdots$$
$$\cdots \longrightarrow [Z, \Omega \mathrm{hofib}(f)]_* \overset{(\Omega j_f)_*}{\longrightarrow} [Z, \Omega E_f]_* \overset{(\Omega p_f)_*}{\longrightarrow} [Z, \Omega Y]_*$$

for any based space Z. Theorem 5.5.2 also says that we have a commutative diagram

$$
\begin{array}{ccccc}
[Z, \Omega^n \mathrm{hofib}(f)]_* & \longrightarrow & [Z, \Omega^n X] & \overset{(\Omega^n f)_*}{\longrightarrow} & [Z, \Omega^n Y]_* \\
\| & & \downarrow{\scriptstyle (\Omega^n i_f)} & & \| \\
[Z, \Omega^n \mathrm{hofib}(f)]_* & \underset{(\Omega^n j_f)}{\longrightarrow} & [Z, \Omega^n E_f] & \underset{(\Omega^n p_f)_*}{\longrightarrow} & [Z, \Omega^n Y]
\end{array}
$$

in which vertical maps are isomorphisms. Thus we may replace $[Z, \Omega^n E_f]_*$ by $[Z, \Omega^n X]_*$ in the above exact sequence. □

By taking $Z = S^0$, we obtain the long exact sequence of homotopy groups.

Corollary 5.8.11. *For a based map $f : X \to Y$, we have a long exact sequence of homotopy groups*

$$\cdots \longrightarrow \pi_n(\mathrm{hofib}(f)) \longrightarrow \pi_n(X) \xrightarrow{f_*} \pi_n(Y) \longrightarrow \pi_{n-1}(\mathrm{hofib}(f)) \longrightarrow \cdots$$

$$\cdots \longrightarrow \pi_1(\mathrm{hofib}(f)) \longrightarrow \pi_1(X) \xrightarrow{f_*} \pi_1(Y).$$

Corollary 5.8.12. *For a based Hurewicz fibration $p : E \to B$, we have the following long exact sequence of homotopy groups*

$$\cdots \longrightarrow \pi_n(F) \longrightarrow \pi_n(E) \xrightarrow{p_*} \pi_n(B) \xrightarrow{\partial_n} \pi_{n-1}(F) \longrightarrow \cdots$$

$$\cdots \longrightarrow \pi_1(F) \longrightarrow \pi_1(E) \xrightarrow{p_*} \pi_1(B).$$

Proof. The exact sequence in Corollary 5.8.11 is

$$\cdots \longrightarrow \pi_n(\mathrm{hofib}(p)) \longrightarrow \pi_n(E) \xrightarrow{p_*} \pi_n(B) \longrightarrow \pi_{n-1}(\mathrm{hofib}(p)) \longrightarrow \cdots$$

$$\cdots \longrightarrow \pi_1(\mathrm{hofib}(p)) \longrightarrow \pi_1(E) \xrightarrow{p_*} \pi_1(B).$$

Since p is a Hurewicz fibration, we have $\mathrm{hofib}(p) \simeq F$ by Corollary 5.5.11. Thus we obtain the long exact sequence. □

When f is an inclusion map, we use the following notation for $\pi_n(\mathrm{hofib}(f))$.

Definition 5.8.13. Let X be a based space and A a subspace containing the base point. The inclusion is denoted by $i : A \hookrightarrow X$. Define $\pi_n(X, A) = \pi_{n-1}(\mathrm{hofib}(i))$. This is called the n-th *relative homotopy group* of the pair (X, Y).

Corollary 5.8.14. *Let X be a based space and A a subspace containing the base point. Then we have the following long exact sequence of homotopy groups*

$$\cdots \longrightarrow \pi_{n+1}(X, A) \xrightarrow{\partial_{n+1}} \pi_n(A) \longrightarrow \pi_n(X) \longrightarrow \pi_n(X, A) \xrightarrow{\partial_n} \cdots$$

$$\cdots \longrightarrow \pi_2(X, A) \xrightarrow{\partial_2} \pi_1(A) \longrightarrow \pi_1(X).$$

There is a more direct description of relative homotopy groups by using the based version of the relative homotopy sets introduced in Definition 4.6.18.

Definition 5.8.15. Let (X, A) and (Y, B) be pairs of topological groups. When X and Y have base points in A and B, respectively, the set of based homotopy classes of based maps of pairs from (X, A) to (Y, B) is denoted by $[(X, A), (Y, B)]_*$.

Remark 5.8.16. By using maps and homotopies of triples of spaces, we may define

$$[(X, A), (Y, B)]_* = [(X, A, *), (Y, B, *)].$$

It should be noted, however, some authors use the notation

$$f : (X; A_1, A_2) \longrightarrow (Y; B_1, B_2)$$

to denote a map $f : X \to Y$ with $f(A_1) \subset B_1$ and $f(A_2) \subset B_2$ even when $A_2 \not\subset A_1$ nor $B_2 \not\subset B_1$.

The following identification justifies the notation $\pi_n(X, A)$.

Lemma 5.8.17. *For a pair of based spaces (X, A), there is a natural bijection*

$$\pi_n(X, A) \cong [(D^n, S^{n-1}), (X, A)]_*.$$

Proof. Let $i : A \hookrightarrow X$ be the inclusion. Define a map

$$\varphi : [(D^n, S^{n-1}), (X, A)]_* \longrightarrow \pi_{n-1}(\mathrm{hofib}(i))$$

as follows. Note that the homotopy fiber $\mathrm{hofib}(i)$ is given by

$$\mathrm{hofib}(i) = \{\omega : [0, 1] \to X \mid \omega(0) \in A, \omega(1) = *\} \subset \mathrm{Map}([0, 1], X).$$

For a based map of pairs $f : (D^n, S^{n-1}) \to (X, A)$, we regard

$$D^n = S^{n-1} \times [0, 1]/(S^{n-1} \times \{1\} \cup \{*\} \times [0, 1]).$$

Under this identification, S^{n-1} on the left hand side correspond to $S^{n-1} \times \{0\}$ in the right hand side. Let

$$p : S^{n-1} \times [0, 1] \longrightarrow S^{n-1} \times [0, 1]/(S^{n-1} \times \{1\} \cup \{*\} \times [0, 1])$$

be the projection and consider the composition.

$$f \circ p : S^{n-1} \times [0, 1] \longrightarrow X.$$

Its adjoint

$$\varphi(f) = \mathrm{ad}(f \circ p) : S^{n-1} \longrightarrow \mathrm{Map}([0, 1], X)$$

satisfies $\varphi(f)(S^{n-1}) \subset \mathrm{hofib}(i)$ and preserves the base points. Thus we obtain a map $\varphi(f) : S^{n-1} \to \mathrm{hofib}(i)$.

If there exists a based homotopy of pairs $f \simeq g$, we may construct a based homotopy $\varphi(f) \simeq \varphi(g)$ by the above construction and we obtain a well-defined map

$$\varphi : [(D^n, S^{n-1}), (X, A)]_* \longrightarrow \pi_{n-1}(\mathrm{hofib}(i)).$$

The process can be reversed to obtain an inverse to φ. $\qquad\qquad\square$

Remark 5.8.18. Under the isomorphism of the above lemma, the map $\partial_n : \pi_n(X, A) \to \pi_{n-1}(A)$ in Corollary 5.8.14 is given by $[f] \mapsto [f|_{S^{n-1}}]$.

Consider the case $(X, A) = (E, F)$ for a fibration $p : E \to B$ with fiber F.

Corollary 5.8.19. *Let* $p : E \to B$ *a Hurewicz fibration with fiber* F. *Then the projection* p *induces an isomorphism*

$$p_* : \pi_n(E, F) \xrightarrow{\cong} \pi_n(B)$$

for all $n \geq 2$.

This is a consequence of Corollary 5.8.12, Corollary 5.8.14 and the following well-known fact, called the five lemma.

Proposition 5.8.20. *Let*

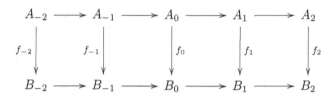

be a commutative diagram of groups and homomorphisms in which the top and bottom rows are exact. Then the following hold:

(1) If f_{-2} *and* f_2 *are surjective and* f_{-1} *and* f_1 *are injective, then* f_0 *is a surjection.*

(2) If f_{-2} *and* f_2 *are injective and* f_{-1} *and* f_1 *are surjective, then* f_0 *is an injection.*

Thus f_{-2}, f_{-1}, f_1, *and* f_2 *are isomorphisms, so is* f_0.

Let us now consider the case when $p : E \to B$ is a Serre fibration. Since S^n has a structure of a CW complex, by Theorem 5.8.9, we still have an exact sequence

$$\pi_n(F) \longrightarrow \pi_n(E) \longrightarrow \pi_n(B)$$

for all n, where F is the fiber over the base point of B. In order to concatenate these sequence into a long exact sequence, we need the following refinement of Corollary 5.8.19.

Proposition 5.8.21. *Let $p : E \to B$ a Serre fibration with fiber F. Then the projection p induces a bijection*

$$p_* : \pi_n(E, F) \xrightarrow{\cong} \pi_n(B)$$

for all $n \geq 1$.

Proof. In order to show that p_* is a surjection, take $f : (D^n, S^{n-1}) \to (B, *)$. By the homeomorphism in Exercise 4.6.7, there exists a quotient map

$$\pi : S^{n-1} \times [0, 1] \longrightarrow D^n$$

such that $\pi(S^{n-1} \times \{1\} \cup \{e_0\} \times [0, 1]) = \{e_0\}$. Thus we obtain a commutative diagram

where $* : S^{n-1} \times \{1\} \to E$ is the constant map to the base point. We obtain the dotted arrow $\tilde{f}$ by the based CHP. Since

$$(p \circ \tilde{f})(S^{n-1} \times \{0\}) = f(\partial S^{n-1}) = *,$$

we have

$$\tilde{f}(S^{n-1} \times \{0\}) \subset p^{-1}(*) = F.$$

Since $\tilde{f}(\{*\} \times [0, 1]) = *$, $\tilde{f}$ induces a map $\tilde{f} : D^n \to E$. By identifying S^{n-1} with $S^{n-1} \times \{0\}$, we obtain a based map

$$\tilde{f} : (D^n, S^{n-1}) \longrightarrow (E, F).$$

The commutativity of the diagram implies that $p_*([\tilde{f}]) = [f]$ and hence p_* is surjective.

To show the injectivity of p_*, suppose $p_*([f]) = 0$. Let $\tilde{f}$ be the lift of f used in the proof of surjectivity and H a homotopy from f to $*$. We construct a homotopy $\tilde{H}$ between $\tilde{f}$ and $*$ by lifting H by CHP. Denote

$$K' = [0, 1] \times \partial[0, 1] \cup \{1\} \times [0, 1]$$

and define a map

$$h : S^{n-1} \times K' \longrightarrow E$$

by

$$h(x, s, t) = \begin{cases} \tilde{f}(x, s), & t = 0 \\ *, & t = 1 \text{ or } s = 1. \end{cases}$$

Then we obtain a commutative diagram

$$
\begin{array}{ccc}
S^{n-1} \times K' & \xrightarrow{\quad h \quad} & E \\
\Big\downarrow & & \Big\downarrow{\scriptstyle p} \\
S^{n-1} \times [0,1] \times [0,1] & \xrightarrow{\pi \times 1_{[0,1]}} D^n \times [0,1] \xrightarrow{\quad H \quad} & B.
\end{array}
$$

By the same argument used in the proof of Theorem 5.4.7, we may find a homeomorphism of pairs

$$([0,1] \times [0,1], K') \cong ([0,1] \times [0,1], [0,1] \times \{0\}),$$

which allows us to apply CHP to this diagram. And we obtain a map

$$\tilde{H} : S^{n-1} \times [0,1] \times [0,1] \longrightarrow E.$$

By Exercise 4.6.7, it induces a map $\tilde{H} : D^n \times [0,1] \to E$, which is a homotopy we wanted. □

Corollary 5.8.22. *Let* $p : E \to B$ *be a Serre fibration. For* $n \geq 1$, *define* $\partial_n : \pi_n(B) \to \pi_{n-1}(F)$ *to be the composition*

$$\pi_n(B) \cong \pi_n(E, F) \xrightarrow{\partial_n} \pi_n(F).$$

Then the sequence

$$\cdots \longrightarrow \pi_n(F) \longrightarrow \pi_n(E) \longrightarrow \pi_n(B) \longrightarrow \pi_{n-1}(F) \longrightarrow \cdots$$
$$\cdots \longrightarrow \pi_1(F) \longrightarrow \pi_1(E) \longrightarrow \pi_1(B).$$

is exact.

By the five lemma (Proposition 5.8.20), we obtain the following analogue of Corollary 5.5.11.

Corollary 5.8.23. *For a Serre fibration* $p : E \to B$ *with* F *the fiber over the base point, we have an isomorphism* $\pi_n(F) \cong \pi_n(\mathrm{hofib}(p))$ *for all* n.

We use the following terminology for those maps that induce isomorphisms of homotopy groups.

Definition 5.8.24. A continuous map $f : X \to Y$ is said to be an n-equivalence if, for any $x \in X$, the induced homomorphisms

$$f_* : \pi_k(X, x) \longrightarrow \pi_k(Y, f(x))$$

satisfy the following conditions:

(1) it is a bijection for any $0 \le k < n$, and
(2) it is a surjection for $k = n$.

A map of pairs $f : (X, A) \to (Y, B)$ is called an n-*equivalence* if the following conditions hold:

(1) The induced map on π_0

$$f_* : \pi_0(X) \longrightarrow \pi_0(Y)$$

satisfies

$$f_*^{-1}(\text{Im}(\pi_0(B) \longrightarrow \pi_0(Y))) = \text{Im}(\pi_0(A) \longrightarrow \pi_0(X)).$$

(2) For any $a \in A$, the induced map

$$f_* : \pi_k(X, A, a) \longrightarrow \pi_k(Y, B, f(a))$$

is

(a) bijective for $0 \le k < n$, and
(b) surjective for $k = n$.

A map is called a *weak homotopy equivalence* if it is an n-equivalence for all n.

When there exists a weak homotopy equivalence from X to Y, we say that they are *weakly homotopy equivalent* and denote $X \underset{w}{\simeq} Y$. We also use the notation $\underset{w}{\simeq}$ to denote the equivalence relation generated by this relation.

Remark 5.8.25. Note that the existence of a weak equivalence is not an equivalence relation.

Before we end this section, we state a useful property of fibrations, which is a corollary to Theorem 5.8.9. Proofs are left to the reader.

Corollary 5.8.26. *Let* $p : E \to B$ *be a fibration and* $f : X \to B$ *a continuous map. If* f *is homotopic to the constant map* $f \simeq *$, *then we* *may deform f into a map whose image is contained in the fiber over $*$. In other words, there exists a map $\tilde{f} : X \to F$ making the diagram*

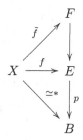

homotopy commutative.

5.9 Cofibrations

In this section, we study the following question:

What makes the path-loop fibration $p : PX \to X$ a fibration?

First note that PX can be written as

$$PX = \{\omega : [0,1] \to X \mid \omega(0) = *\}$$
$$= \mathrm{Map}_*([0,1], X).$$

Here we regard 0 as the base point of $[0,1]$. We also have the following description of X

$$X = \{x \in X\}$$
$$= \{f : \{*\} \to X\}$$
$$= \{f : \{0,1\} \to X \mid f(0) = *\}$$
$$= \{f : \partial[0,1] \to X \mid f(0) = *\}$$
$$= \mathrm{Map}_*(\partial[0,1], X).$$

Lemma 5.9.1. *Under the above identification, p is the map induced by the restriction to $\partial[0,1]$. In other words, we have a commutative diagram*

$$
\begin{array}{ccc}
PX & \xrightarrow{\quad p \quad} & X \\
\| & & \| \\
\mathrm{Map}_*([0,1], X) & \xrightarrow{\quad i^* \quad} & \mathrm{Map}_*(\partial[0,1], X),
\end{array}
$$

where $i : \partial[0,1] \to [0,1]$ is the inclusion map, and i^* is the map given by post-composing i or the restriction to $\partial[0,1]$.

Furthermore the fiber of p can be written as

$$\Omega X = \{\omega : [0,1] \to X \mid \omega(0) = \omega(1) = *\}$$
$$= \mathrm{Map}_*([0,1]/\partial[0,1], X).$$

The above argument implies that the map $p : PX \to X$ is defined by the inclusion map

$$i : \partial[0,1] \hookrightarrow [0,1] \tag{5.13}$$

and is independent of X. The fact that $p : PX \to X$ is a fibration should follow from some property of the map (5.13). Let us be a bit more general and consider a condition on the inclusion map $i : A \hookrightarrow X$ of based spaces under which the induced map

$$i^* : \mathrm{Map}_*(X, Y) \longrightarrow \mathrm{Map}_*(A, Y)$$

is a based fibration for any based space Y.

Suppose that the map $\mathrm{Map}_*(X, Y) \to \mathrm{Map}_*(A, Y)$ is a based fibration. Then for any based space Z and based maps

$$H : Z \times [0,1] \longrightarrow \mathrm{Map}_*(A, Y)$$
$$f : Z \longrightarrow \mathrm{Map}_*(X, Y)$$

making the diagram

$$
\begin{array}{ccc}
Z \times \{0\} & \xrightarrow{\ f\ } & \mathrm{Map}_*(X,Y) \\
\downarrow & & \downarrow{\scriptstyle i^*} \\
Z \times [0,1] & \xrightarrow{\ H\ } & \mathrm{Map}_*(A,Y)
\end{array}
$$

commutative, there exists a based homotopy $\widetilde{H} : Z \times [0,1] \to \mathrm{Map}_*(X, Y)$ making the diagram

$$
\begin{array}{ccc}
Z \times \{0\} & \xrightarrow{\ f\ } & \mathrm{Map}_*(X,Y) \\
\downarrow & {\scriptstyle \widetilde{H}}\nearrow & \downarrow{\scriptstyle i^*} \\
Z \times [0,1] & \xrightarrow{\ H\ } & \mathrm{Map}_*(A,Y)
\end{array}
\tag{5.14}
$$

commutative.

Let us take adjoints and define maps

$$f' : X \times Z \longrightarrow Y \qquad \text{by } f'(x,z) = f(z)(x)$$
$$H' : A \times Z \times [0,1] \longrightarrow Y \quad \text{by } H'(a,z,t) = H(z,t)(a)$$
$$\widetilde{H}' : X \times Z \times [0,1] \longrightarrow Y \quad \text{by } \widetilde{H}'(x,z,t) = \widetilde{H}(z,t)(x),$$

respectively. Then the commutativity of the diagram (5.14) is equivalent to

$$H' = \widetilde{H}'|_{A \times Z \times [0,1]}$$
$$\widetilde{H}'(x, z, 0) = f'(x, z).$$

In order to find conditions on X and A, we introduce new maps

$f'' : X \longrightarrow \mathrm{Map}_*(Z, Y)$ by $f''(x)(z) = f'(x, z)$

$H'' : A \times [0, 1] \longrightarrow \mathrm{Map}_*(Z, Y)$ by $H''(a, t)(z) = H'(a, z, t)$

$\widetilde{H}'' : X \times [0, 1] \longrightarrow \mathrm{Map}_*(Z, Y)$ by $\widetilde{H}''(x, t)(z) = \widetilde{H}'(x, z, t)$.

Then the above conditions are equivalent to

$$\widetilde{H}''|_{A \times [0,1]} = H''$$
$$\widetilde{H}''(x, 0) = f''(x).$$

The above argument can be summarized as follows.

Proposition 5.9.2. *For a based space X and its subspace A containing the base point, let $i : A \hookrightarrow X$ be the inclusion map. Then the induced map*

$$i^* : \mathrm{Map}_*(X, Y) \longrightarrow \mathrm{Map}_*(A, Y)$$

is a based fibration for any based space Y if and only if, for any based space Z, a based map and a based homotopy

$$H : A \times [0, 1] \longrightarrow \mathrm{Map}_*(Z, Y)$$
$$f : X \times [0, 1] \longrightarrow \mathrm{Map}_*(Z, Y)$$

satisfying $H|_{A \times \{0\}} = f|_A$, there exists a based homotopy $\widetilde{H} : X \times [0, 1] \to \mathrm{Map}_*(Z, Y)$ such that

$$\widetilde{H}|_{A \times [0,1]} = H$$
$$\widetilde{H}|_{X \times \{0\}} = f.$$

This observation leads to the following definition.

Definition 5.9.3. An inclusion map $i : A \hookrightarrow X$ is said to have the *homotopy extension property*, or *HEP* for short, with respect to a space Z, if for any maps

$$H : A \times [0, 1] \longrightarrow Z$$
$$f : X \longrightarrow Z$$

with $H|_{A \times \{0\}} = f|_A$, there exists a homotopy $\widetilde{H} : X \times [0, 1] \to Z$ such that

$$\widetilde{H}|_{A \times [0,1]} = H$$
$$\widetilde{H}|_{X \times \{0\}} = f.$$

By requiring all spaces, maps, and homotopy to be based, we obtain the notion of the *based homotopy extension property*.

Remark 5.9.4. The homotopy extension property can be expressed by the following diagram.

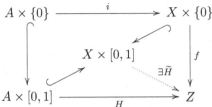

The HEP says that, when the outside square is commutative, there exists a homotopy $\widetilde{H}$ making the diagram commutative.

Theorem 5.9.5. *For a based space X and a subspace A containing a base point, the inclusion map $i : A \hookrightarrow X$ induces a based fibration*

$$i^* : \mathrm{Map}_*(X, Y) \longrightarrow \mathrm{Map}_*(A, Y)$$

for any based space Y if and only if $i : A \hookrightarrow X$ has the based HEP for any based space Z.

Proof. We have already seen that $i^* : \mathrm{Map}_*(X, Y) \to \mathrm{Map}_*(A, Y)$ is a based fibration if and only if i has the based HEP for any space of the form $\mathrm{Map}_*(Z, Y)$. On the other hand, any based space can be expressed as $\mathrm{Map}_*(Z, Y)$, since $\mathrm{Map}_*(\{0, 1\}, Y) = Y$. $\qquad\square$

Definition 5.9.6. The inclusion map $i : A \hookrightarrow X$ is called a *cofibration* if i has the HEP for any space. The quotient space X/A is called the *cofiber* of i.

When all spaces and maps are based, we obtain the notion of *based cofibration*.

Corollary 5.9.7. *For a based space X and a subspace A containing the base point the inclusion map $i : A \hookrightarrow X$ induces a*

$$i^* : \mathrm{Map}_*(X, Y) \longrightarrow \mathrm{Map}_*(A, Y)$$

based fibration for any based space Y if and only if i is a based cofibration. In this case, the fiber of i^* over the base point is $\mathrm{Map}_*(X/A, Y)$.

Proof. The base point of $\mathrm{Map}_*(A, Y)$ is the constant map to the base point of Y. Thus the fiber over the base point is given by

$$(i^*)^{-1}(\{*\}) = \{f \in \mathrm{Map}_*(X, Y) \mid f \circ i = *\}$$
$$= \{f \in \mathrm{Map}_*(X, Y) \mid f|_A = *\}$$
$$= \{f \in \mathrm{Map}(X, Y) \mid f(a) = * \text{ for any } a \in A\}$$
$$= \mathrm{Map}_*(X/A, Y).$$

$\qquad\square$

Example 5.9.8. An important example of a cofibration is the inclusion $A \hookrightarrow X$ of a subcomplex A in a CW complex X. $\square$

Cofibrations are closely related to NDR pairs used in §4.10. See Strøm's paper [Strøm (1966)] for a proof.

Theorem 5.9.9. *For a topological space X and its closed subspace A, the inclusion $A \hookrightarrow X$ is a cofibration if and only if the pair (X, A) is an NDR pair.*

Corollary 5.9.10. *For a CW complex X and its subcomplex A, the pair (X, A) is an NDR pair.*

In mathematics, the prefix "co-" is used to denote a dual concept. The meaning of duality depends on the context, but it often means that it is obtained from the original concept by reversing the direction of arrows (maps). We use the term cofibration since it has properties dual to fibrations. We should be able to obtain properties of cofibrations by reversing arrows in theorems on fibrations.

5.10 Duality between Fibrations and Cofibrations

In order to compare fibrations and cofibrations, let us recall important properties of fibrations we have proved so far.

(1) The pullback of a fibration along a continuous map is again a fibration.

(2) Any continuous map can be replaced by a fibration up to homotopy.

(3) Any continuous map $f : X \to Y$ gives rise to a homotopy fiber sequence

$$\cdots \longrightarrow \Omega^n \mathrm{hofib}(f) \longrightarrow \Omega^n X \longrightarrow \Omega^n Y \longrightarrow \Omega^{n-1} \mathrm{hofib}(f) \longrightarrow \cdots$$
$$\cdots \longrightarrow \Omega \mathrm{hofib}(f) \longrightarrow \Omega X \longrightarrow \Omega Y \longrightarrow \mathrm{hofib}(f) \longrightarrow X \longrightarrow Y.$$

(4) Let $p : E \to B$ be a based fibration with fiber F. Then for any based space X and an integer $n \geq 1$, the sequence

$$[X, \Omega^n F]_* \xrightarrow{(\Omega^n i)_*} [X, \Omega^n E]_* \xrightarrow{(\Omega^n p)_*} [X, \Omega^n B]_*$$

is exact.

(5) For a based map $f : X \to Y$ and a based space Z, the sequence

$$\cdots \longrightarrow [Z, \Omega^n \mathrm{hofib}(f)]_* \longrightarrow [Z, \Omega^n X]_* \longrightarrow [Z, \Omega^n Y]_*$$
$$\longrightarrow [Z, \Omega^{n-1} \mathrm{hofib}(f)]_* \longrightarrow \cdots$$
$$\cdots \longrightarrow [Z, \Omega \mathrm{hofib}(f)]_* \longrightarrow [Z, \Omega X]_* \longrightarrow [Z, \Omega Y]_*$$

induced by the homotopy fiber sequence of f is an exact sequence of groups.

Among these properties, (5) follows from (4) and (5). Let us consider other four properties to see if dual statements hold for cofibrations.

For (1), we first need to find a dual notion to pullback.

Definition 5.10.1. For continuous maps

$$f : X \longrightarrow Y$$
$$g : X \longrightarrow Z,$$

define $Y \amalg Z$ as the quotient map

$$Y \cup_X Z = Y \amalg Z / \sim,$$

where the relation $\sim$ is the equivalence relation generated by $f(x) \sim g(x)$ for $x \in X$. The compositions $Y \hookrightarrow Y \amalg Z \to Y \cup_X Z$ and $Z \hookrightarrow Y \amalg Z \to Y \cup_X Z$ are denoted by

$$f_*(g) : Y \longrightarrow Y \cup_X Z$$
$$g_*(f) : Z \longrightarrow Y \cup_X Z,$$

respectively. Note that we have a commutative diagram

$$
\begin{array}{ccc}
X & \xrightarrow{\ f\ } & Y \\
{\scriptstyle g}\big\downarrow & & \big\downarrow{\scriptstyle f_*(g)} \\
Z & \xrightarrow[g_*(f)]{} & Y \cup_X Z.
\end{array}
$$

This is called the *pushout* of Y and Z over X by f and g.

Proposition 5.10.2. *Let* $i : A \hookrightarrow X$ *be a cofibration, for any continuous map* $f : A \to Y$, $i_*(f)$ *is a cofibration.*

Proof. In order to show that $i_*(f)$ is a cofibration, consider a commutative diagram

$$
\begin{array}{ccc}
Y \times \{0\} & \xrightarrow{\;i_*(f)\;} & (X \cup_A Y) \times \{0\} \\
\downarrow & & \downarrow{\scriptstyle g} \\
Y \times [0,1] & \xrightarrow{\quad H \quad} & Z.
\end{array}
$$

By composing the pushout diagram, we obtain a commutative diagram

$$
\begin{array}{ccccc}
A & =\!=\!=\!=\!= & A \times \{0\} & \xrightarrow{\;\;i\;\;} & X \\
 & & \downarrow{\scriptstyle f} & & \downarrow{\scriptstyle f_*(i)} \\
 & & Y \times \{0\} & \xrightarrow{\;i_*(f)\;} & (X \cup_A Y) \times \{0\} \\
\downarrow & & \downarrow & & \downarrow{\scriptstyle g} \\
A \times [0,1] & \xrightarrow{\;f \times 1_{[0,1]}\;} & Y \times [0,1] & \xrightarrow{\quad H \quad} & Z.
\end{array}
$$

Since i is a cofibration, we obtain a map $G : X \times [0,1] \to Z$ by applying HEP to the outer square. Define $\widetilde{H} : (X \cup_A Y) \times [0,1] \to Z$ by

$$
\widetilde{H}([u], t) = \begin{cases} G(u,t), & u \in X \\ H(u,t), & u \in Y. \end{cases}
$$

This is well defined, since if $u = f(a)$ for $a \in A$, we have

$$
H(u,t) = H(f(a), t) = G(a,t)
$$

by the defining diagram of G. It is straightforward to verify that this homotopy satisfies the conditions for HEP. $\square$

Remark 5.10.3. Note that this proof is obtained by reversing arrows in the proof of Proposition 5.4.2.

Remark 5.10.4. Note also that the pushout is defined as quotient space of the union of two spaces, while the pullback is define as a subspace of the product of two spaces. In general, we have the following correspondence is regarded as a part of the duality in spaces and continuous maps:

$$
X \times Y \Longleftrightarrow X \amalg Y
$$
$$
X \wedge Y \Longleftrightarrow X \vee Y
$$
$$
\text{subspace} \Longleftrightarrow \text{quotient space}
$$
$$
\text{pullback} \Longleftrightarrow \text{pushout}
$$
$$
\text{fibration} \Longleftrightarrow \text{pushout}
$$
$$
\vdots
$$

Another important pair of dual notions is injection and surjection. In the definition of cofibration, we assumed that $i : A \to X$ is an inclusion. On the other hand, we did not assume that the projection of a fibration $p : E \to B$ is a surjection. However, if the base space B is arcwise connected and E is nonempty, we may prove that p is surjective by using CHP with respect to $[0, 1]$.

Let us consider (2) next. As is the case of fibrations, any continuous map can be deformed into a cofibration.

Definition 5.10.5. For a continuous map $f : X \to Y$, define

$$Z_f = Y \amalg X \times [0, 1]/_\sim,$$

where $\sim$ is the equivalence relation generated by

$$f(x) \sim (x, 0)$$

for $x \in X$. This is called the *mapping cylinder* of f.

Define maps

$$i_f : X \hookrightarrow Z_f$$
$$r_f : Z_f \longrightarrow Y$$
$$j_f : Y \hookrightarrow Z_f$$

by

$$i_f(x) = [(x, 1)]$$
$$r_f([x, t]) = f(x) \quad (x, t) \in X \times [0, 1]$$
$$r_f([y]) = y \quad y \in Y$$
$$j_f(y) = [y].$$

The mapping cylinder Z_f of f is a space obtained by attaching the cylinder $X \times [0, 1]$ to Y via f as is shown in Figure 5.10.

The following is an analogue of Theorem 5.5.2.

Theorem 5.10.6. *For any continuous map $f : X \to Y$ the following hold.*

(1) The map $i_f : X \to Z_f$ is a cofibration.

(2) The map f factors as $f = r_f \circ i_f$

$$(5.15)$$

and we have $i_f \simeq j_f \circ f$.

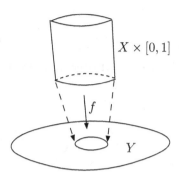

Fig. 5.10   mapping cylinder

(3) The maps r_f and j_f are homotopy inverse to each other and homotopies H and H' for $r_f \circ j_f \simeq 1_Y$ and $j_f \circ r_f \simeq 1_{Z_f}$ can be chosen to make the diagram

$$
\begin{array}{ccc}
X \times [0,1] & \xrightarrow{\mathrm{pr}_1} & X \\
\downarrow{\scriptstyle f \times 1} & & \downarrow{\scriptstyle f} \\
Y \times [0,1] & \xrightarrow{H} & Y
\end{array}
\qquad
\begin{array}{ccc}
X \times [0,1] & \xrightarrow{\mathrm{pr}_1} & X \\
\downarrow{\scriptstyle i_f \times 1} & & \downarrow{\scriptstyle i_f} \\
Z_f \times [0,1] & \xrightarrow{H'} & Z_f
\end{array}
$$

commutative and homotopy commutative, respectively.

Proof. Let us first show that i_f is a cofibration. It is induced from the inclusion $X \times \{0\} \hookrightarrow X \times [0,1]$ and can be regarded as an inclusion of a subspace. Suppose a commutative diagram

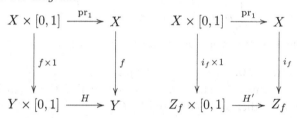

is given. A homotopy $\widetilde{H} : Z_f \times [0,1] \to Z$ can be defined by

$$\widetilde{H}([y], t) = \varphi(y)$$

for $y \in Y$ and by

$$\widetilde{H}([(x,s)],t) = \begin{cases} H(x,t-2(1-s)) & t \geq 2(1-s) \\ \varphi(x,\frac{2s}{2-t}) & t \leq 2(1-s) \end{cases}$$

for $(x,s) \in X \times [0,1]$. It is left to the reader to verify the following:

(1) $\widetilde{H}$ is well-defined.
(2) $\widetilde{H}|_{X\times[0,1]} = H$.
(3) $\widetilde{H}|_{Z_f \times \{0\}} = \varphi$.
(4) The diagram (5.15) is commutative.
(5) $j_f \circ f \simeq i_f$.

Let us show that r_f and j_f are homotopy inverse to each other. Since $r_f \circ j_f = 1_Y$, it remains to find a homotopy for $j \circ r \simeq 1_{Z_f}$. Since

$$\begin{cases} (j_f \circ r_f)([y]) = j(y) = y, & y \in Y \\ (j_f \circ r_f)([x,s]) = j_f(f(x)) = f(x), & (x,s) \in X \times [0,1] \end{cases},$$

we define

$$\begin{cases} H'([y],t) = y, & y \in Y \\ H'([(x,s)],t) = (x,st), & (x,s) \in X \times [0,1] \end{cases}$$

then this is a homotopy between $j \circ r$ and 1_{Z_f}. $\square$

Definition 5.10.7. We say the map $i_f : X \hookrightarrow Z_f$ is obtained by *replacing f by a cofibration*. The cofiber of i_f is called the mapping cone or the homotopy cofiber and is denoted by C_f.

The quotient map $Z_f \to C_f$ is denoted by q_f.

A direct description of C_f is given as follows.

Lemma 5.10.8. *Let* $CX = X \times [0,1]/X \times \{0\}$ *be the cone[8] on* X. *Then we have*

$$C_f = (Y \amalg CX)/\sim,$$

where $\sim$ *is the equivalence relation generated by* $(x,0) \sim f(x)$.

In other words, the mapping cone of f is obtained by attaching the cone of X to Y via f, as is shown in Figure 5.11.

Let us consider the case of based maps. In the case of fibrations, the mapping track construction can be used without a change to replace a based map by a based fibration. In the case of cofibrations, we need to modify the definition of the mapping cylinder to make i_f into a based map. For

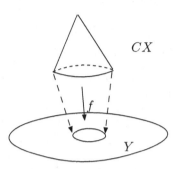

Fig. 5.11   mapping cone

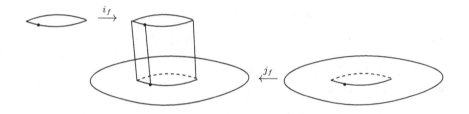

Fig. 5.12   base points and i_f and j_f

example, if we choose $(*, 1)$ as a base point of Z_f so that i_f is a based map, then $j_f : Y \hookrightarrow Z_f$ does not preserve base points, as is shown in Figure 5.12.

In order to avoid this problem, we identify $(*, 1)$ and the base point of Y together with points in the interval between these two points.

Definition 5.10.9. For a based map $f : X \to Y$, define
$$\widetilde{Z}_f = Z_f / \{*\} \times [0, 1].$$
This is called the *reduced mapping cylinder* of f. And
$$\widetilde{C}_f = C_f / \{*\} \times [0, 1]$$
is called the *reduced mapping cone* of f. The quotient map $\widetilde{Z}_f \to \widetilde{C}_f$ is also denoted by q_f.

---

[8]Definition 4.4.8.

In particular, when f is the identity map, we define

$$\widetilde{C}X = CX/\{*\} \times [0,1].$$

This is called the *reduced cone* of X.

It is straightforward to modify the proof of Theorem 5.10.6 for based maps. The proof is omitted.

Theorem 5.10.10. *For a based map $f : X \to Y$, Theorem 5.10.6 holds if we replace Z_f by $\widetilde{Z}_f$ and maps and homotopies by based ones.*

As is the case of fibrations, we should be able to obtain a sequence of cofibrations by iterating this "replacing with cofibration" construction.

Let $f : X \to Y$ be a based map. By Theorem 5.10.10, we obtain a sequence

$$X \xrightarrow{f} Y \xrightarrow{j_f} \widetilde{C}_f$$

is which $\widetilde{C}_f$ is the (reduced) homotopy cofiber of f.

Lemma 5.10.11. *The map $j_f : Y \to \widetilde{C}_f$ is a based cofibration.*

Proof. We show that j_f has the based HEP for any space Z. Suppose a commutative diagram

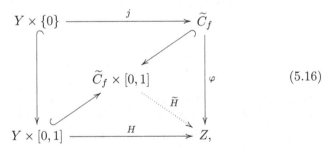

$$(5.16)$$

is given. By the definition of $\widetilde{C}_f$, we may rewrite $\widetilde{C}_f \times [0,1]$ as follows

$$\widetilde{C}_f \times [0,1] = (Y \times [0,1] \amalg X \times [0,1] \times [0,1]) /_\sim,$$

where $\sim$ is the equivalence relation generated by

$$(f(x), t) \sim (x, 0, t)$$
$$(*, t) \sim *$$

for $x \in X$ and $t \in [0,1]$.

Under this identification, define $\widetilde{H} : \widetilde{C}_f \times [0,1] \to Z$ by

$$\widetilde{H}([u]) = \begin{cases} H(f(x), t - s), & \exists x \in X \text{ s.t. } u = (x, s, t) \text{ and } s \geq t \\ \varphi([(x, s - t)]), & \exists x \in X \text{ s.t. } u = (x, s, t) \text{ and } s \leq t \\ H(y, t), & \exists y \in Y \text{ s.t. } u = (y, t). \end{cases}$$

The domains of the first and the second maps overlap when $s = t$, in which case we have

$$H(f(x), 0) = \varphi(j(x)) = \varphi([(x, 0)]).$$

There is also an overlap of domains in the first and the third maps when $s = 0$, but their values coincide. Furthermore, since both H and φ preserve base points, $\{*\} \times [0,1]$ is mapped to $\{*\}$. Thus $\widetilde{H}$ is well-defined and continuous.

The $\widetilde{H}$ is defined in such a way it makes the diagram (5.16) commutative. $\qquad\square$

Thanks to this lemma, we do not need to replace j_f by a cofibration. In order to describe its cofiber, we need the following construction.

Definition 5.10.12. For a topological space X, define

$$\Sigma X = CX / X \times \{1\}.$$

This is called the *suspension* of X. When X is a based space, define

$$\widetilde{\Sigma}X = \widetilde{C}X / X \times \{1\} = \Sigma X / \{*\} \times [0,1].$$

This is called the *reduced suspension* of X.

Remark 5.10.13. Since

$$\widetilde{\Sigma}X = X \times [0,1] / (X \times \partial[0,1] \cup \{*\} \times [0,1]),$$

the homeomorphism $[0,1]/\partial[0,1] \cong S^1$ induces a homeomorphism

$$\widetilde{\Sigma}X \cong X \wedge S^1.$$

Hence iterated suspensions can be written as

$$\widetilde{\Sigma}^n X \cong X \wedge \underbrace{S^1 \wedge \cdots \wedge S^1}_{n} \cong X \wedge S^n.$$

Note that we used Lemma 5.7.5 for the last homeomorphism.

Lemma 5.10.14. *For a based map* $f : X \to Y$, *the cofiber of* j_f *is* $\widetilde{\Sigma}X$.

Proof. Since j_f is the map which embeds Y into the bottom of $\widetilde{C}_f$, we have

cofiber of $j_f = \widetilde{C}_f/Y$

$$= ((Y \amalg (X \times [0,1]))/(X \times \{1\} \cup \{*\} \times [0,1]))/{\sim})/Y$$
$$= X \times [0,1]/X \times \{0,1\} \cup \{*\} \times [0,1]$$
$$= \widetilde{\Sigma}X.$$

$\square$

As is the case of fibrations, the next step is a bit complicated. We first replace j_f by a cofibration and obtain a diagram

$$
\begin{array}{ccccc}
Y & \longrightarrow & \widetilde{Z}_{j_f} & \xrightarrow{\ q_{f_f}\ } & \widetilde{C}_{j_f} \\
\downarrow{\scriptstyle \|} & & \downarrow{\scriptstyle \simeq} & & \downarrow{\scriptstyle \simeq?} \\
Y & \xrightarrow{\ j_f\ } & \widetilde{C}_f & \longrightarrow & \widetilde{\Sigma}X.
\end{array}
\tag{5.17}
$$

In the diagram (5.17), the middle vertical map is a homotopy equivalence. In order to obtain a homotopy equivalence between cofibers, we need a concept which is dual to fiber homotopy equivalence.

Definition 5.10.15. Let

$$i : A \hookrightarrow X$$
$$i' : A \hookrightarrow X'$$

be cofibrations. We say i and i' are *cofiber homotopy equivalent* if there exist map f and g making the diagram

$$
\begin{array}{ccc}
A =\!=\!= A \\
i \downarrow \quad\quad \downarrow i' \\
X \xrightarrow{\ f\ } X'
\end{array}
\qquad
\begin{array}{ccc}
A =\!=\!= A \\
i' \downarrow \quad\quad \downarrow i \\
X' \xrightarrow{\ g\ } X
\end{array}
$$

commutative and that there exist homotopies H and H' for $g \circ f \simeq 1_X$ and $f \circ g \simeq 1_{X'}$ making the diagram

$$
\begin{array}{ccc}
A \times [0,1] \xrightarrow{\ \mathrm{pr}_1\ } A \\
\downarrow \quad\quad\quad \downarrow \\
X \times [0,1] \xrightarrow{\ H\ } X
\end{array}
\qquad
\begin{array}{ccc}
A \times [0,1] \xrightarrow{\ \mathrm{pr}_1\ } A \\
\downarrow \quad\quad\quad \downarrow \\
X' \times [0,1] \xrightarrow{\ H'\ } X'
\end{array}
$$

commutative.

Maps f and g are called *cofiber homotopy equivalences*.

The following fact is dual to Corollary 5.4.11 and can be proved by "reversing arrows". The proof is left to the reader.

Proposition 5.10.16. *If cofibrations $i : A \hookrightarrow X$ and $i' : A \hookrightarrow X'$ are cofiber homotopy equivalent to each other, their cofibers are based homotopy equivalent.*

Thus we see that the map i_{j_f} in the diagram (5.17) induces a homotopy equivalence $\overline{i_{j_f}} : \widetilde{C}_{j_f} \xrightarrow{\simeq} \widetilde{\Sigma}X$.
The composition

$$\widetilde{C}_f \xrightarrow{i_{j_f}} \widetilde{Z}_{j_f} \xrightarrow{q_{j_f}} \widetilde{C}_{j_f}$$

is a cofibration and its cofiber is

$$\widetilde{C}_{j_f} / \widetilde{C}_f = \widetilde{\Sigma}Y.$$

Let us denote the quotient map $\widetilde{C}_{j_f} \to \widetilde{\Sigma}Y$ by π_f and consider the composition

$$\widetilde{\Sigma}X \xrightarrow{\overline{i_{j_f}}} \widetilde{C}_{j_f} \xrightarrow{\pi_f} \widetilde{\Sigma}Y.$$

As is the case of fibrations, this map is not homotopic to $\widetilde{\Sigma}f$.

Definition 5.10.17. For a based map $f : X \to Y$, define maps

$$\widetilde{\Sigma}f, -\widetilde{\Sigma}f : \widetilde{\Sigma}X \longrightarrow \widetilde{\Sigma}Y$$

by

$$(\widetilde{\Sigma}f)([x,t]) = [f(x),t]$$
$$(-\widetilde{\Sigma}f)([x,t]) = [f(x),1-t].$$

Lemma 5.10.18. *For a based map $f : X \to Y$, we have $\pi_f \circ \overline{i_{j_f}} \simeq -\widetilde{\Sigma}f$.*

Thus we obtain a sequence of based spaces and based maps

$$X \xrightarrow{f} Y \xrightarrow{j_f} \widetilde{C}_f \longrightarrow \widetilde{\Sigma}X \xrightarrow{-\widetilde{\Sigma}f} \widetilde{\Sigma}Y$$

in which

(1) $\widetilde{C}_f$ is the homotopy cofiber of f.
(2) $\widetilde{\Sigma}X$ the cofiber of j.
(3) $\widetilde{\Sigma}Y$ is homotopy equivalent to the cofiber of $\widetilde{C}_f \to \widetilde{\Sigma}X$.

Definition 5.10.19. A sequence of based spaces and based maps

$$X \xrightarrow{f} Y \xrightarrow{g} W$$

is called a *homotopy cofiber sequence* if the homotopy cofiber of f is homotopy equivalent to W by g.

By iterating this construction, we obtain the following sequence. The proof is omitted.

Theorem 5.10.20. *For a based map $f : X \to Y$, there exists a sequence of based spaces and based maps*

$$X \xrightarrow{f} Y \xrightarrow{j} \widetilde{C}_f \longrightarrow \widetilde{\Sigma} X \xrightarrow{-\widetilde{\Sigma} f} \widetilde{\Sigma} Y \xrightarrow{-\widetilde{\Sigma} j} \widetilde{\Sigma} \widetilde{C}_f \longrightarrow \cdots$$

$$\cdots \longrightarrow \widetilde{\Sigma}^{n-1} \widetilde{C}_f \longrightarrow \widetilde{\Sigma}^n X \xrightarrow{(-1)^n \widetilde{\Sigma}^n f} \widetilde{\Sigma}^n Y \xrightarrow{(-1)^n \widetilde{\Sigma}^n j} \widetilde{\Sigma}^n \widetilde{C}_f \longrightarrow \cdots$$

in which any consecutive three terms is a homotopy cofiber sequence.

Definition 5.10.21. The sequence in the above theorem is called the *dual Puppe sequence* of f.

Let us consider (4), i.e. an analogue of Theorem 5.8.9, which is a relation among based homotopy sets in which Ω^n appears in the range. The construction of the dual Puppe sequence suggests that we should replace Ω^n by $\widetilde{\Sigma}^n$ and switch domains and ranges to obtain a corresponding result for cofibrations.

Theorem 5.10.22. *For a based cofibration $A \hookrightarrow X$ and a based space Z, the sequence*

$$\left[\widetilde{\Sigma}^n (X/A), Z\right]_* \xrightarrow{(\widetilde{\Sigma}^n q)^*} \left[\widetilde{\Sigma}^n X, Z\right]_* \xrightarrow{(\widetilde{\Sigma}^n i)^*} \left[\widetilde{\Sigma}^n A, Z\right]_*$$

is an exact sequence of groups for $n \geq 1$, where $q : X \to X/A$ is the quotient map.

Remark 5.10.23. The maps $(\widetilde{\Sigma}^n q)^*$ and $(\widetilde{\Sigma}^n i)^*$ are given by recomposing $\widetilde{\Sigma}^n q$ and $\widetilde{\Sigma}^n i$, respectively. See Definition 5.8.2.

Before we go into the proof of Theorem 5.10.22, we need to make the homotopy set $\left[\widetilde{\Sigma} X, Y\right]_*$ into a group for any based sets X and Y.

Lemma 5.10.24. *For based spaces X and Y, the based homotopy set $[\widetilde{\Sigma} X, Y]_*$ has a structure of a group.*

Proof. By Theorem 5.7.6, we obtain a homeomorphism

$$\mathrm{Map}_*(X \wedge Z, Y) \cong \mathrm{Map}_*(X, \mathrm{Map}_*(Z, Y)).$$

When $Z = S^1$, we have

$$\mathrm{Map}_*(\widetilde{\Sigma}X, Y) = \mathrm{Map}_*(X \wedge S^1, Y) \cong \mathrm{Map}_*(X, \mathrm{Map}_*(S^1, Y)) = \mathrm{Map}_*(X, \Omega Y)$$

by Remark 5.10.13. By taking homotopy classes, we obtain a bijection

$$\left[\widetilde{\Sigma}X, Y\right]_* \cong [X, \Omega Y]_*.$$

By Corollary 5.8.3, the loop product makes $[X, \Omega Y]_*$ into a group and the above bijection can be used to define a group structure on $\left[\widetilde{\Sigma}X, Y\right]_*$. $\square$

This proof of Lemma 5.10.24 suggests that Theorem 5.10.22 also follows from the corresponding result for fibrations. In fact, it is the case.

Proof of Theorem 5.10.22. By Corollary 5.9.7, when $i : A \hookrightarrow X$ is a based cofibration, the map induced by i

$$i^* : \mathrm{Map}_*(X, Z) \longrightarrow \mathrm{Map}_*(A, Z)$$

is a based fibration with fiber $\mathrm{Map}_*(X/A, Z)$. Thus by Corollary 5.8.12, we obtain an exact sequence

$$\pi_n(\mathrm{Map}_*(X/A, Z)) \longrightarrow \pi_n(\mathrm{Map}_*(X, Z)) \longrightarrow \pi_n(\mathrm{Map}_*(A, Z)).$$

Theorem 5.7.6 allows us to rewrite

$$\begin{aligned}
\pi_n(\mathrm{Map}_*(Y, Z)) &= [S^n, \mathrm{Map}_*(Y, Z)]_* \\
&= [S^n \wedge Y, Z]_* \\
&\cong [Y \wedge S^n, Z]_* \\
&\cong [\widetilde{\Sigma}^n Y, Z]_*.
\end{aligned}$$

Note that we used a homeomorphism in Remark 5.10.13. And we obtain an exact sequence

$$\pi_n(\mathrm{Map}_*(X/A, Z)) \longrightarrow \pi_n(\mathrm{Map}_*(X, Z)) \longrightarrow \pi_n(\mathrm{Map}_*(A, Z))$$

$$\left\| \qquad\qquad\qquad \right\| \qquad\qquad\qquad \left\| \right.$$

$$\left[\widetilde{\Sigma}^n(X/A), Z\right]_* \longrightarrow \left[\widetilde{\Sigma}^n X, Z\right]_* \longrightarrow \left[\widetilde{\Sigma}^n A, Z\right]_*.$$

$$\square$$

We obtain a long exact sequence by combining Theorem 5.10.20 and Theorem 5.10.22.

Corollary 5.10.25. *For a based map* $f : X \to Y$ *and a based space* Z, *we have an exact sequence of groups*

$$\cdots \longrightarrow \left[\widetilde{\Sigma}^n \widetilde{C}_f, Z\right]_* \longrightarrow \left[\widetilde{\Sigma}^n Y, Z\right]_* \longrightarrow \left[\widetilde{\Sigma}^n X, Z\right]_* \longrightarrow \cdots$$

$$\cdots \longrightarrow \left[\widetilde{\Sigma}^2 X, Z\right]_* \longrightarrow \left[\widetilde{\Sigma} \widetilde{C}_f, Z\right]_* \longrightarrow \left[\widetilde{\Sigma} Y, Z\right]_* \longrightarrow \left[\widetilde{\Sigma} X, Z\right]_* .$$

We conclude this section by the following fact, which is a dual to Corollary 5.8.26.

Proposition 5.10.26. *Let* $i : A \hookrightarrow X$ *be a cofibration and* $f : X \to Y$ *a continuous map. If the composition*

$$A \hookrightarrow X \longrightarrow Y$$

is homotopic to the constant map, the there exists a continuous map

$$\tilde{f} : X/A \longrightarrow Y$$

making the diagram

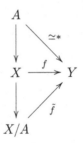

homotopy commutative.

Proof. Let

$$H : A \times [0,1] \longrightarrow Y$$

be a homotopy from $f \circ i$ and the constant map $*$. Since i is a cofibration, there exists a homotopy $\widetilde{H}$ making the diagram

$$
\begin{array}{ccc}
A \times \{0\} & \xrightarrow{\quad j \quad} & X \times \{0\} \\
\downarrow & & \downarrow f \\
& X \times [0,1] & \\
\downarrow & \quad \exists \widetilde{H} & \downarrow \\
A \times [0,1] & \xrightarrow{\quad H \quad} & Y
\end{array}
$$

commutative.

Define a map $\tilde{f} : X/A \to Y$ by
$$\tilde{f}([x]) = \tilde{H}(x, 1).$$
This is well defined and continuous for $\tilde{H}(A \times \{1\}) = H(A \times \{1\}) = \{*\}$. Furthermore $\tilde{H}$ is a homotopy between f and the composition $X \to X/A \xrightarrow{\tilde{f}}$, which implies that $\tilde{f}$ satisfies the requirement. □

This fact is often used in the following form.

Corollary 5.10.27. *Let $f : X \to Y$ and $g : Y \to Z$ be continuous maps. If the composition $g \circ f$ is homotopic to the constant map, then g has an extension $\tilde{g} : C_f \longrightarrow Z$.*

In other words, we have the following homotopy commutative diagram

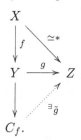

5.11 Quasifibrations

We introduced Serre fibrations and Hurewicz fibrations as generalizations of fiber bundles. In this section, we introduce a further generalization. Let us begin with a motivation.

In the proof of the classification theorem of principal bundles, the key is a construction of a principal bundle whose total space has trivial homotopy groups. We used the long exact sequence of homotopy groups[9] in the construction of universal bundles in §4.9. In general, it is quite hard to construct a contracting homotopy by hand. On the other hand, lots of techniques have been developed to compute homology and homotopy groups, one of which is the long exact sequence of homotopy groups.

A generalization of fibrations which only requires the existence of a long exact sequence of homotopy groups would be useful. Such a generalization is called a quasifibration, which is the subject of this section.

The notion of quasifibrations was introduced by the paper [Dold and Thom (1958)] and basic properties are already studied in the paper, which

---

[9]Corollary 4.8.12.

makes this paper still one of the fundamental references for quasifibrations. Unfortunately, however, the paper is written in German. The reader is recommended to use [May (1990)] as a reference.

In the first half of this section, we review basic properties of quasifibrations based on these two papers. In the second half, two applications, i.e. Milgram's construction of universal bundles for topological monoids and infinite symmetric products, are exposed. Proofs are basically omitted.

Recall that if $p : E \to B$ is a Hurewicz fibration and B is arcwise connected, the fiber $F_x = p^{-1}(x)$ and the homotopy fiber $\mathrm{hofib}_x(p)$ over x are homotopy equivalent[10] $F_x \simeq \mathrm{hofib}_x(p)$ for any $x \in B$. This homotopy equivalence is given by the map $i : F_x \to \mathrm{hofib}_x(p)$ defined by $i(y) = (y, c_x)$, where c_x is the constant loop at x. If p is a Serre fibration, the map i may not be a homotopy equivalence but it induces an isomorphism on homotopy groups[11]

$$i_* : \pi_*(F_x) \xrightarrow{\cong} \pi_*(\mathrm{hofib}_x(p)).$$

Recall from Definition 5.8.24 that such a map is called a weak homotopy equivalence. It should be a reasonable idea to use this property to generalize Serre fibrations.

Definition 5.11.1. A continuous map $p : E \to B$ is called a *quasifibration* if

(1) p is surjective, and
(2) the map $p : (E, p^{-1}(x)) \to (B, x)$ is a weak homotopy equivalence for any $x \in B$.

As is the case of fibrations, B and E are called the *base space* and the *total space*, respectively. For $x \in B$, $p^{-1}(x)$ is called the *fiber* over x.

Example 5.11.2. Any Serre fibration is a quasifibration.   □

Corollary 5.11.3. *Let* $p : E \to B$ *be a quasifibration. Then for any* $x \in B$ *and* $y \in F = p^{-1}(x)$, *we have the following long exact sequence*

$$\cdots \longrightarrow \pi_n(F, y) \longrightarrow \pi_n(E, y) \longrightarrow \pi_n(B, x) \longrightarrow \pi_{n-1}(F, y) \longrightarrow \cdots$$
$$\cdots \longrightarrow \pi_2(B, x) \longrightarrow \pi_1(F, y) \longrightarrow \pi_1(E, y) \longrightarrow \pi_1(B, x).$$

The following is an example of a quasifibration appeared in [Dold and Thom (1958)] which is not a fibration.

---

[10]Corollary 5.5.11.
[11]Corollary 5.8.23.

Example 5.11.4. Let p be the map from the region E in $\mathbb{R}^2$ to $B = [-1, 1]$ given by Figure 5.13. More precisely, we may use

$$E = [-1, 0] \times [-2, -1] \cup \{0\} \times [-2, 2] \cup [0, 1] \times [1, 2]$$

and $p : E \to B$ is the projection onto the first coordinate.

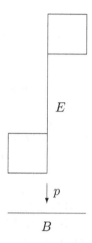

Fig. 5.13   a quasifibration which is not a fibration

This map does not have the CHP. For example, let $\ell : [-1, 1] \to B$ be the identity map regarded as a path in B. This path does not have a continuous lift and thus p is not a Serre fibration. This is a quasifibration, since E, B, and arbitrary fibers are contractible. $\square$

The reader might expect that properties of fibrations hold for quasifibrations after a slight modification. In fact, Corollary 5.11.3 is one of such properties. Unfortunately, however, the pullback of a quasifibration may not be a quasifibration.

Example 5.11.5. Define a subspace E of $\mathbb{R}^3$ by

$$E = ([-1, 1] \times [-1, 1] \times [0, 1] \setminus [-1, 1] \times [0, 1] \times \{1\})$$
$$\cup \left\{ (x, y, yz + 1) \,\middle|\, x^2 + z^2 = 1, -1 \le x \le 1, 0 \le y \le 1, z \ge 0 \right\}.$$

See Figure 5.14.

Let $B = [-1, 1] \times [-1, 1] \times [0, 1]$ and $p : E \to B$ be the map which collapses the round roof. This is a quasifibration is both E and B are contractible and all fibers are single points.

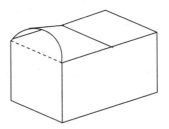

Fig. 5.14   quasifibration cannot be restricted

On the other hand, the restriction of p to $[-1,1] \times \{1\} \times [0,1]$, i.e. the pullback along the inclusion $i : [-1,1] \times \{1\} \times [0,1] \hookrightarrow B$ is not a quasifibration any more. To see this, define $B' = [-1,1] \times \{1\} \times [0,1]$. The restriction of p to B' is denoted by $p' : E' \to B'$. If this is a quasifibration, $\pi_*(E',p'^{-1}(a)) \cong \pi_*(B')$ for any $a \in B'$. Since the fiber $p'^{-1}(a)$ is a single point and B' is contractible, we should have $\pi_*(E') = 0$ by the long exact sequence of homotopy groups. However $E' \simeq S^1$ and thus $\pi_1(E') \neq 0$. Therefore p' is not a quasifibration. $\square$

Since we may not always restrict a quasifibration, we use the following terminology.

Definition 5.11.6. Let $p : E \to B$ be a continuous map. A subset $U \subset B$ is called *distinguished* if the restriction $p|_{p^{-1}(U)} : p^{-1}(U) \to U$ is a quasifibration.

The pullback is a quite useful way of constructing new fiber bundles or fibrations from known fiber bundles or fibrations. We need an alternative way of constructing new quasifibrations or showing a map to be a quasifibration.

One of the most useful methods already appeared in the paper by Dold and Thom. The idea is to glue quasifibrations to obtain a new quasifibration.

Theorem 5.11.7. *A continuous map $p : E \to B$ is a quasifibration, if there exists an open covering $\mathcal{U}$ of B satisfying the following conditions:*

(1) For any $U, V \in \mathcal{U}$, $U \cap V \in \mathcal{U}$.
(2) Each $U \in \mathcal{U}$ is distinguished, i.e. the restriction $p|_U : p^{-1}(U) \to U$ is a quasifibration.

See [Dold and Thom (1958)] or May's paper [May (1990)] for a proof. The following criterion is also due to Dold and Thom. Although the conditions appear to be quite complicated, it turns out to be quite useful for practical applications.

Theorem 5.11.8 (Dold-Thom Criterion). *Let* $p : E \to B$ *a continuous map. Suppose B has a filtration $\{F_n B\}_{n \geq 0}$ satisfying the following conditions:*

(1) B *has the weak topology with respect to* $\{F_n B\}_{n \geq 0}$.
(2) *For each* $n > 0$, $(F_n B, F_{n-1} B)$ *is an NDR pair.*
(3) $F_0 B$ *is distinguished.*
(4) *For each* $n > 0$, *any open subset* $U \subset F_n B \backslash F_{n-1} B$ *is distinguished.*
(5) *For each* $n > 0$, *there exists an open neighborhood* U_n *of* $F_{n-1} B$ *in* $F_n B$ *and maps*

$$h : U_n \times [0, 1] \longrightarrow U_n$$

$$H : p^{-1}(U_n) \times [0, 1] \longrightarrow p^{-1}(U_n)$$

satisfying the following conditions:
  (a) h *is a deformation retract of* U_n *onto* $F_{n-1} B$.
  (b) H *is a deformation retract of* $p^{-1}(U_n)$ *onto* $p^{-1}(F_{n-1} B)$.
  (c) (H_1, h_1) *is a fiber-preserving map.*
  (d) *For each* $b \in U_n$, $H_1 : p^{-1}(b) \to p^{-1}(h_1(b))$ *is a weak homotopy equivalence.*

Then each $F_n B$ *is distinguished and* p *is a quasifibration.*

This is proved by induction on n by using Theorem 5.11.7 and the following fact. See May's paper for details.

Lemma 5.11.9. *Let* $p : E \to B$ *be a continuous map and* $A \subset B$ *a distinguished subspace. If there exist homotopies*

$$h : U \times [0, 1] \longrightarrow U \tag{5.18}$$

$$H : p^{-1}(U) \times [0, 1] \longrightarrow p^{-1}(U) \tag{5.19}$$

such that

(1) h *is a deformation retract of* U *onto* A,
(2) H *is a deformation retract of* $p^{-1}(U)$ *onto* $p^{-1}(A)$,
(3) (H, h) *is a fiber-preserving map, and*
(4) *for each* $b \in B$, $H_1 : p^{-1}(b) \to p^{-1}(h_1(p))$ *is a weak homotopy equivalence,*

then p *is a quasifibration.*

We conclude this section with applications of the Dold-Thom criterion.

Example 5.11.10. Let M be a topological monoid.[12] In Milgram's construction[13] of universal bundles for topological groups, we did not use inverses. The construction can be applied to topological monoids and we obtain topological spaces EM and BM and a continuous map $p : EM \to BM$. Although p may not be a fiber bundle if M is not a topological group, Milgram proved that this map is a quasifibration under a mild condition by using the Dold-Thom criterion (Theorem 5.11.8).

Note that spaces EM and BM are equipped with filtrations $\{E_n M\}$ and $\{B_n M\}$ by definition. It is easy to verify the conditions (1) and (3). We have already verified (2) in the proof of Theorem 4.10.19 under the condition that $(M, \{e\})$ is an NDR pair. The map (4.18) also defines a homeomorphism

$$\varphi_n : E_n M \setminus E_{n-1} M \xrightarrow{\cong} (B_n M \setminus B_{n-1} M) \times M$$

for any topological monoid M.

It remains to prove (5). We need to require that $(M, \{e\})$ is a *strong NDR pair* in the sense that it has such an NDR representation (h, u) that $u^{-1}([0,1))$ is preserved by the homotopy h. When $(M, \{e\})$ is a strong NDR pair, the construction in the proof of Theorem 4.10.19 gives us strong NDR representations $(\tilde{h}_n, \tilde{u}_n)$ and (h_n, u_n) of $(E_n M, E_{n-1} M)$ and $(B_n M, B_{n-1} M)$, respectively. Define $U_n = u_n^{-1}([0,1))$. The restriction of h_n defines a homotopy $h : U_n \times [0,1] \to U_n$, which is a deformation retraction of U_n onto $B_{n-1} M$. The restriction of $\tilde{h}_n$ is a deformation retraction

$$H : p^{-1}(U_n) \times [0,1] \longrightarrow p^{-1}(U_n)$$

of $p^{-1}(U_n)$ onto $p^{-1}(B_{n-1} M) = E_{n-1} M$. And (H_1, h_1) is a fiber-preserving map by construction. □

Example 5.11.11. Nowadays Milgram's construction is regarded as a special case of the following construction.

For a topological monoid M, a space X with a right action of M and a space Y with a left action of M, define $B_n(X, M, Y) = X \times M^n \times Y$. Then the collection $B_*(X, M, Y) = \{B_n(X, M, Y)\}_{n \geq 0}$ has a structure of a simplicial space[14] and is called the *two-sided bar construction*. See Example A.3.6. When $X = * = Y = M$ or $Y = *$, the geometric realizations are $|B_*(*, M, M)| = EM$ and $|B_*(*, M, *)| = BM$.

---

[12]Definition 3.4.20.
[13]Definition 4.10.11.
[14]Definition A.3.7.

The argument in Example 5.11.10 can be used to prove that the map

$$|B_*(X, M, M)| \longrightarrow |B_*(X, M, *)|$$

induced by the collapsing map $M \to *$ is a quasifibration, when $(M, \{e\})$ is a strong NDR pair.

See Appendix A.3 for a summary of simplicial techniques. □

Dold and Thom introduced quasifibrations in order to study *infinite symmetric products*.

Definition 5.11.12. Let Σ_n denote the symmetric group of n letters. For a topological space X, define a right action $\mu : X^n \times \Sigma_n \to X^n$ of Σ_n on the n-fold product X^n by

$$\mu(x_1, \ldots, x_n; \sigma) = (x_{\sigma(1)}, \ldots, x_{\sigma(n)})$$

for $\sigma \in \Sigma_n$ and $(x_1, \ldots, x_n) \in X^n$.

The *n-fold symmetric product* of X is defined to be the quotient space

$$\mathrm{SP}^n(X) = X^n / {}_{\Sigma_n}.$$

The element of $\mathrm{SP}^n(X)$ represented by $(x_1, \cdots, x_n) \in X^n$ is denoted by $[x_1, \cdots, x_n]$.

For each n, define an inclusion $\mathrm{SP}^n(X) \hookrightarrow \mathrm{SP}^{n+1}(X)$ by $[x_1, \ldots, x_n] \mapsto [x_1, \ldots, x_n, *]$ and define

$$\mathrm{SP}^\infty(X) = \bigcup_{n=1}^\infty \mathrm{SP}^n(X).$$

It is topologized by the weak topology with respect to $\{\mathrm{SP}^n(X)\}_{n \geq 1}$. This is called the *infinite symmetric product* of X.

Example 5.11.13. It is an important observation by Dold and Thom that the infinite symmetric product relates homotopy groups and homology groups. Namely, for a based space X with a nondegenerate base point, we have a natural isomorphism

$$\pi_*(\mathrm{SP}^\infty(X)) \cong \widetilde{H}_*(X; \mathbb{Z}),$$

where the right hand side is the reduced homology groups with coefficients in $\mathbb{Z}$.

This is a consequence of the uniqueness of homology theories due to Eilenberg and Steenrod [Eilenberg and Steenrod (1952)]. In other words,

$$\pi_* \circ \mathrm{SP}^\infty : \mathbf{Top} \longrightarrow \mathbf{GradeAbel}$$

is a functor from the category of topological spaces to that of graded Abelian groups satisfying the Eilenberg-Steenrod axioms.

Among the axioms, the existence of a long exact sequence

$$\cdots \longrightarrow \pi_n(\mathrm{SP}^\infty(A)) \longrightarrow \pi_n(\mathrm{SP}^\infty(X)) \longrightarrow \pi_n(\mathrm{Sp}^\infty(X/A))$$
$$\longrightarrow \pi_{n-1}(\mathrm{SP}^\infty(A)) \longrightarrow \cdots$$

follows the fact that the map

$$\mathrm{SP}^\infty(X) \longrightarrow \mathrm{SP}^\infty(X/A)$$

induced by the quotient map $X \to X/A$ is a quasifibration with fiber $\mathrm{SP}^\infty(A)$ if (X, A) is an NDR pair.

Dold and Thom proved this fact by using Theorem 5.11.8. The proof is analogous to the case of Example 5.11.10. Instead of repeating the argument, we show that this is a special case of Example 5.11.11 following [Tamaki (2013a,b)].

It is a fundamental property of NDR pairs that the collapsing map $X \cup CA \to (X \cup CA)/CA = X/A$ is a homotopy equivalence. We also have a homotopy equivalence $X \simeq X \cup A \times [0, 1]$ that is compatible with the above homotopy equivalence. Thus it suffices to show that the map

$$\mathrm{SP}^\infty(X \cup A \times [0, 1]) \longrightarrow \mathrm{SP}^\infty(X \cup CA) \qquad (5.20)$$

induced by the collapsing map $X \cup A \times [0, 1] \to X \cup CA$ is a quasifibration. An interesting fact is that these spaces can be obtained by the two-sided bar construction.

For any based space X, the infinite symmetric product $\mathrm{SP}^\infty(X)$ can be made into a topological monoid by the concatenation

$$[x_1, \ldots, x_m] \cdot [y_1, \ldots, y_n] = [x_1, \ldots, x_m, y_1, \ldots, y_n].$$

It is elementary to construct homeomorphisms

$$\mathrm{SP}^\infty(X \cup A \times [0, 1]) \cong |B_*(\mathrm{SP}^\infty(X), \mathrm{SP}^\infty(A), \mathrm{SP}^\infty(A))|$$
$$\mathrm{SP}^\infty(X \cup CA) \cong |B_*(\mathrm{SP}^\infty(X), \mathrm{SP}^\infty(A), *)|$$

and thus (5.20) is a quasifibration by Example 5.11.11.   □

Chapter 6

Postscript

We have seen throughout this book that fiber bundles and more general classes of fibrations are closely related to continuous deformations of maps and spaces, namely *homotopy*. Ever since its discovery in topology, the notion of homotopy has been making profound impact on the development of various fields in mathematical sciences. We now have a research field called "homotopy theory".

The term "homotopy theory" used to mean the study of homotopy types of topological spaces, especially CW complexes. We have seen in this book that homotopy theory of this type is useful in the study of fiber bundles. On the other hand, tools and ideas discovered in topology, such as homology and cohomology, have been imported into other fields, which resulted in discoveries of analogous notions of homotopy in such fields. For example, a categorical structure called triangulated category was independently discovered in algebraic geometry and topology. Nowadays the meaning of "homotopy theory" is not clear.

The author choose to close this book by his personal view on "homotopy theory".

6.1 What is Homotopy Theory?

In this book, the notion of homotopy first appeared in Definition 2.2.13 as continuous deformations of paths. It was extended to homotopy between continuous maps in §4.3 as continuous deformations of maps.

On the other hand, we introduced homotopy equivalence in Definition 5.3.13 as a relation between topological spaces. The meaning is, however, yet to be discussed. It was defined formally replacing "=" in the definition of homeomorphism by "≃". The term "homotopy" gives a wrong impression that, if two spaces are homotopy equivalent, then one can be "continuously deformed" into another. Unfortunately, however, such an interpretation is not possible. In the case of maps, a homotopy H between $f, g : X \to Y$ gives rise to a family of maps $\{h_t = H|_{X \times \{t\}}\}$ parameterized by $t \in [0, 1]$. The map h_t can be interpreted as the stage at time t during a continuous deformation of f into g. In the case of spaces, we do not have such a deformation of spaces.

Note that even in the case of maps, it is not easy to discuss the continuity with respect to the parameter if we regard a homotopy as a family of maps. It was Brouwer who found the current definition of homotopy as a $H : X \times [0, 1] \to Y$ in 1911. The product topology on $X \times [0, 1]$ allows us to discuss the continuity of homotopy. It is possible to define a parameterized family of spaces as a path in a moduli space or a map $\pi : E \to [0, 1]$, but these constructions work only in limited cases.

The notion of homotopy equivalence was introduced by Hurewicz in [Hurewicz (1935)]. It seems quite difficult to come up with the definition of homotopy equivalence from the viewpoint of continuous deformations. It required a big change of viewpoint. It seems that Hurewicz actually had an idea of homotopy category, although the language of categories and functors was yet to be discovered in Hurewicz' era.

Definition 6.1.1. The category whose objects are topological spaces and whose set of maps from X to Y is $[X, Y]$ is denoted by ho($\mathbf{Top}$) and is called the *homotopy category* of the category of topological spaces.

Then a homotopy equivalence from X to Y is nothing but an isomorphism in the homotopy category ho($\mathbf{Top}$). Besides the definition of homotopy equivalence, we have introduced new definitions by replacing $\cong$ with $\simeq$. We have also seen that many properties of fiber bundles still hold for fibrations if we replace $\cong$ with $\simeq$. Such definitions and theorems can be stated in much simpler ways if we use the homotopy category.

The importance of categorical viewpoint and the usefulness of homotopy category were made clear when homology theories were axiomatized by Eilenberg and Steenrod [Eilenberg and Steenrod (1952)]. Homology groups were originally introduced by Poincaré at the end of 19th century. From the viewpoint of Eilenberg and Steenrod, a homology theory is a family of functors

$$H_n : \mathbf{Top} \longrightarrow \mathbf{Abel}$$

which assigns an Abelian group $H_n(X)$ to a topological space X and each integer n satisfying a certain set of axioms. The homotopy invariance is a part of the axioms and we obtain a functor

$$H_n : \mathrm{ho}(\mathbf{Top}) \longrightarrow \mathbf{Abel}. \tag{6.1}$$

Hurewicz defined not only homotopy equivalence but also homotopy groups,[1] with which the notion of weak homotopy equivalence[2] was introduced by J.H.C. Whitehead[3] [Whitehead (1939)]. Henry Whitehead also discovered the following remarkable fact.

Theorem 6.1.2. *If CW complexes[4] X and Y are weakly homotopy equivalent, they are homotopy equivalent.*

The reader might wonder why this is a remarkable fact. This is remarkable, since showing a map to be a weak homotopy equivalence is much easier than showing it to be a homotopy equivalence. This is why we used long exact sequence of homotopy groups in §4.8. This special property made CW complexes into a main target of study in algebraic topology.

Another remarkable discovery of Henry Whitehead is the following relation between homology groups and homotopy groups.

Theorem 6.1.3. *If a continuous map $f : X \to Y$ is a weak homotopy equivalence, then the induced map*

$$f_* : H_n(X) \longrightarrow H_n(Y)$$

is an isomorphism for all n. Furthermore if X and Y are simply connected, the converse holds.

Homology groups are defined in terms of an algebraic structure called chain complexes, which allows us to come up with various techniques of computing homology groups. Because of Theorem 6.1.3 and the computability of homology groups, weak homotopy equivalence is now considered as a fundamental relationship among topological spaces.

Theorem 6.1.3 also suggests to define a homotopy category $\mathrm{ho}_w(\mathbf{Top})$ based on weak equivalences, with which we regard homology groups as a family of functors

$$H_n : \mathrm{ho}_w(\mathbf{Top}) \longrightarrow \mathbf{Abel}.$$

However, weak homotopy equivalence is not a relation among maps, but a relation among spaces. We need a new idea to define $\mathrm{ho}_w(\mathbf{Top})(X, Y)$. This is one of the origins of abstract homotopy theory.

---

[1] Definition 4.6.5.

[2] Definition 5.8.24.

[3] Called Henry Whitehead in the rest of this book to distinguish from another famous homotopy theorist George Whitehead.

[4] Henry Whitehead actually proved the case of finite simplicial complexes.

6.2 Many Kinds of Homotopies

It should be noted that, in the study of homology groups, a notion of homotopy for maps between chain complexes was discovered. In order to discuss the meaning of homotopy, let us recall the definition of chain complex and chain homotopy.

Definition 6.2.1. (1) A sequence of Abelian groups and homomorphisms

$$C : \cdots \longrightarrow C_{n+1} \overset{\partial_{n+1}}{\longrightarrow} C_n \overset{\partial_n}{\longrightarrow} C_{n-1} \longrightarrow \cdots$$

is called a *chain complex* if $\partial_n \circ \partial_{n+1} = 0$ for all n.

(2) For a chain complex C, the group $H_n(C) = \operatorname{Ker} \partial_n / \operatorname{Im} \partial_{n+1}$ is called the n-th *homology group* of C.

(3) A *chain map* from a chain complex C to another D is a sequence of homomorphisms $f_n : C_n \to D_n$ which commute with ∂_n.

(4) A *chain homotopy* between chain maps $f, g : C \to D$ is a family of homomorphisms $\Phi_n : C_n \to D_{n+1}$ satisfying

$$\partial_{n+1} \circ \Phi_n + \Phi_{n-1} \circ \partial_n = f_n - g_n$$

for all n.

We say f and g are *chain homotopic* if there exists a chain homotopy between them and we write $f \simeq g$.

(5) We say two chain complexes C and D are *chain homotopy equivalent* and denote $C \simeq D$ if there exist chain maps $f : C \to D$ and $g : D \to C$ such that $g \circ f \simeq 1_C$ and $f \circ g \simeq 1_D$.

(6) The category of chain complexes and chain maps is denoted by **Ch**.

The reader might wonder why such a relation is called "homotopy". It looks very much different from the homotopy of continuous maps. A relation between the homotopy of continuous maps and the homotopy of chain maps is given by the singular chain complex functor.

Theorem 6.2.2. *For a topological space X, let $S_n(X)$ be the free Abelian group generated by $\operatorname{Map}(\Delta^n, X)$. Then $S_*(X) = \{S_n(X)\}_{n \geq 0}$ can be made into a chain complex and defines a functor*

$$S_* : \mathbf{Top} \longrightarrow \mathbf{Ch}.$$

$S_*(X)$ *is called the* singular chain complex *of X. Furthermore if $f, g : X \to Y$ are homotopic, the induced maps $S_*(f), S_*(g) : S_*(X) \to S_*(Y)$ are chain homotopic.*

If we define the homotopy category ho($\mathbf{Ch}$) of chain complexes by using chain homotopy, we obtain a functor.

$$S_* : \mathrm{ho}(\mathbf{Top}) \longrightarrow \mathrm{ho}(\mathbf{Ch}).$$

It is also easy to see that, for chain complexes C and D, if $C \simeq D$ then $H_n(C) \cong H_n(D)$. In other words, we have functors

$$H_n : \mathrm{ho}(\mathbf{Ch}) \longrightarrow \mathbf{Abel}.$$

The composition

$$\mathrm{ho}(\mathbf{Top}) \xrightarrow{S_*} \mathrm{ho}(\mathbf{Ch}) \xrightarrow{H_n} \mathbf{Abel}$$

is called the *singular homology theory* and is known to satisfy the Eilenberg-Steenrod axioms. In particular, Theorem 6.1.3 holds for the singular homology theory.

We also have an analogue of weak homotopy equivalence for chain complexes.

Definition 6.2.3. A chain map $f : C \to D$ is called a *quasi-isomorphism* if $f_* : H_n(C) \to H_n(D)$ is an isomorphism for all n.

If we could define a category ho$_w$($\mathbf{Ch}$) of chain complexes in which quasi-isomorphisms are made into isomorphisms, Theorem 6.1.3 implies that the singular chain complex functor induces a functor

$$S_* : \mathrm{ho}_w(\mathbf{Top}) \longrightarrow \mathrm{ho}_w(\mathbf{Ch}).$$

So far we have seen two kinds of homotopies for maps between topological spaces. The category of chain complexes also have two kinds of homotopies, if we regard quasi-isomorphism as an analogue of weak homotopy equivalence. We have also seen that, if we could define homotopy categories with respect to such homotopies, statements of important properties of functors between them can be simplified. A homotopy relation is also introduced for maps between simplicial spaces in Definition A.3.11 in Appendix A.3.

It was D. Quillen who noticed the existence of many kinds of homotopies in various fields of mathematical sciences and tried to build a unified theory of homotopies and homotopy categories.

6.3 Framework of Homotopy Theory

What Quillen tried to say in his book "Homotopical Algebra" [Quillen (1967)] is, from my personal point of view, that homotopy equivalences can be characterized by their relations with fibrations and cofibrations. He axiomatized the properties of weak homotopy equivalences, fibrations, and cofibrations and introduced the notion of model categories.

Quillen's axioms have been refined in various ways since its appearance. Here we use a formulation based on weak factorization systems due to Joyal and Tierney which appeared in an Appendix of their paper [Joyal and Tierney (2007)].

Definition 6.3.1. A *weak factorization system* in a category X is a a pair of subcategories (A, B) satisfying the following conditions:

(1) Any morphism f in X can be factors as $f = p \circ i$ by a morphism i in A and a morphism p in B.
(2) A morphism $i : A \to X$ in X belongs to A if and only if i has the *right lifting property* with respect to all morphisms $p : E \to B$ in B. In other words, given a commutative diagram

$$
\begin{array}{ccc}
A & \longrightarrow & E \\
{\scriptstyle i}\downarrow & \nearrow & \downarrow{\scriptstyle p} \\
X & \longrightarrow & B
\end{array}
\qquad (6.2)
$$

there exists a morphism $X \to E$ making the triangles commutative.
(3) A morphism $p : E \to B$ in X belongs to B if and only if p has the *left lifting property* with respect to all morphisms $i : A \to X$ in A. Namely, for a commutative diagram (6.2), there exists a morphism $X \to E$ making the triangles commutative.

Definition 6.3.2. A *model structure* on a category X is a triple (W, C, F) of subcategories satisfying the following conditions:

(1) Let $f : X \to Y$ and $g : Y \to Z$ be composable morphisms in X. If any two of morphisms f, g, $g \circ f$ belong to W, then so is the third.
(2) The pair $(C \cap W, F)$ is a weak factorization system.
(3) The pair $(C, W \cap F)$ is also a weak factorization system.

When a category X is equipped with a model structure, it is called a *model category*.

Morphisms in the subcategories W, C, and F are called *weak equivalences*, *cofibrations*, and *fibrations*.

Those who are interested in model categories are recommended to read the exposition [Dwyer and Spaliński (1995)] by Dwyer and Spalinski first and then study Hovey's book [Hovey (1999)] and Hirschhorn's book [Hirschhorn (2003)].

One of the most fundamental examples is the category **Top** of topological spaces. Quillen proved that **Top** can be made into a model category by defining weak equivalences to be weak homotopy equivalences and fibrations to be Serre fibrations. Strøm proved in [Strøm (1972)] that **Top** can be made into a model category by requiring fibrations to be Hurewicz fibrations, cofibrations to be inclusions of NDR pairs, and weak equivalences to be homotopy equivalences.

Another important example already discovered by Quillen is the category $\mathbf{Ch}^{+}(A)$ of bounded below chain complexes in an Abelian category A having enough projectives. Weak equivalences are given by quasi-isomorphisms. Fibrations are surjections. And cofibrations are injections $f : C \to D$ such that Coker f consists of projective objects in A in each degree. There is another model structure on the category of chain complexes in which weak equivalences are chain homotopy equivalences. This is proved by Golasiński and Gromadzki [Golasiński and Gromadzki (1982)].

On the other hand, another attempt had been already made to construct a homotopy category of chain complexes before Quillen. Grothendieck (and Verdier) introduced the notion of *derived category* by formally inverting quasi-isomorphisms. Grothendieck also introduced a generalization of derived categories, called *triangulated categories*, which now plays an indispensable role in algebraic geometry and representation theory. See Neeman's book [Neeman (2001)] for details.

The derived category of an Abelian category is defined by formally adding inverses for quasi-isomorphisms in the category of chain complexes. More generally we have the following construction.

Definition 6.3.3. Let C be a category and $\mathcal{W}$ a class of morphisms in C. A *localization* of C with respect to $\mathcal{W}$ is a pair of the category $C[\mathcal{W}^{-1}]$ and a functor

$$L : C \longrightarrow C[\mathcal{W}^{-1}]$$

satisfying the following conditions:

(1) For any $f \in \mathcal{W}$, $L(f)$ is an isomorphism.
(2) Given a functor $F : C \to D$ which sends all morphisms in $\mathcal{W}$ to

isomorphisms, there exists a unique functor $\widetilde{F}$ making the diagram

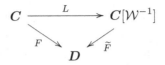

commutative.

See the book [Gabriel and Zisman (1967)] by Gabriel and Zisman, for more details on localizations of categories. With this definition, the homotopy categories of **Top** and **Ch** with respect to weak homotopy equivalences and quasi-isomorphisms appeared in §6.2 can be defined by

$$\mathrm{ho}_w(\mathbf{Top}) = \mathbf{Top}[\mathcal{W}_{\mathbf{Top}}^{-1}]$$
$$\mathrm{ho}_w(\mathbf{Ch}) = \mathbf{Ch}[\mathcal{W}_{\mathbf{Ch}}^{-1}],$$

where $\mathcal{W}_{\mathbf{Top}}$ and $\mathcal{W}_{\mathbf{Ch}}$ are the classes of weak homotopy equivalences and quasi-isomorphisms, respectively. We may also define the derived category $D(A)$ of an Abelian category A by

$$D(A) = \mathrm{ho}_w(\mathbf{Ch}(A)) = \mathbf{Ch}(A)[\mathcal{W}_{\mathbf{Ch}(A)}^{-1}],$$

where $\mathcal{W}_{\mathbf{Ch}(A)}$ is the class of quasi-isomorphisms.

It should be noted that it is not easy to grasp morphisms in $D(A)$. Another problem is that the collection of all morphisms between two objects might become huge. We usually localize the homotopy category $\mathrm{ho}(\mathbf{Ch}(A))$ with respect to chain homotopies and then localize it with respect to quasi-isomorphisms to defined the derived category.

The reader might have noticed that we do not need fibrations nor cofibrations in order to define derived categories. We simply added formal inverses to weak equivalences. In fact, many categorical structures for which "homotopy category" can be defined have been introduced.

- *dg category* (*differential graded category*) [Bondal and Kapranov (1990)].
- A_∞-*category* [Bespalov *et al.* (2008)].
- *Spectral category* (According to [Blumberg and Mandell (2012)], this is due to Jeff Smith.)
- *Waldhausen category* [Waldhausen (1978, 1979, 1985)].
- *Category with fibrant objects* [Brown (1973)], its refinements [Anderson (1978); Cisinski (2003, 2008, 2010)], and its dual.
- *Weak fibration category* [Barnea and Schlank (2016)] and its dual.

- *Semimodel category* [Spitzweck (2001)].
- *Cartan-Eilenberg category* [Guillén *et al.* (2010)].
- *Homotopical category* [Dwyer *et al.* (2004)].
- *Relative category* [Barwick and Kan (2012)].
- $(\infty, 1)$-*category*.

The last one is one of the most popular approaches besides model categories. It is based on the idea that homotopies can be encoded in higher categorical structures. Several models, such as *quasicategory*, *complete Segal space*, and *Segal category* have been proposed based on simplicial methods. See Lurie's book [Lurie (2009)] for details.

Quillen introduced the notion of model categories as a framework which unifies various kinds of homotopies. Unfortunately, a lot of homotopies which cannot be handled by model category have been discovered, which lead to many kinds of "categories with homotopy".

This is because people begin to realize the existence and the importance of the notion of "homotopy" in a wide range of research fields. For example, Kontsevich and Soibelman write as follows in Introduction to an unpublished book on deformation theory [Kontsevich and Soibelman (unknown)] by quoting Gel′fand.

Any area of mathematics is a kind of deformation theory.

The author do not know what Gel′fand means by deformation theory, but it should include continuous deformations. As Gel′fand's words predict, many kinds of "theory of homotopy" will continue to be discovered. The author hopes that this book will be of some help to understand such theories of homotopies.

Appendix A

Related Topics

A.1 The Meaning of Compact-Open Topology

The compact-open topology introduced in §3.4 is complicated and its meaning is not clear. Unfortunately(?) it has lots of useful properties, which forces us to use the compact-open topology when we need to topologize spaces of continuous maps. In order to get familiar with the compact-open topology, we discuss its meaning in this appendix.

Let us recall the following fact from calculus.

Theorem A.1.1. *Any continuous map $f : [a, b] \to \mathbb{R}$ has a maximum and a minimum.*

This fact can be proved by the fact that $f([a, b])$ is a compact subset of $\mathbb{R}$, which is bounded and closed. The same argument applies to continuous functions on compact spaces.

Theorem A.1.2. *If X is compact, any continuous map $f : X \to \mathbb{R}$ has a maximum and a minimum.*

Corollary A.1.3. *If X is compact, for continuous functions $f, g : X \to \mathbb{R}$ the function $x \mapsto |f(x) - g(x)|$ has a maximum.*

It is straightforward to verify that it defines a metric on the set of continuous functions $\mathrm{Map}(X, \mathbb{R})$.

Theorem A.1.4. *Suppose X is a compact space. For $f, g \in \mathrm{Map}(X, \mathbb{R})$, define*

$$d(f, g) = \max_{x \in X} |f(x) - g(x)|.$$

Then d is a metric on $\mathrm{Map}(X, \mathbb{R})$.

291

Remark A.1.5. This metric is associated with the *norm* on $\mathrm{Map}(X, \mathbb{R})$ defined by $\|f\| = \max_{x \in X} |f(x)|$. This is called the *sup norm*.

The sup norm is also defined on $C(X) = \mathrm{Map}(X, \mathbb{C})$ by using the norm of $\mathbb{C}$. This is known to have a structure of a C^*-*algebra*.

Note that we only used the fact that $\mathbb{R}$ is a metric space.

Theorem A.1.6. *Let X be a compact topological space and (Y, δ) a metric space. For $f, g \in \mathrm{Map}(X, Y)$, define*
$$d(f, g) = \max_{x \in X} \delta(f(x), g(x)).$$
Then d is a metric on $\mathrm{Map}(X, Y)$.

This is a much more accessible way of defining a topology on $\mathrm{Map}(X, Y)$ than the compact-open topology, but it only works when X is compact and Y is a metric space.

This topology on $\mathrm{Map}(X, Y)$ is known as the topology of uniform convergence in analysis and generalized to noncompact spaces as follows.

Definition A.1.7. Let X be a topological space and (Y, δ) a metric space. For $f \in \mathrm{Map}(X, Y)$ and $\varepsilon > 0$, define
$$U(f; \varepsilon) = \{g \in \mathrm{Map}(X, Y) \mid \delta(f(x), g(x)) < \varepsilon \text{ for any } x \in X\}.$$
The topology on $\mathrm{Map}(X, Y)$ defined by using $\{U(f; \varepsilon)\}_f$ as open basis is called the *topology of uniform convergence*.

Remark A.1.8. A sequence $\{f_n\}_{n \in \mathbb{N}} \subset \mathrm{Map}(X, Y)$ is said to *converge to f uniformly*, if for any $\varepsilon > 0$, there exists an integer N such that, for any $x \in X$ and $n > N$, $\delta(f(x), f_n(x)) < \varepsilon$.

When X is not compact, the maximum $\max_{x \in X} \delta(f(x), g(x))$ might not exist. The following variation is more practical.

Definition A.1.9. Let X be a topological space and (Y, δ) a metric space. For each compact subset $K \subset X$ equip $\mathrm{Map}(K, Y)$ with the topology of uniform convergence. The weakest topology on $\mathrm{Map}(X, Y)$ with which the restriction map
$$i_K^* : \mathrm{Map}(X, Y) \longrightarrow \mathrm{Map}(K, Y)$$
is continuous for any compact subset $K \subset X$ is called the *topology of compact convergence*.

In other words, it is the topology defined by
$$\left\{(i_K^*)^{-1}(U_\varepsilon^K(f)) \mid K \subset X \text{ compact}, f \in \mathrm{Map}(K, Y), \varepsilon > 0\right\}$$
as an open basis.

Theorem A.1.10. *If X is locally compact Hausdorff and Y is a metric space, the compact-open topology and the topology of compact convergence coincide.*

Proof. For any compact set $K \subset X$ and an open set $V \subset Y$, let us show that $W(K, V)$ is open in the topology of compact convergence.

For $f \in W(K, V)$, there exists $\varepsilon_x > 0$ and $U_{\varepsilon_x}(f(x)) \subset V$, since $f(x) \in V$ for any $x \in K$ and V is open. Here $U_{\varepsilon_x}(f(x))$ is the ε-neighborhood of $f(x)$ in Y. Then

$$f(K) \subset \bigcup_{x \in K} U_{\frac{\varepsilon_x}{2}}(f(x)).$$

By the compactness of K, we may choose a finite number of points to cover $f(K)$

$$f(K) \subset \bigcup_{i=1}^{n} U_{\frac{\varepsilon_{x_i}}{2}}(f(x_i)). \tag{A.1}$$

Define $\varepsilon = \min_i \varepsilon_{x_i}$ and $U = (i_K^*)^{-1}(U_{\frac{\varepsilon}{2}}^K(i_K^*(f)))$, where $U_{\frac{\varepsilon}{2}}^K(i_K^*(f))$ is the $\frac{\varepsilon}{2}$-neighborhood of $i_K^*(f)$ in $\mathrm{Map}(K, Y)$ by the sup norm. Then we have $f \in U \subset W(K, V)$. This can be verified as follows. For $g \in U$ and $x \in K$, we have $\delta(f(x), g(x)) < \frac{\varepsilon}{2}$. By (A.1), there exists an i such that $\delta(f(x), f(x_i)) < \frac{\varepsilon_{x_i}}{2}$ for any $x \in K$ and we have

$$\delta(g(x), f(x_i)) < \delta(g(x), f(x)) + \delta(f(x), f(x_i))$$
$$< \frac{\varepsilon}{2} + \frac{\varepsilon_{x_i}}{2}$$
$$< \frac{\varepsilon_{x_i}}{2} + \frac{\varepsilon_{x_i}}{2}$$
$$= \varepsilon_{x_i},$$

which implies $g(x) \in U_{\varepsilon_{x_i}}(f(x_i)) \subset V$, and $g(K) \subset V$. By the definition of the topology of compact convergence, U is open in the topology of compact convergence. Thus $W(K, V)$ open in the compact convergence topology.

Conversely, let us show that any open set $U \subset \mathrm{Map}(X, Y)$ in the topology of compact convergence is open in the compact-open topology.

For $f \in U$, by the definition of the topology of compact convergence, there exists a compact subset K and $\varepsilon > 0$ such that $(i_K^*)^{-1}\left(U_\varepsilon^K(i_K^*(f))\right) \subset U$. On the other hand, $K \subset \bigcup_{x \in K} f^{-1}(U_{\frac{\varepsilon}{3}}(f(x)))$. Since X is locally compact, for each $x \in K$, there exists an open neighborhood V_x of x in X such that $V_x \subset f^{-1}(U_{\frac{\varepsilon}{3}}(f(x)))$ and $\overline{V_x}$ is compact. By the compactness of

K we may choose a finite number from the open covering $K \subset \bigcup_{x \in K} V_x$ to cover K

$$K \subset \bigcup_{i=1}^{n} V_{x_i}.$$

Define

$$U_i = U_{\frac{\varepsilon}{3}}(f(x_i))$$
$$K_i = \overline{V_{x_i}} \cap K.$$

Then, for any $g \in \bigcap_{i=1}^{n} W(K_i, U_i)$, we have $i_K^*(g) \in U_{\varepsilon}^K(i_K^*(f))$. In fact, since $K = \bigcup_{i=1}^{n} K_i$, for any $x \in K$, there exists an i such that $x \in K_i$. Then

$$\delta(f(x), g(x)) \leq \delta(f(x), f(x_i)) + \delta(f(x_i), g(x))$$
$$< \frac{\varepsilon_K}{3} + \delta(f(x_i), g(x)).$$

By the inclusion $V_{x_i} \subset f^{-1}(U_{\frac{\varepsilon}{3}}(f(x_i)))$, we have $\overline{V_{x_i}} \subset U_{\frac{2\varepsilon}{3}}(f(x_i))$ and thus

$$\delta(f(x), g(x)) < \frac{\varepsilon}{3} + \frac{2\varepsilon}{3} = \varepsilon.$$

Thus we have

$$i_K^* \left(\bigcap_{i=1}^{n} W(K_i, U_i) \right) \subset U_{\varepsilon}^K(i_K^*(f))$$

and U is open in the compact-open topology. $\square$

Another important and useful fact which justifies the use of the compact-open topology is the following.

Theorem A.1.11. *For topological spaces X and Y, define the evaluation map* ev : $\mathrm{Map}(X, Y) \times X \to Y$ *by* $\mathrm{ev}(f, x) = f(x)$.

If X is locally compact, the compact-open topology is the weakest topology on $\mathrm{Map}(X, Y)$ which makes ev *continuous.*

See Kelly's book [Kelley (1975)], for example.

We leave the following two facts to the reader as exercise problems.

Exercise A.1.12. Regard the set $[n] = \{1, 2, \ldots, n\}$ as a topological space by the compact-open topology. Define a map $e : \mathrm{Map}([n], X) \to X^n$ by $e(f) = (f(1), f(2), \ldots, f(n))$. Prove that this is a homeomorphism.

Exercise A.1.13. Recall that we defined a topology on $M_n(\mathbb{R})$ by identifying it with $\mathbb{R}^{n^2}$ in §3.3. Similarly the set $M_{m,n}(\mathbb{R})$ of $m \times n$ real matrices can be identified with $\mathbb{R}^{mn}$. On the other hand, any $A \in M_{m,n}(\mathbb{R})$ defines a linear map $A : \mathbb{R}^n \to \mathbb{R}^m$, which allows us to regard $M_{m,n}(\mathbb{R}) \subset \mathrm{Map}(\mathbb{R}^n, \mathbb{R}^m)$.

Show that the topology on $M_{m,n}(\mathbb{R})$ induced from the compact-open topology on $\mathrm{Map}(\mathbb{R}^n, \mathbb{R}^m)$ agrees with the metric topology of $\mathbb{R}^{mn}$.

A.2 Vector Bundles

Maps of fiber bundles are defined in §4.1 when the structure groups and fibers are the same. For practical applications, however, we sometimes need maps between fiber bundles even when structure groups or fibers are different. For example, we need such maps for vector bundles.[5]

In the first half of this section, we discuss maps between vector bundles as a complement for §4.1. In the latter half, important constructions of vector bundles from known vector bundles are explained.

In the rest of this section, we assume that vector bundles are real vector bundles of finite rank. Analogous results hold for complex vector bundles under appropriate modifications.

Let us begin with the definition of maps between vector bundles.

Definition A.2.1. Let $\xi = (p : E \to B)$ and $\xi' = (p' : E' \to B')$ be vector bundles of rank m and n, respectively. A *map of vector bundles* from ξ to ξ' is a fiber-preserving map

$$f = (\tilde{f}, f) : (E, B) \longrightarrow (E', B')$$

satisfying the following condition: for local trivializations $\varphi_\alpha : p^{-1}(U_\alpha) \to U_\alpha \times \mathbb{R}^m$ and $\psi_\beta : p'^{-1}(V_\beta) \to V_\beta \times \mathbb{R}^n$ of ξ and ξ', respectively, satisfying $U_\alpha \cap f^{-1}(V_\beta) \neq \emptyset$, the map

$$L^f_{\alpha\beta} : U_\alpha \cap f^{-1}(V_\beta) \longrightarrow \mathrm{Map}(\mathbb{R}^m, \mathbb{R}^n)$$

in Definition 4.1.2 takes values in the set of linear maps $\mathrm{Hom}_{\mathbb{R}}(\mathbb{R}^m, \mathbb{R}^n)$ from $\mathbb{R}^m$ to $\mathbb{R}^n$. Namely there exists a continuous map $\bar{L}^f_{\alpha\beta}$ making the diagram

$$
\mathrm{Hom}_{\mathbb{R}}(\mathbb{R}^m, \mathbb{R}^n) \lhook\joinrel\longrightarrow \mathrm{Map}(\mathbb{R}^m, \mathbb{R}^n)
$$
$$
\exists \bar{L}^f_{\alpha\beta} \qquad U_\alpha \cap f^{-1}(V_\beta) \qquad L^f_{\alpha\beta}
$$

commutative.

[5] Definition 3.5.24.

When $B = B'$ and $f = 1_B$, it is called a *homomorphism of vector bundles.*

Remark A.2.2. In some books, a map or a homomorphism of vector bundles is defined as a fiber-preserving map $\tilde{f}$ which is a "linear map on each fiber". However, the fiber $p^{-1}(x)$ over $x \in B$ and the fiber $p'^{-1}(f(x))$ over $f(x)$ are nothing more than topological spaces. We need to use local trivializations to make them into $\mathbb{R}^m$ and $\mathbb{R}^n$ in order to discuss the linearity of maps.

The set $\mathrm{Hom}_{\mathbb{R}}(\mathbb{R}^m, \mathbb{R}^n)$ appeared in the above definition itself has a structure of vector space.

Definition A.2.3. For finite dimensional real vector spaces V and W, define

$$\mathrm{Hom}_{\mathbb{R}}(V, W) = \{f : V \to W \mid \text{linear map}\}.$$

The sum and the scalar multiplication of linear maps make it into a real vector space. In particular, when $W = \mathbb{R}$, it is denoted by $V^* = \mathrm{Hom}_{\mathbb{R}}(V, \mathbb{R})$ and is called the *dual vector space* of V.

Given a basis $\mathcal{B} = \{v_1, \ldots, v_n\}$ of V, define $v_1^*, \ldots, v_n^* \in V^*$ by

$$v_i^*(v_j) = \begin{cases} 1, & i = j \\ 0, & i \neq j \end{cases}.$$

Then $\mathcal{B}^* = \{v_1^*, \ldots, v_n^*\}$ is a basis for V^*. This is called the *dual basis* of $\mathcal{B}$.

The definition of inner products can be extended to vector bundles as a homomorphism of vector bundles. In order to state the definition, we first need to define the direct sum of vector bundles. For vector spaces V and W, the direct sum $V \oplus W$ is defined to be the direct product $V \times W$ as a set. We know from Proposition 3.5.28 that the direct product of fiber bundles is again a fiber bundle. When $p : E \to B$ and $p' : E' \to B'$ are vector bundles of rank m and n, respectively, the product

$$p \times p' : E' \times E \longrightarrow B \times B'$$

is a vector bundle with fiber $\mathbb{R}^m \times \mathbb{R}^n$. Let us denote this bundle by $\xi \times \xi'$. Proposition 3.5.28 also says that the structure group is $\mathrm{GL}_m(\mathbb{R}) \times \mathrm{GL}_n(\mathbb{R})$. This bundle $\xi \times \xi'$ is a vector space of rank $m + n$, since we may regard $\mathrm{GL}_m(\mathbb{R}) \times \mathrm{GL}_n(R)$ as a subgroup of $\mathrm{GL}_{m+n}(\mathbb{R})$ under the correspondence

$$(A, B) \longmapsto \begin{pmatrix} A & O \\ O & B \end{pmatrix}.$$

In order to obtain a vector bundle over B, we take a pullback along the diagonal map.

Definition A.2.4. Let $\xi = (p : E \to B)$ and $\xi' = (p' : E' \to B)$ be vector bundles over the same base space B. The pullback of $\xi \times \xi'$ along the diagonal map $\Delta : B \to B \times B$ is denoted by $\xi \oplus \xi'$ and is called the *direct sum* or the *Whitney sum* of ξ and ξ'. The total space of this bundle is denoted by $E \oplus_B E'$ or $E \times_B E'$.

Remark A.2.5. The total space of the direct sum of two vector bundles can be also defined by the pullback diagram

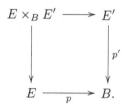

Note also that if ξ and ξ' are vector bundles of rank m and n, respectively, then $\xi \oplus \xi'$ is a vector bundle of rank $m + n$ by Theorem 4.2.3.

Definition A.2.6. Let $\xi = (p : E \to B)$ be a vector bundle of rank n. Denote the trivial bundle $B \times \mathbb{R} \to B$ by $\mathbb{R}_B$. An *inner product* or a *metric* on ξ is a homomorphism of vector bundles

$$g : \xi \oplus \xi \longrightarrow \mathbb{R}_B$$

such that the map on each fiber gives rise to an inner product $\mathbb{R}^n \times \mathbb{R}^n \to \mathbb{R}$ when translated by local trivializations.

Proposition A.2.7. *Any vector bundle over a paracompact Hausdorff space has an inner product.*

Proof. Let $\xi = (p : E \to B)$ be a vector bundle of rank n over a paracompact Hausdorff space B. Let $\mathcal{U} = \{U_\alpha\}_{\alpha \in A}$ be an open covering of B such that ξ has a local trivialization $\varphi_\alpha : p^{-1}(U_\alpha) \to U_\alpha \times \mathbb{R}^n$ on each U_α.

Choose an inner product on $\mathbb{R}^n$. Then it induces an inner product on each restriction $\xi|_{U_\alpha} = (p^{-1}(U_\alpha) \to U_\alpha)$ by the local trivialization φ_α.

Since B is paracompact Hausdorff there exists a partition of unity subordinate to the open covering $\mathcal{U}$ by Theorem 4.3.19. As we have done in the proof of Theorem 4.3.7, we can glue inner products on $\xi|_{U_\alpha}$ to obtain an inner product on ξ by using the partition of unity. $\square$

When a finite dimensional vector space has an inner product, the Gram-Schmidt orthonormalization allows us to transform an arbitrary basis into an orthonormal basis. By applying this process to vector bundles of rank n, we may reduce the structure group to the orthogonal group $O(n)$.

Theorem A.2.8. *Let ξ be a vector bundle of rank n over a paracompact Hausdorff space. Then there exists a vector bundle ξ' which is isomorphic to ξ whose structure group can be reduced to $O(n)$.*

Proof. We construct ξ' by Theorem 3.5.31. Let $\{\varphi_\alpha : p^{-1}(U_\alpha) \to U_\alpha \times \mathbb{R}^n\}_{\alpha \in A}$ be a local trivialization of ξ. Coordinate transformations are denoted by $\{\Phi^{\alpha\beta} : U_\alpha \cap U_\beta \to \mathrm{GL}_n(\mathbb{R})\}$. By Proposition A.2.7, we obtain an inner product $g : \xi \oplus \xi \to \mathbb{R}_B$ on ξ, which induces a map $\bar{L}^g_{\alpha\beta}$ making the diagram

$$\mathrm{Hom}_\mathbb{R}(\mathbb{R}^n \times \mathbb{R}^n, \mathbb{R}) \lhook\joinrel\xrightarrow{\hspace{3cm}} \mathrm{Map}(\mathbb{R}^n \times \mathbb{R}^n, \mathbb{R})$$

$$\bar{L}^g_{\alpha\beta} \nwarrow \qquad \nearrow L^g_{\alpha\beta}$$

$$U_\alpha \cap U_\beta$$

commutative. Thus we obtain an inner product $\bar{L}^g_{\alpha\beta}(x) : \mathbb{R}^n \times \mathbb{R}^n \to \mathbb{R}$ on $\mathbb{R}^n$ for each x. Note that, although this inner product depends on x, the $\mathbb{R}^n$'s in the domain and the range of $\Phi^{\alpha\beta}(x)$ correspond to the fiber over x and has the same inner product.

Let $\{e_1, \ldots, e_n\}$ be a basis of $\mathbb{R}^n$. Apply the Gram-Schmit orthonormalization to the basis $\{\Phi^{\alpha\beta}(x)(e_1), \ldots, \Phi^{\alpha\beta}(x)(e_n)\}$ with respect to the inner product $L^g_{\alpha\beta}(x)$ to obtain an orthonormal basis $\{v_1(x), \ldots, v_n(x)\}$. Let $\Psi^{\alpha\beta}(x) = (v_1(x), \ldots, v_n(x))$ be the matrix obtain from this orthonormal basis. Then this is an orthogonal matrix. Thus we obtain a map

$$\Psi^{\alpha\beta} : U_\alpha \cap U_\beta \longrightarrow O(n).$$

This is a continuous map since the Gram-Schmit orthonormalization is defined in terms of the addition, the inner product, and the scalar multiplication. By Theorem 3.5.31, we obtain a vector bundle ξ' with structure group $O(n)$, which is isomorphic to ξ. $\qquad\qquad\square$

Definition A.2.4 is a "globalization" of the direct sum of vector spaces. Other than the direct sum, there are many ways to construct new vector spaces. Let us recall some of them in order to globalize them.

Definition A.2.9. Let V and W be finite dimensional vector spaces. Suppose a vector space $V \otimes W$ and a bilinear map $p : V \times W \to V \otimes W$ are given. If this bilinear map satisfies the following universal property, $V \otimes W$ is called the *tensor product* of V and W: for any vector space U and a bilinear map $f : V \times W \to U$, there exists a unique linear map $\bar{f}$ making the diagram

$$
\begin{array}{ccc}
V \times W & \xrightarrow{\ f\ } & U \\
{\scriptstyle p}\big\downarrow & \nearrow & \\
& \exists! \bar{f} & \\
V \otimes W &
\end{array}
$$

commutative.

The image of $(\boldsymbol{v}, \boldsymbol{w}) \in V \times W$ by p is denoted by $\boldsymbol{v} \otimes \boldsymbol{w}$.

Remark A.2.10. Note that elements of $V \otimes W$ are, in general, linear combinations of elements of the form $\boldsymbol{v} \otimes \boldsymbol{w}$.

By the defining universal property of tensor product, given linear maps $f : V \to V'$ and $g : W \to W'$, there exists a linear map

$$
f \otimes g : V \otimes W \longrightarrow V' \otimes W'
$$

making the diagram

$$
\begin{array}{ccc}
V \times W & \xrightarrow{\ f \times g\ } & V' \times W' \\
{\scriptstyle p}\big\downarrow & & \big\downarrow {\scriptstyle p} \\
V \otimes W & \xrightarrow[\ f \otimes g\ ]{} & V' \otimes W'
\end{array}
$$

commutative.

Definition A.2.11. For a vector space V, define a right action of the symmetric group Σ_k of k letters on the k-fold tensor product $V^{\otimes k} = \underbrace{V \otimes \cdots \otimes V}_{k}$ by

$$
(\boldsymbol{v}_1 \otimes \cdots \otimes \boldsymbol{v}_k)\sigma = \operatorname{sgn}(\sigma)\boldsymbol{v}_{\sigma(1)} \otimes \cdots \otimes \boldsymbol{v}_{\sigma(k)}.
$$

The quotient vector space of $V^{\otimes k}$ by this action is denoted by

$$
\wedge^k(V) = V^{\otimes k} / \left\{ x \in V^{\otimes k} \,\middle|\, x - x\sigma \right\}.
$$

This is called the *k-fold exterior power* of V.

For $\boldsymbol{v}_1, \ldots, \boldsymbol{v}_k \in V$, the equivalence class represented by $\boldsymbol{v}_1 \otimes \cdots \otimes \boldsymbol{v}_k$ is denoted by $\boldsymbol{v}_1 \wedge \cdots \wedge \boldsymbol{v}_k$.

Remark A.2.12. The 2-fold exterior power $\wedge^2(V)$ is the vector space obtained from $V \otimes V$ by requiring the relation $\boldsymbol{v} \wedge \boldsymbol{v}' = -\boldsymbol{v}' \wedge \boldsymbol{v}$ for $\boldsymbol{v}, \boldsymbol{v}' \in V$. In particular,

$$\boldsymbol{v} \otimes \boldsymbol{v} = 0.$$

Thus when V has a basis $\boldsymbol{v}_1, \ldots, \boldsymbol{v}_n$,

$$\boldsymbol{v}_1 \otimes \boldsymbol{v}_2, \boldsymbol{v}_1 \otimes \boldsymbol{v}_3, \ldots, \quad \boldsymbol{v}_1 \otimes \boldsymbol{v}_n,$$
$$\boldsymbol{v}_2 \otimes \boldsymbol{v}_3, \ldots, \quad \boldsymbol{v}_2 \otimes \boldsymbol{v}_n,$$
$$\vdots$$
$$\boldsymbol{v}_{n-1} \otimes \boldsymbol{v}_n$$

is a basis of $\wedge^2(V)$ and $\dim \wedge^2 V = \binom{n}{2}$.

More generally, when $n = \dim V$, the dimension of the k-fold exterior power is given by $\dim \wedge^k V = \binom{n}{k}$.

As is the case of inner product, these operations on vector spaces can be "globalized" to vector bundles. Let us first define the tensor product of vector bundles. We first need to define the exterior tensor product, which correspond to $\xi \times \xi'$ when we defined the direct sum $\xi \oplus \xi'$.

Definition A.2.13 (Tensor Product). Let $\xi = (E \to B)$ and $\xi' = (E' \to B'$ be vector bundles of rank m and n with coordinate transformations $\{\Phi^{\alpha\alpha'} : U_\alpha \cap U_{\alpha'} \to \mathrm{GL}_m(\mathbb{R})\}$ and $\{\Psi^{\beta\beta'} : V_\beta \cap V_{\beta'} \to \mathrm{GL}_n(\mathbb{R})\}$, respectively. Define a map

$$\Phi^{\alpha\alpha'} \otimes \Psi^{\beta\beta'} : (U_\alpha \times V_\beta) \cap (U_{\alpha'} \times V_{\beta'}) = (U_\alpha \cap U_{\alpha'}) \times (V_\beta \cap V_{\beta'}) \longrightarrow \mathrm{GL}_{mn}(\mathbb{R})$$

by the composition

$$(U_\alpha \cap U_{\alpha'}) \times (V_\beta \cap V_{\beta'}) \xrightarrow{\Phi^{\alpha\alpha'} \times \Psi^{\beta\beta'}} \mathrm{GL}_m(\mathbb{R}) \times \mathrm{GL}_n(\mathbb{R}) \xrightarrow{\otimes} \mathrm{GL}_{mn}(\mathbb{R}),$$

where $\otimes : \mathrm{GL}_m(\mathbb{R}) \times \mathrm{GL}_n(\mathbb{R}) \to \mathrm{GL}_{mn}(\mathbb{R})$ is given by assigning the composition

$$\mathbb{R}^{mn} \cong \mathbb{R}^m \otimes \mathbb{R}^n \xrightarrow{A \otimes B} \mathbb{R}^m \otimes \mathbb{R}^n \cong \mathbb{R}^{mn}$$

to $A \in \mathrm{GL}_m(\mathbb{R})$ and $B \in \mathrm{GL}_n(\mathbb{R})$.

These maps satisfy the condition of Theorem 3.5.31. The vector bundle obtained by gluing the trivial bundles $\{(U_\alpha \times V_\beta) \times \mathbb{R}^m \otimes \mathbb{R}^n\}_{\alpha \in A, \beta \in B}$ is denoted by $E \widehat{\otimes} E'$ and is called the *external tensor product* of E and E'.

When $B = B'$, the pullback of $\xi \widehat{\otimes} \xi'$ along the diagonal map $\Delta : B \to B \times B$ is denoted by $\xi \otimes \xi'$ and called the *tensor product* of ξ and ξ'.

Definition A.2.14 (Hom Bundle). Let $\xi = (E \to B)$ and $\xi' = (E' \to B)$ be vector bundles over the same base space B of rank m and n, respectively. We may assume that local trivializations are defined on a common open covering by taking a subdivision if necessary. Let $\{\Phi^{\alpha\alpha'} : U_\alpha \cap U_{\alpha'} \to \mathrm{GL}_m(\mathbb{R})\}$ and $\{\Psi^{\alpha\alpha'} : U_\alpha \cap U_{\alpha'} \to \mathrm{GL}_n(\mathbb{R})\}$ be coordinate transformations of ξ and ξ', respectively. Then the family of maps obtained by the composition

$$^t\Phi^{\alpha\alpha'} \otimes \Psi^{\alpha\alpha'} : U_\alpha \cap U_{\alpha'} \overset{\Phi^{\alpha\alpha'} \times \Psi^{\alpha\alpha'}}{\longrightarrow} \mathrm{GL}_m(\mathbb{R}) \times \mathrm{GL}_n(\mathbb{R})$$

$$\overset{^t(-) \times 1}{\longrightarrow} \mathrm{GL}_m(\mathbb{R}) \times \mathrm{GL}_n(\mathbb{R}) \overset{\otimes}{\longrightarrow} \mathrm{GL}_{mn}(\mathbb{R})$$

satisfies the conditions of Theorem 3.5.31.

The vector bundle obtained by gluing $U_\alpha \times \mathrm{Hom}_\mathbb{R}(\mathbb{R}^m, \mathbb{R}^n)$ by these maps is denoted by $\mathrm{Hom}(E, E')$ and is called the *Hom bundle*.

When E' is the 1-dimensional trivial bundle $\mathbb{R}_B$, We denote $E^* = \mathrm{Hom}(E, \mathbb{R}_B)$ and call it the *dual vector bundle* of E.

The use of transpose in the definition of Hom bundle is justified by the following fact.

Exercise A.2.15. Let $f : V \to W$ be a linear map between finite dimensional vector spaces and $\mathcal{B}_V = \{v_1, \ldots, v_m\}$ and $\mathcal{B}_W = \{w_1, \ldots, w_n\}$ bases of V and W, respectively. Define a linear map

$$f^* : W^* \longrightarrow V^*$$

by $f^*(\varphi) = \varphi \circ f$ for $\varphi \in W^*$.

Show that, if A a matrix representing f with respect to $\mathcal{B}_V$ and $\mathcal{B}_W$, the matrix representing f^* with respect to $\mathcal{B}_W^*$ and $\mathcal{B}_V^*$ is ^tA.

Definition A.2.16 (Exterior Product). Let ξ be a vector bundle of rank n over B. Define a vector bundle $\wedge^k \xi$ over B as follows.

Define a map

$$\wedge^k : \mathrm{GL}_n(\mathbb{R}) \longrightarrow \mathrm{GL}_{\binom{n}{k}}(\mathbb{R}) \tag{A.2}$$

by assigning the map

$$A^{\wedge k} : \wedge^k \mathbb{R}^n \longrightarrow \wedge^k \mathbb{R}^n$$

induced by $A^{\otimes k} : (\mathbb{R}^n)^{\otimes k} \to (\mathbb{R}^n)^{\otimes k}$ to $A \in \mathrm{GL}_n(\mathbb{R})$.

When $\{\Phi^{\alpha\beta} : U_\alpha \cap U_\beta \to \mathrm{GL}_n(\mathbb{R})\}$ is the coordinate transformations of ξ, the vector bundle obtained by using the composition

$$U_\alpha \cap U_\beta \overset{\Phi^{\alpha\beta}}{\longrightarrow} \mathrm{GL}_n(\mathbb{R}) \overset{\wedge^k}{\longrightarrow} \mathrm{GL}_{\binom{n}{k}}(\mathbb{R})$$

as coordinate transformations is denoted by $\wedge^k \xi$. This is called the *k-fold exterior power* of ξ.

In the rest of this section, we discuss classification of vector bundles and K-theory.

Definition A.2.17. The set of isomorphism classes of real vector bundles of rank n over a topological space X is denoted by $\mathrm{Vect}_n(X)$. The isomorphism classes of complex vector bundles of rank n is denoted by $\mathrm{Vect}_n^{\mathbb{C}}(X)$.

In order to make these sets closed under direct sums and tensor products, we define

$$\mathrm{Vect}(X) = \coprod_{n=0}^{\infty} \mathrm{Vect}_n(X)$$

$$\mathrm{Vect}^{\mathbb{C}}(X) = \coprod_{n=0}^{\infty} \mathrm{Vect}_n^{\mathbb{C}}(X).$$

Lemma A.2.18. *Both* $\mathrm{Vect}(X)$ *and* $\mathrm{Vect}^{\mathbb{C}}(X)$ *become a commutative monoid under the direct sum of vector bundles.*

The K-theory is defined by making this commutative monoid into an Abelian group by adding inverses.

Definition A.2.19. For a commutative monoid $(M, +, 0)$, define a relation $\sim$ on $M \times M$ by

$$(m, n) \sim (m', n') \iff m + n' = m' + n.$$

The set of equivalence classes is denoted by

$$G(M) = (M \times M)/{\sim}$$

and is called the *group completion* of M.

Example A.2.20. When M is the monoid $\mathbb{Z}_{\geq 0}$ of nonnegative integers by addition, we have $G(\mathbb{Z}_{\geq 0}) = \mathbb{Z}$. $\square$

Definition A.2.21. The group completions of $\mathrm{Vect}(X)$ and $\mathrm{Vect}^{\mathbb{C}}(C)$ are denoted by $KO(X)$ and $K(X)$ and called the *real K-theory* and the *complex K-theory* of X, respectively.

By definition vector bundles of rank n have $\mathrm{GL}_n(\mathbb{R})$ and $\mathrm{GL}_n(\mathbb{C})$ as structure groups. Thus the classification of vector bundles is reduced to the classification of principal $\mathrm{GL}_n(\mathbb{R})$-bundles and principal $\mathrm{GL}_n(\mathbb{C})$-bundles. When the base space is paracompact Hausdorff, the structure groups can be reduced to $O(n)$ and $U(n)$ by Theorem A.2.8 (and its complex version) and we have bijections

$$\text{Vect}_n(X) \cong P_{O(n)}(X)$$
$$\text{Vect}_n^{\mathbb{C}}(X) \cong P_{U(n)}(X).$$

The problem is now reduced to the classification of principal $O(n)$-bundles and principal $U(n)$-bundles. By Theorem 4.5.21, we obtain the following descriptions of $\text{Vect}(X)$ and $\text{Vect}^{\mathbb{C}}(X)$.

Corollary A.2.22. *For a CW complex X, we have natural bijections*

$$\text{Vect}_n(X) \cong [X, BO(n)]$$
$$\text{Vect}_n^{\mathbb{C}}(X) \cong [X, BU(n)]$$

and

$$\text{Vect}(X) \cong \left[X, \coprod_{n=0}^{\infty} BO(n)\right] \tag{A.3}$$

$$\text{Vect}^{\mathbb{C}}(X) \cong \left[X, \coprod_{n=0}^{\infty} BU(n)\right].$$

In order to obtain a description of K-theory in terms of homotopy sets, we need a "homotopy theoretic group completion". We first note that the map $\oplus : O(m) \times O(n) \to O(m+n)$ in Definition 4.9.17 is a homomorphism and induces a continuous map

$$BO(m) \times BO(n) \cong B(O(m) \times O(n)) \xrightarrow{B\oplus} BO(m+n)$$

by Theorem 4.10.26 and Theorem 4.10.28. By the associativity of $\oplus$, these maps make $\coprod_{n=0}^{\infty} BO(n)$ into a topological monoid, which makes the homotopy set $[X, \coprod_{n=0}^{\infty} BO(n)]$ into a monoid and (A.3) into an isomorphism of monoids.

Thus we may apply Milgram's construction in §4.10 to $\coprod_{n=0}^{\infty} BO(n)$ to obtain the classifying space $B\left(\coprod_{n=0}^{\infty} BO(n)\right)$. It turns out that the group completion correspond to $\Omega B(-)$.

Theorem A.2.23. *For a CW complex X, we have natural isomorphisms of Abelian groups*

$$KO(X) \cong \left[X, \Omega B\left(\coprod_{n=0}^{\infty} BO(n)\right)\right]$$

$$K(X) \cong \left[X, \Omega B\left(\coprod_{n=0}^{\infty} BU(n)\right)\right].$$

On the other hand, we have a sequence

$$BO(1) \longrightarrow BO(2) \longrightarrow \cdots \longrightarrow BO(n) \longrightarrow BO(n+1) \longrightarrow \cdots$$

induced by the inclusions $O(n) \hookrightarrow O(n+1)$ in Definition 4.9.15. By taking the "limit", we define

$$BO = \operatorname*{colim}_n BO(n).$$

Then it is known that there is a homotopy equivalence

$$\Omega B \left(\coprod_{n=0}^{\infty} BO(n) \right) \simeq BO \times \mathbb{Z}$$

which is compatible with Hopf space structures. Thus we have an isomorphism of Abelian groups

$$KO(X) \cong [X, BO \times \mathbb{Z}].$$

We also have

$$K(X) \cong [X, BU \times \mathbb{Z}].$$

Furthermore the tensor product defines multiplications on $KO(X)$ and $K(X)$ making them into a ring. The tensor product also defines multiplications on $BO \times \mathbb{Z}$ and $BU \times \mathbb{Z}$ and the above isomorphisms are isomorphisms of rings.

More details on K-theory can be found in [Atiyah (1989)] and [Karoubi (1978)], for example.

A.3　Simplicial Techniques

The structure appeared in Milgram's construction of universal bundles in §4.10 resembles that of a simplicial complex. Such a structure is now generalized as simplicial objects in a category and plays very important roles in homotopy theory.

This section is a summary of basic techniques used in the study of simplicial objects. Proofs are basically omitted. The reader is recommend to refer to May's book [May (1992)], Curtis' exposition [Curtis (1971)], the book by Goerss and Jardine [Goerss and Jardine (2009)], and Friedman's exposition [Friedman (2012)]. The monograph by Dwyer [Dwyer and Henn (2001)] on classifying spaces is also useful.

Let us begin with classical simplicial complexes and their geometric realizations.

Definition A.3.1. An *abstract simplicial complex* K with *vertex set* V is a family of finite subsets of V satisfying the following condition

$$\sigma \in K \text{ and } \tau \subset \sigma \Longrightarrow \tau \in K.$$

Define $F(K) = K \setminus \{\emptyset\}$ and regard it as a partially ordered set (poset, for short). This is called the *face poset* of K.

An element σ of $F(K)$ is called an *n-simplex* if it consists of $n+1$ points.

When V is equipped with a partial order in such a way that each simplex of K is totally ordered, K is called an *ordered simplicial complex*. An n-simplex $\sigma = \{v_0, \ldots, v_n\}$ of an ordered simplicial complex is represented by $\sigma = [v_0, \ldots, v_n]$ with $v_0 < \cdots < v_n$.

There is another kind of simplicial complexes used in elementary textbooks on topology. Such a simplicial complex consists of geometric simplices in Euclidean spaces and called an *Euclidean simplicial complex*. By a *geometric simplex*, we mean a subspace of an Euclidean space defined as the convex hull of a finite number of points. Since a geometric simplex can be reconstructed from its vertices as the convex hull, the set of vertices of simplices has essentially the same information as the Euclidean simplicial complex, which justifies the definition of abstract simplicial complexes.

In the case of a Euclidean simplicial complex K, the geometric realization $\|K\|$ can be simply defined as the union of simplices. On the other hand, it is not clear how to define a geometric realization of an abstract simplicial complex. When the vertex set V is finite, we may embed V in an Euclidean space $\mathbb{R}^N$ in such a way that elements of V are affinely independent. By taking the convex hulls of the images of simplices, we obtain a subspace of $\mathbb{R}^N$, which can be called a geometric realization of K. However, the definition depends on the embedding of V in $\mathbb{R}^N$ and only works for finite simplicial complexes. If K is an order simplicial complex, we have an alternative way. We may glue standard simplices in Definition 4.10.6 by using the ordering of vertices of each simplex.

Definition A.3.2. Let K be an ordered simplicial complex. The set of n-simplices is denoted by K_n. The map given by removing the i-th vertex is denoted by $d_i : K_n \to K_{n-1}$. This is called a *face operator*. Define

$$\|K\| = \left(\coprod_{n=0}^{\infty} K_n \times \Delta^n \right) \Big/ \underset{d}{\sim},$$

where the relation $\underset{d}{\sim}$ is the equivalence relation generated by

$$(d_i(\sigma), t) \underset{d}{\sim} (\sigma, d^i(t))$$

and the map $d^i : \Delta^{n-1} \to \Delta^n$ is given by

$$d^i(t_0, \ldots, t_{n-1}) = (t_0, \ldots, t_{i-1}, 0, t_i, \ldots, t_{n-1}).$$

The space $\|K\|$ is called the *geometric realization of K*.

In this definition, each K_n is regarded as a topological space by the discrete topology. The definition can be applied without a change when K_n is a more general topological space.

Definition A.3.3. A *semisimplicial space* $X_\bullet$ consists of a sequence of topological spaces $\{X_n\}_{n=0}^\infty$ and continuous maps $\{d_i : X_n \to X_{n-1}\}_{i=0}^n$ satisfying the condition that, for $i < j$,

$$d_i \circ d_j = d_{j-1} \circ d_i. \tag{A.4}$$

The maps d_i are called *face operators*.

For a semisimplicial space $X_\bullet$, its *geometric realization* $\|X_\bullet\|$ is defined by

$$\|X_\bullet\| = \left(\coprod_{n=0}^\infty K_n \times \Delta^n \right)\Big/_{\underset{d}{\sim}},$$

where $\underset{d}{\sim}$ is the same as the relation used in the geometric realization of ordered simplicial complexes.

Remark A.3.4. Semisimplicial spaces are often called Δ-*spaces* in the literature, for example, in [Rourke and Sanderson (1971a,b)]. We choose to use the terminology used in the more recent literature such as [Ebert and Randal-Williams (2019)].

Remark A.3.5. The relation (A.4) used in the definition of semisimplicial spaces can be interpreted as the relation that arise when two vertices of a simplex are removed.

Example A.3.6. Let G be a topological monoid. Suppose G acts on topological spaces X and Y from the right and the left, respectively. Define a space $B_n(X, G, Y)$ by

$$B_n(X, G, Y) = X \times G^n \times Y$$

and maps

$$d_i : B_n(X, G, Y) \longrightarrow B_{n-1}(X, G, Y)$$

by

$$d_i(x; g_1, \ldots, g_n; y) = \begin{cases} (\mu_X(x, g_1); g_2, \ldots, g_n; y), & i = 0 \\ (x; g_1, \ldots, \mu_G(g_i, g_{i+1}), \ldots, g_n; y), & 1 \leq i \leq n - 1 \\ (x; g_1, \ldots, g_{n-1}; \mu_Y(g_n, y)), & i = n, \end{cases}$$

where μ_X, μ_Y, and μ_G are actions of G on X, Y, and G itself. Then the collection $B_\bullet(X, G, Y) = \{B_n(X, G, Y)\}_{n \geq 0}$ is a semisimplicial space.

Consider the case when $X = *$ and $Y = G$. The geometric realization is

$$\|B_\bullet(*, G, G)\| = \left(\coprod_{n=0}^{\infty} \{*\} \times G^n \times G \times \Delta^n \right) \Big/ \underset{d}{\sim} .$$

Since

$$d_i(*; g_1, \ldots, g_n; g) = \begin{cases} (*; g_2, \ldots, g_n; g), & i = 0 \\ (*; g_1, \ldots, g_i g_{i+1}, \ldots, g_n; g), & 1 \leq i \leq n - 1 \\ (*; g_1, \ldots, g_{n-1}; g_n g), & i = n, \end{cases}$$

the first three relations used in Milgram's construction of universal bundles (Definition 4.10.11) appear as defining relations of $\|B_*(*, G, G)\|$. Thus Milgram's construction EG is a quotient space of $\|B_*(*, G, G)\|$ by certain relations defined in terms of the unit of G. $\qquad \square$

Definition A.3.7. A *simplicial space* $X_\bullet$ consists of a sequence of topological spaces $\{X_n\}_{n=0}^{\infty}$ and continuous maps $d_i : X_n \to X_{n-1}$ $(0 \leq i \leq n)$ and $s_i : X_n \to X_{n+1}$ $(0 \leq i \leq n)$ satisfying the following conditions:

(1) the maps d_i make $\{X_n\}$ into a semisimplicial space,
(2) $d_i \circ s_j = s_{j-1} \circ d_i$ for $i < j$,
(3) $d_j \circ s_j = 1 = d_{j+1} \circ s_j$,
(4) $d_i \circ s_j = s_j \circ d_{i-1}$ for $i > j + 1$, and
(5) $s_i \circ s_j = s_{j+1} \circ s_i$ for $i \leq j$.

The maps s_i are called *degeneracy operators*.

The *geometric realization* is defined by

$$|X_\bullet| = \|X_\bullet\|/\underset{s}{\sim},$$

where the relation $\underset{s}{\sim}$ is the equivalence relation generated by $(s_i(x), t) \underset{s}{\sim} (x, s^i(t))$.

For simplicial spaces $X_\bullet$ and $Y_\bullet$, a *map of simplicial spaces* $f_\bullet : X_\bullet \to Y_\bullet$ is a sequence of continuous maps $\{f_n : X_n \to Y_n\}_{n \geq 0}$ that commute with d_i and s_i. The map induced by map of simplicial spaces is denoted by

$$|f_\bullet| : |X_\bullet| \longrightarrow |Y_\bullet|.$$

Example A.3.8. For a topological space Y, define $c_n(Y) = Y$ for all n. The face and degeneracy operators are defined by $d_i = 1_Y$ and $s_i = 1_Y$. This simplicial space $c_\bullet(Y)$ is called the *constant simplicial space* generated by Y. It is easy to see that we have a homeomorphism $|c_\bullet(Y)| \cong Y$.   □

Example A.3.9. In Example A.3.6, define maps

$$s_i : B_n(X, G, Y) \longrightarrow B_{n+1}(X, G, Y)$$

by

$$s_i(x; g_1, \ldots, g_n; y) = (x; g_1, \ldots, g_{i-1}, e, g_i, \ldots, g_n; y),$$

where e is the unit of G. Then we have a simplicial space $B_\bullet(X, G, Y)$. This is called the *two-sided bar construction*.

By definition, we have

$$|B_\bullet(*, G, *)| = BG$$
$$|B_\bullet(*, G, G)| = EG$$

and the projection $p_G : EG \to BG$ in Milgram's construction agrees with the map

$$|B_\bullet(*, G, G)| \longrightarrow |B_\bullet(*, G, *)|$$

induced from the map of simplicial spaces $B_\bullet(*, G, G) \to B_\bullet(*, G, *)$ defined by $G \to *$.   □

Now we can use general theory of simplicial spaces to study Milgram's construction. We first consider products of simplicial spaces.

Theorem A.3.10. *For simplicial spaces $X_\bullet$ and $Y_\bullet$, define the product $X_\bullet \times Y_\bullet$ by $(X_\bullet \times Y_\bullet)_n = X_n \times Y_n$. The face and degeneracy operators are defined by products of those of X and Y. Then the projections onto the first and the second factors induce a homeomorphism*

$$|X_\bullet \times Y_\bullet| \cong |X_\bullet| \times |Y_\bullet|.$$

See Theorem 14.3 of May's book [May (1992)] for a proof.

Now Theorem 4.10.28 is a corollary to this fact.

Proof of Theorem 4.10.28. For topological groups (or monoids) G and H, we have an isomorphism of simplicial spaces

$$B_\bullet(*, G \times H, *) \cong B_\bullet(*, G, *) \times B_\bullet(*, H, *).$$

Thus we obtain a homeomorphism $B(G \times H) \cong BG \times BH$ by Theorem A.3.10. $\qquad\square$

In Corollary 4.10.25, we proved that EG is ∞-connected by using Proposition 4.10.23. We can show that EG is contractible by constructing a simplicial homotopy.

Definition A.3.11. Let $f_\bullet, g_\bullet : X_\bullet \to Y_\bullet$ be maps of simplicial spaces. A *simplicial homotopy* from $f_\bullet$ to $g_\bullet$ is a family of maps

$$h_{n,i} : X_n \longrightarrow Y_{n+1}$$

indexed by nonnegative integers n and $0 \le i \le n$ satisfying the following relations

$$d_0 \circ h_{n,0} = f_n$$
$$d_{n+1} \circ h_{n,n} = g_n$$

$$d_i \circ h_{n,j} = \begin{cases} h_{n-1,j-1} \circ d_i & \text{if } i < j \\ d_i \circ h_{n,i-1}, & \text{if } i = j > 0 \\ h_{n-1,j} \circ d_{i-1} & \text{if } i > j+1 \end{cases}$$

$$s_i \circ h_{n,j} = \begin{cases} h_{n+1,j+1} \circ s_i & \text{if } i \le j \\ h_{n+1,j} \circ s_{i-1} & \text{if } i > j. \end{cases}$$

When there exists a simplicial homotopy between $f_\bullet$ and $g_\bullet$, we write $f_\bullet \underset{s}{\simeq} g_\bullet$.

Remark A.3.12. The first two conditions correspond to the conditions $H|_{X \times \{0\}} = f$ and $H|_{X \times \{1\}} = g$ in the usual homotopy. The third and the fourth conditions correspond to the conditions it defines a map of simplicial spaces

$$h_\bullet : X_\bullet \times I_\bullet \longrightarrow Y_\bullet$$

where $I_\bullet$ is the simplicial space defined by

$$I_n = \{\alpha : \{0 < \cdots < n\} \to \{0 < 1\} \mid \text{order preserving map}\}.$$

By a homeomorphism $|I| \cong [0,1]$ and Theorem A.3.10, we obtain the following.

Proposition A.3.13. *Let* $f_\bullet, g_\bullet : X_\bullet \to Y_\bullet$ *be simplicial maps between simplicial spaces. If* $f_\bullet \underset{s}{\simeq} g_\bullet$, *then* $|f_\bullet| \simeq |g_\bullet|$.

Thanks to this fact, in order to show that the geometric realization of a simplicial space $X_\bullet$ is contractible, it suffices to construct a simplicial homotopy between the identity map on $X_\bullet$ and the collapsing map to a point $x_0 \in X_0$. More generally, we have the following fact.

Definition A.3.14. An *augmentation* of a simplicial space $X_\bullet$ is a continuous map $d_0 : X_0 \to X_{-1}$ satisfying $d_0 \circ d_0 = d_0 \circ d_1$.

Theorem A.3.15. *Let* $X_\bullet$ *be a simplicial space equipped with an* augmentation $d_0 : X_0 \to X_{-1}$. *Suppose there exists a family of continuous maps* $s_{n+1} : X_n \to X_{n+1}$ *for* n *satisfying the following conditions*

$$d_i \circ s_{n+1} = \begin{cases} s_n \circ d_i, & i \le n \\ 1, & i = n+1 \end{cases}$$

$$s_i \circ s_{n+1} = s_{n+2} \circ s_i, \ \forall i \le n+1.$$

Then $|X_\bullet|$ *is homotopy equivalent to* X_{-1}. *In particular, when* $X_{-1} = \{*\}$, $|X_\bullet|$ *is contractible.*

Proof. Define $\varepsilon_n : X_n \to X_{-1}$ by

$$\varepsilon_n = \underbrace{d_0 \circ \cdots \circ d_0}_{n+1}$$

by using the augmentation $d_0 : X_0 \to X_{-1}$. Then it defines a map of simplicial spaces $\varepsilon_\bullet : X_\bullet \to c_\bullet(X_{-1})$.

We also obtain a map of simplicial spaces $\xi_\bullet : c_\bullet(X_{-1}) \to X_\bullet$ by

$$\xi_n = s_n \circ s_{n-1} \circ \cdots \circ s_1 \circ s_0.$$

Since $\varepsilon_n \circ \xi_n = 1_{X_{-1}}$ for all n, it suffices to construct a simplicial homotopy $h = \{h_{n,i} : X_n \to X_{n+1}\}$ between $\xi_\bullet \circ \varepsilon_\bullet$ and $1_{X_\bullet}$.

Such a simplicial homotopy can be defined by

$$h_{n,i} = \begin{cases} s_{n+1} \circ \cdots \circ s_{i+1} \circ d_{i+1} \circ \cdots \circ d_n, & i < n \\ s_{n+1}, & i = n. \end{cases}$$

$\square$

Remark A.3.16. The maps s_{n+1} appeared in the proof of Theorem A.3.15 are called *upper extra degeneracies*. We may also use maps $s_{-1} : X_n \to X_{n+1}$ satisfying

$$d_i \circ s_{-1} = \begin{cases} s_{-1} \circ d_{i-1}, & i > 0 \\ 1 & i = 0 \end{cases}$$

$$s_i \circ s_{-1} = s_{-1} \circ s_{i-1}, \ \forall i \geq 0$$

as such "extra degeneracies". These are called *lower extra degeneracies*. See §3.3 of Barr's monograph [Barr (2002)] for details.

Example A.3.17. Consider $EG = |B_\bullet(*, G, G)|$. Define $B_{-1}(*, G, G) = *$. Then the collapsing map $d_0 : B_0(*, G, G) = G \to * = B_{-1}(*, G, G)$ is an augmentation.

For each n, define $s_{n+1} : B_n(*, G, G) \to B_{n+1}(*, G, G)$ by

$$s_{n+1}(*; g_1, \ldots, g_n; g_{n+1}) = (*; g_1, \ldots, g_n, g_{n+1}; e).$$

Then these are upper extra degeneracies and we obtain a homotopy equivalence

$$EG \simeq B_{-1}(*, G, G) = *$$

by Theorem A.3.15. □

Bibliography

Abe, E. (1980). *Hopf algebras, Cambridge Tracts in Mathematics*, Vol. 74 (Cambridge University Press, Cambridge), ISBN 0-521-22240-0, translated from the Japanese by Hisae Kinoshita and Hiroko Tanaka.

Adams, J. F. (1960). On the non-existence of elements of Hopf invariant one, *Ann. of Math. (2)* **72**, pp. 20–104.

Anderson, D. W. (1978). Fibrations and geometric realizations, *Bull. Amer. Math. Soc.* **84**, 5, pp. 765–788.

Atiyah, M. and Segal, G. (2004). Twisted K-theory, *Ukr. Mat. Visn.* **1**, 3, pp. 287–330, math/0407054.

Atiyah, M. F. (1989). *K-theory*, 2nd edn., Advanced Book Classics (Addison-Wesley Publishing Company, Advanced Book Program, Redwood City, CA), ISBN 0-201-09394-4, notes by D. W. Anderson.

Barnea, I. and Schlank, T. M. (2016). A projective model structure on pro-simplicial sheaves, and the relative étale homotopy type, *Adv. Math.* **291**, pp. 784–858, doi:10.1016/j.aim.2015.11.014, 1109.5477, https://doi.org/10.1016/j.aim.2015.11.014.

Barr, M. (2002). *Acyclic models, CRM Monograph Series*, Vol. 17 (American Mathematical Society, Providence, RI), ISBN 0-8218-2877-0.

Barwick, C. and Kan, D. M. (2012). Relative categories: another model for the homotopy theory of homotopy theories, *Indag. Math. (N.S.)* **23**, 1-2, pp. 42–68, doi:10.1016/j.indag.2011.10.002, 1011.1691, http://dx.doi.org/10.1016/j.indag.2011.10.002.

Baum, P. F., Hajac, P. M., Matthes, R., and Szymanski, W. (2007). Noncommutative Geometry Approach to Principal and Associated Bundles, math/0701033.

Bespalov, Y., Lyubashenko, V., and Manzyuk, O. (2008). *Pretriangulated A_∞-categories, Proceedings of Institute of Mathematics of NAS of Ukraine. Mathematics and its Applications*, Vol. 76 (Natsīonal'na Akademīya Nauk Ukraïni, Īnstitut Matematiki, Kiev), ISBN 978-966-02-4861-8.

Blumberg, A. J. and Mandell, M. A. (2012). Localization theorems in topological Hochschild homology and topological cyclic homology, *Geom. Topol.* **16**, 2, pp. 1053–1120, doi:10.2140/gt.2012.16.1053, 0802.3938, http://dx.doi.org/10.2140/gt.2012.16.1053.

Boardman, J. M. and Vogt, R. M. (1968). Homotopy-everything H-spaces, *Bull. Amer. Math. Soc.* **74**, pp. 1117–1122.

Boardman, J. M. and Vogt, R. M. (1973). *Homotopy invariant algebraic structures on topological spaces*, Lecture Notes in Mathematics, Vol. 347 (Springer-Verlag, Berlin).

Bondal, A. I. and Kapranov, M. M. (1990). Framed triangulated categories, *Mat. Sb.* **181**, 5, pp. 669–683.

Brouwer, L. E. J. (1976). *Collected works, Vol. 2* (North-Holland Publishing Co., Amsterdam), geometry, analysis, topology and mechanics, Edited by Hans Freudenthal.

Brown, K. S. (1973). Abstract homotopy theory and generalized sheaf cohomology, *Trans. Amer. Math. Soc.* **186**, pp. 419–458, doi:10.2307/1996573, https://doi.org/10.2307/1996573.

Cairns, S. S. (1934). On the triangulation of regular loci, *Ann. of Math. (2)* **35**, 3, pp. 579–587, doi:10.2307/1968752, http://dx.doi.org/10.2307/1968752.

Cannas da Silva, A. (2001). *Lectures on symplectic geometry*, Lecture Notes in Mathematics, Vol. 1764 (Springer-Verlag, Berlin), ISBN 3-540-42195-5, https://doi.org/10.1007/978-3-540-45330-7.

Cartan, H. (1951). La transgression dans un groupe de Lie et dans un espace fibré principal, in *Colloque de topologie (espaces fibrés), Bruxelles, 1950* (Georges Thone, Liège), pp. 57–71.

Cisinski, D.-C. (2003). Images directes cohomologiques dans les catégories de modèles, *Ann. Math. Blaise Pascal* **10**, 2, pp. 195–244, http://ambp.cedram.org/item?id=AMBP_2003__10_2_195_0.

Cisinski, D.-C. (2008). Propriétés universelles et extensions de Kan dérivées, *Theory Appl. Categ.* **20**, 17, pp. 605–649.

Cisinski, D.-C. (2010). Catégories dérivables, *Bull. Soc. Math. France* **138**, 3, pp. 317–393, doi:10.24033/bsmf.2592, https://doi.org/10.24033/bsmf.2592.

Curtis, E. B. (1971). Simplicial homotopy theory, *Advances in Math.* **6**, pp. 107–209, https://doi.org/10.1016/0001-8708(71)90015-6.

Dehn, M. and Heegaard, P. (1907). *Analysis Situs, Enzyclopädie der mathematischen Wissenschaften*, Vol. III 1 AB 3 (Teubner, Leipzig).

Dieudonné, J. (1989). *A history of algebraic and differential topology. 1900–1960* (Birkhäuser Boston Inc., Boston, MA), ISBN 0-8176-3388-X.

Dold, A. (1963). Partitions of unity in the theory of fibrations, *Ann. of Math. (2)* **78**, pp. 223–255.

Dold, A. and Thom, R. (1958). Quasifaserungen und unendliche symmetrische Produkte, *Ann. of Math. (2)* **67**, pp. 239–281.

Dugundji, J. (1978). *Topology* (Allyn and Bacon Inc., Boston, Mass.), ISBN 0-205-00271-4.

Dwyer, W. G. and Henn, H.-W. (2001). *Homotopy theoretic methods in group cohomology*, Advanced Courses in Mathematics—CRM Barcelona (Birkhäuser Verlag, Basel), ISBN 3-7643-6605-2.

Dwyer, W. G., Hirschhorn, P. S., Kan, D. M., and Smith, J. H. (2004). *Homotopy limit functors on model categories and homotopical categories, Mathemat-*

ical Surveys and Monographs, Vol. 113 (American Mathematical Society, Providence, RI), ISBN 0-8218-3703-6.

Dwyer, W. G. and Spaliński, J. (1995). Homotopy theories and model categories, in *Handbook of algebraic topology* (North-Holland, Amsterdam), pp. 73–126, doi:10.1016/B978-044481779-2/50003-1, http://dx.doi.org/10.1016/B978-044481779-2/50003-1.

Ebert, J. and Randal-Williams, O. (2019). Semisimplicial spaces, *Algebr. Geom. Topol.* **19**, 4, pp. 2099–2150, doi:10.2140/agt.2019.19.2099, 1705.03774, https://doi.org/10.2140/agt.2019.19.2099.

Eilenberg, S. and Steenrod, N. (1952). *Foundations of algebraic topology* (Princeton University Press, Princeton, New Jersey).

Fomenko, A. T., Fuchs, D. B., and Gutenmacher, V. L. (1986). *Homotopic topology* (Akadémiai Kiadó (Publishing House of the Hungarian Academy of Sciences), Budapest), ISBN 963-05-3544-0, translated from the Russian by K. Mályusz.

Fox, R. H. (1945). On topologies for function spaces, *Bull. Amer. Math. Soc.* **51**, pp. 429–432, https://doi.org/10.1090/S0002-9904-1945-08370-0.

Friedman, G. (2012). Survey article: an elementary illustrated introduction to simplicial sets, *Rocky Mountain J. Math.* **42**, 2, pp. 353–423, doi: 10.1216/RMJ-2012-42-2-353, 0809.4221, http://dx.doi.org/10.1216/RMJ-2012-42-2-353.

Gabriel, P. and Zisman, M. (1967). *Calculus of fractions and homotopy theory*, Ergebnisse der Mathematik und ihrer Grenzgebiete, Band 35 (Springer-Verlag New York, Inc., New York).

Gajer, P. (1997). Geometry of Deligne cohomology, *Invent. Math.* **127**, 1, pp. 155–207, doi:10.1007/s002220050118, https://doi.org/10.1007/s002220050118.

Goerss, P. G. and Jardine, J. F. (2009). *Simplicial homotopy theory*, Modern Birkhäuser Classics (Birkhäuser Verlag, Basel), ISBN 978-3-0346-0188-7, https://doi.org/10.1007/978-3-0346-0189-4, reprint of the 1999 edition [MR1711612].

Golasiński, M. and Gromadzki, G. (1982). The homotopy category of chain complexes is a homotopy category, *Colloq. Math.* **47**, 2, pp. 173–178, doi: 10.4064/cm-47-2-173-178, https://doi.org/10.4064/cm-47-2-173-178.

Grothendieck, A. (2003). *Revêtements étales et groupe fondamental (SGA 1)*, Documents Mathématiques (Paris) [Mathematical Documents (Paris)], 3 (Société Mathématique de France, Paris), ISBN 2-85629-141-4, math/0206203, séminaire de géométrie algébrique du Bois Marie 1960–61. [Algebraic Geometry Seminar of Bois Marie 1960-61], Directed by A. Grothendieck, With two papers by M. Raynaud, Updated and annotated reprint of the 1971 original [Lecture Notes in Math., 224, Springer, Berlin; MR0354651 (50 #7129)].

Guillén, F., Navarro, V., Pascual, P., and Roig, A. (2010). A Cartan-Eilenberg approach to homotopical algebra, *J. Pure Appl. Algebra* **214**, 2, pp. 140–164, doi:10.1016/j.jpaa.2009.04.009, 0707.3704, http://dx.doi.org/10.1016/j.jpaa.2009.04.009.

Hironaka, H. (1975). Triangulations of algebraic sets, in *Algebraic geometry (Proc. Sympos. Pure Math., Vol. 29, Humboldt State Univ., Arcata, Calif., 1974)* (Amer. Math. Soc., Providence, R.I.), pp. 165–185.

Hirschhorn, P. S. (2003). *Model categories and their localizations, Mathematical Surveys and Monographs*, Vol. 99 (American Mathematical Society, Providence, RI), ISBN 0-8218-3279-4.

Hovey, M. (1999). *Model categories, Mathematical Surveys and Monographs*, Vol. 63 (American Mathematical Society, Providence, RI), ISBN 0-8218-1359-5.

Hu, S.-t. (1964). *Elements of general topology* (Holden-Day Inc., San Francisco, Calif.).

Hurewicz, W. (1935). Homotopie, homologie und lokaler zusammenhang, *Fund. Math.* **25**, 1, pp. 467–485.

Hurewicz, W. (1955). On the concept of fiber space, *Proc. Nat. Acad. Sci. U. S. A.* **41**, pp. 956–961.

Husemoller, D. (1994). *Fibre bundles, Graduate Texts in Mathematics*, Vol. 20, 3rd edn. (Springer-Verlag, New York), ISBN 0-387-94087-1, doi:10.1007/978-1-4757-2261-1, http://dx.doi.org/10.1007/978-1-4757-2261-1.

Joyal, A. and Tierney, M. (2007). Quasi-categories vs Segal spaces, in *Categories in algebra, geometry and mathematical physics, Contemp. Math.*, Vol. 431 (Amer. Math. Soc., Providence, RI), pp. 277–326, math/0607820.

Karoubi, M. (1978). *K-theory* (Springer-Verlag, Berlin), ISBN 3-540-08090-2, an introduction, Grundlehren der Mathematischen Wissenschaften, Band 226.

Kelley, J. L. (1975). *General topology, Graduate Texts in Mathematics*, Vol. 27 (Springer-Verlag, New York).

Komatsu, A., Nakaoka, M., and Sugawara, M. (1967). *Topology I* (Iwanami Shoten, Tokyo, Japan), in Japanese.

Kontsevich, M. and Soibelman, Y. (unknown). Deformation Theory. I, http://www.math.ksu.edu/~soibel/Book-vol1.ps.

Kudo, T. and Araki, S. (1956a). On $H_*(\Omega^N(S^n); \mathbb{Z}_2)$, *Proc. Japan Acad.* **32**, pp. 333–335.

Kudo, T. and Araki, S. (1956b). Topology of H_n-spaces and H-squaring operations, *Mem. Fac. Sci. Kyūsyū Univ. Ser. A.* **10**, pp. 85–120.

Lefschetz, S. and Whitehead, J. H. C. (1933). On analytical complexes, *Trans. Amer. Math. Soc.* **35**, 2, pp. 510–517, doi:10.2307/1989779, http://dx.doi.org/10.2307/1989779.

Lenstra, H. W. (2008). Galois theory for schemes, http://websites.math.leidenuniv.nl/algebra/GSchemes.pdf.

Liu, Y. M. (1978). A necessary and sufficient condition for the producibility of CW-complexes, *Acta Math. Sinica* **21**, 2, pp. 171–175.

Lundell, A. T. and Weingram, S. (1969). *Topology of CW-Complexes* (Van Nostrand Reinhold, New York).

Lurie, J. (2009). *Higher topos theory, Annals of Mathematics Studies*, Vol. 170 (Princeton University Press, Princeton, NJ), ISBN 978-0-691-14049-0, doi:10.1515/9781400830558, http://dx.doi.org/10.1515/9781400830558.

Mac Lane, S. (1998). *Categories for the working mathematician, Graduate Texts in Mathematics*, Vol. 5, 2nd edn. (Springer-Verlag, New York), ISBN 0-387-98403-8.

May, J. P. (1972). *The geometry of iterated loop spaces* (Springer-Verlag, Berlin), lectures Notes in Mathematics, Vol. 271.

May, J. P. (1990). Weak equivalences and quasifibrations, in *Groups of self-equivalences and related topics (Montreal, PQ, 1988), Lecture Notes in Math.*, Vol. 1425 (Springer, Berlin), pp. 91–101, doi:10.1007/BFb0083834, http://dx.doi.org/10.1007/BFb0083834.

May, J. P. (1992). *Simplicial objects in algebraic topology*, Chicago Lectures in Mathematics (University of Chicago Press, Chicago, IL), ISBN 0-226-51181-2, reprint of the 1967 original.

May, J. P. (1999). *A concise course in algebraic topology*, Chicago Lectures in Mathematics (University of Chicago Press, Chicago, IL), ISBN 0-226-51182-0.

Milgram, R. J. (1967). The bar construction and abelian H-spaces, *Illinois J. Math.* **11**, pp. 242–250, http://projecteuclid.org/euclid.ijm/1256054662.

Milnor, J. (1956). Construction of universal bundles. I, *Ann. of Math. (2)* **63**, pp. 272–284, doi:10.2307/1969609, http://dx.doi.org/10.2307/1969609.

Milnor, J. (1963). *Morse theory*, Based on lecture notes by M. Spivak and R. Wells. Annals of Mathematics Studies, No. 51 (Princeton University Press, Princeton, N.J.).

Milnor, J. W. and Moore, J. C. (1965). On the structure of Hopf algebras, *Ann. of Math. (2)* **81**, pp. 211–264, doi:10.2307/1970615, http://dx.doi.org/10.2307/1970615.

Milnor, J. W. and Stasheff, J. D. (1974). *Characteristic classes* (Princeton University Press, Princeton, N.J.), annals of Mathematics Studies, No. 76.

Mimura, M. and Toda, H. (1991). *Topology of Lie groups. I, II, Translations of Mathematical Monographs*, Vol. 91 (American Mathematical Society, Providence, RI), ISBN 0-8218-4541-1, translated from the 1978 Japanese edition by the authors.

Munkres, J. R. (2000). *Topology*, 2nd edn. (Prentice-Hall Inc., Englewood Cliffs, N.J.).

Neeman, A. (2001). *Triangulated categories, Annals of Mathematics Studies*, Vol. 148 (Princeton University Press, Princeton, NJ), ISBN 0-691-08685-0; 0-691-08686-9.

Oliva, W. M. (2002). *Geometric mechanics, Lecture Notes in Mathematics*, Vol. 1798 (Springer-Verlag, Berlin), ISBN 3-540-44242-1, https://doi.org/10.1007/b84214.

Quillen, D. G. (1967). *Homotopical algebra*, Lecture Notes in Mathematics, No. 43 (Springer-Verlag, Berlin).

Ravenel, D. C. and Wilson, W. S. (1980). The Morava K-theories of Eilenberg-Mac Lane spaces and the Conner-Floyd conjecture, *Amer. J. Math.* **102**, 4, pp. 691–748, doi:10.2307/2374093, http://dx.doi.org/10.2307/2374093.

Riehl, E. (2016). *Category Theory in Context (Aurora: Dover Modern Math Originals)* (Dover Publications), ISBN 9780486809038.

Rothenberg, M. and Steenrod, N. E. (1965). The cohomology of classifying spaces of H-spaces, *Bull. Amer. Math. Soc.* **71**, pp. 872–875.

Rourke, C. P. and Sanderson, B. J. (1971a). $\triangle$-sets. I. Homotopy theory, *Quart. J. Math. Oxford Ser. (2)* **22**, pp. 321–338.

Rourke, C. P. and Sanderson, B. J. (1971b). $\triangle$-sets. II. Block bundles and block fibrations, *Quart. J. Math. Oxford Ser. (2)* **22**, pp. 465–485.

Serre, J.-P. (1951). Homologie singulière des espaces fibrés. Applications, *Ann. of Math. (2)* **54**, pp. 425–505, doi:10.2307/1969485, http://dx.doi.org/10.2307/1969485.

Spitzweck, M. (2001). Operads, Algebras and Modules in General Model Categories, math/0101102.

Stasheff, J. D. (1963). Homotopy associativity of H-spaces. I, II, *Trans. Amer. Math. Soc.* **108** *(1963)*, *275-292; ibid.* **108**, pp. 293–312.

Steenrod, N. E. (1967). A convenient category of topological spaces, *Michigan Math. J.* **14**, pp. 133–152, http://projecteuclid.org/euclid.mmj/1028999711.

Strøm, A. (1966). Note on cofibrations, *Math. Scand.* **19**, pp. 11–14, doi:10.7146/math.scand.a-10791, https://doi.org/10.7146/math.scand.a-10791.

Strøm, A. (1972). The homotopy category is a homotopy category, *Arch. Math. (Basel)* **23**, pp. 435–441, doi:10.1007/BF01304912, https://doi.org/10.1007/BF01304912.

Sweedler, M. E. (1969). *Hopf algebras*, Mathematics Lecture Note Series (W. A. Benjamin, Inc., New York).

Tamaki, D. (2013a). Twisting Segal's K-homology theory, in *Noncommutative geometry and physics. 3, Keio COE Lect. Ser. Math. Sci.*, Vol. 1 (World Sci. Publ., Hackensack, NJ), pp. 197–235, doi:10.1142/9789814425018_0007, http://dx.doi.org/10.1142/9789814425018_0007.

Tamaki, D. (2013b). Two-sided bar constructions for partial monoids and applications to K-homology theory, in *Noncommutative geometry and physics. 3, Keio COE Lect. Ser. Math. Sci.*, Vol. 1 (World Sci. Publ., Hackensack, NJ), pp. 177–195, doi:10.1142/9789814425018_0006, http://dx.doi.org/10.1142/9789814425018_0006.

Tamaki, D. (2018). Cellular stratified spaces, in *Combinatorial and toric homotopy, Lect. Notes Ser. Inst. Math. Sci. Natl. Univ. Singap.*, Vol. 35 (World Sci. Publ., Hackensack, NJ), pp. 305–435, 1609.04500.

Waldhausen, F. (1978). Algebraic K-theory of generalized free products. I, II, *Ann. of Math. (2)* **108**, 1, pp. 135–204.

Waldhausen, F. (1979). Algebraic K-theory of topological spaces. II, in *Algebraic topology, Aarhus 1978 (Proc. Sympos., Univ. Aarhus, Aarhus, 1978)*, *Lecture Notes in Math.*, Vol. 763 (Springer, Berlin), pp. 356–394.

Waldhausen, F. (1985). Algebraic K-theory of spaces, in *Algebraic and geometric topology (New Brunswick, N.J., 1983)*, *Lecture Notes in Math.*, Vol. 1126 (Springer, Berlin), pp. 318–419, doi:10.1007/BFb0074449, http://dx.doi.org/10.1007/BFb0074449.

Whitehead, G. W. (1978). *Elements of homotopy theory, Graduate Texts in Mathematics*, Vol. 61 (Springer-Verlag, New York), ISBN 0-387-90336-4.

Whitehead, J. H. C. (1939). Simplicial Spaces, Nuclei and m-Groups, *Proc. London Math. Soc. (2)* **45**, 4, pp. 243–327, doi:10.1112/plms/s2-45.1.243, https://doi.org/10.1112/plms/s2-45.1.243.

Whitehead, J. H. C. (1948). Note on a theorem due to Borsuk, *Bull. Amer. Math. Soc.* **54**, pp. 1125–1132, doi:10.1090/S0002-9904-1948-09138-8, https://doi.org/10.1090/S0002-9904-1948-09138-8.

Yokota, I. (2009). Exceptional Lie groups, 0902.0431.

Index

K-theory, 302
n-connected, 141
n-equivalence, 254
n-fold loop space, 236
n-universal, 159

abstract simplicial complex, 305
action, 26, 57
adjoint, 51
annulus, 2
antipode, 211
augmentation, 310

base point, 10, 18
base point preserving map, 142
base space, 7, 33, 200, 274
based cofibration, 258
based covering homotopy property, 233
based covering homotopy theorem, 155
based fibration, 234
based homotopy, 142
based homotopy set, 142
based map, 142
based space, 18
boundary, 133
bundle isomorphism, 102
bundle map, 101

Cayley, 47

cell, 133
cell complex, 133
cell decomposition, 132
cellular approximation theorem, 152
cellular map, 153
chain complex, 284
chain homotopy, 284
chain map, 284
characteristic class, 171
characteristic map, 133
CHP, 200
classification theorem, 140, 141
classifying space, 141, 190
closed subgroup, 47
closure finite, 138
cofiber, 258
cofiber homotopy equivalence, 268
cofiber homotopy equivalent, 268
cofibration, 258, 286
compact-open topology, 52
complex K-theory, 302
complex projective space, 72
cone, 132
conjugation, 58
connecting homomorphism, 163
constant simplicial space, 308
contractible, 126
contraction, 126
coordinate transformation, 41
covering homotopy extension theorem, 153

321

covering homotopy property, 200
covering homotopy theorem, 122
covering space, 7
covering transformation, 19
covering transformation group, 193
cross section, 85
CW complex, 138
cylinder, 1

deformation retract, 124
deformation retraction, 124
degeneracy operator, 307
diameter, 13
dihedral group, 60
dimension, 140
direct sum, 297
distinguished, 276
Dold-Thom criterion, 277
dual basis, 296
dual Puppe sequence, 270
dual vector bundle, 301
dual vector space, 296

Eilenberg-MacLane space, 191
end point, 10
equivariant, 102
evaluation map, 294
exact, 161
exact sequence, 161
exterior power, 299
exterior tensor product, 300

face operator, 305
face poset, 305
faithful, 89
fiber, 7, 33, 200, 274
fiber bundle, 33
fiber homotopic, 217
fiber homotopy, 216
fiber homotopy equivalent, 217
fiber map, 33
fiber product, 113, 214
fiber-preserving map, 100, 216
fiberwise homotopic, 217
fiberwise homotopy, 216
fiberwise homotopy equivalent, 217

fibration, 200, 286
finite complex, 140
folding map, 147
free action, 79
fundamental group, 25

general linear group, 45
geometric realization, 306
Grassmannian manifold, 173
Graves, 47
group, 21
group completion, 302

H-space, 210
HEP, 257
Hom bundle, 301
homology, 284
homomorphism, 48
homotopic, 115
homotopy, 14, 115, 309
homotopy associativity, 210
homotopy category, 282
homotopy class, 25
homotopy cofiber, 264
homotopy cofiber sequence, 270
homotopy commutative, 210
homotopy equivalence, 213
homotopy equivalent, 213
homotopy extension property, 257
homotopy fiber, 225
homotopy fiber sequence, 235
homotopy group, 143
homotopy inverse, 213
homotopy long exact sequence, 167
homotopy set, 140
Hopf algebra, 211
Hopf bundle, 37, 63, 92
Hopf space, 210
Hurewicz fibration, 200

induced homomorphism, 160
induced map, 244
infinite dimensional real projective
    space, 171
infinite dimensional sphere, 168
infinite symmetric product, 279

initial point, 10
inner product, 297
isomorphic, 48
isomorphism, 48
isomorphism of covering spaces, 19
isotropy subgroup, 79

join, 178

Legesgue number, 13
Legesgue's Lemma, 13
lens space, 72
lift, 12
local cross section, 85
local trivialization, 34
localization, 287
locally arcwise connected, 20
locally compact, 53
locally connected, 20
locally contractible, 20
locally countable, 158
locally finite, 158
locally simply connected, 20
long exact sequence, 161
loop, 10
loop product, 21, 206
loop space, 18, 204

map of covering spaces, 19
map of vector bundles, 295
mapping cone, 264
mapping cylinder, 262
mapping track, 224
metric, 297
model category, 286
model structure, 286
monodromy, 27
monoid, 56
Moore loop space, 212
Möbius band, 4, 35, 43, 88, 95

NDR pair, 183
NDR representation, 183
nondegenerate base point, 183
normal, 117
numerable covering, 122

octonion, 47
orbit, 68
ordered simplicial complex, 305
orthogonal group, 45

paracompact, 123
partition of unity, 121
path, 10
path space, 202
path-loop fibration, 202
pinching map, 147
pointed space, 18
principal bundle, 87
projection, 7, 33, 200
pullback, 108, 214
pullback diagram, 214
Puppe sequence, 236
pushout, 260

quasifibration, 274
quotient map, 70
quotient space, 68
quotient topology, 66

rank, 64
real K-theory, 302
real projective space, 71
real Stiefel manifold, 172
reduced, 65
reduced cone, 145, 266
reduced mapping cone, 265
reduced mapping cylinder, 265
reduced suspension, 267
regular, 53
relative homotopy group, 249
restriction, 113
retract, 124
retraction, 124
rotation group, 59

semilocally simply connected, 30
semisimplicial space, 306
Serre fibration, 200
simplicial homotopy, 309
simplicial space, 307
simply connected, 15

skeleton, 133
smash product, 237
standard simplex, 180
strong NDR pair, 278
structure group, 42, 63
structure map, 22
subcomplex, 140
subgroup, 47
sup norm, 292
suspension, 267
symmetric product, 279
symmetrized compact-open topology,
  55

tangent bundle, 64
tensor product, 299
topological group, 44
topological monoid, 56
topology of compact convergence, 292
topology of uniform convergence, 292
torus, 3
total space, 33, 200, 274

transitive, 73
triangulated category, 287
trivial action, 26
trivial bundle, 35
trivial covering space, 9
trivial fibration, 223
two-sided bar construction, 278, 308

unitary group, 46
universal bundle, 141
universal covering, 30
Urysohn's Lemma, 117

vector bundle, 64
vertex set, 305

weak equivalence, 286
weak factorization system, 286
weak homotopy equivalence, 254
weak topology, 138, 169
wedge sum, 145
Whitney sum, 297

Printed in the United States
by Baker & Taylor Publisher Services